U0218222

物联网开发与应用丛书

物联网系统

综合开发与应用

廖建尚 杨尚森 潘必超 / 编著

電子工業出版社
Publishing House of Electronics Industry
北京•BEIJING

内 容 简 介

本书基于短距离无线通信技术（ZigBee、BLE、Wi-Fi）和长距离无线通信技术（LoRa、NB-IoT、LTE），详细阐述物联网系统中传感器的驱动开发技术、客户端（Web端和Android端）开发技术，由浅入深地分析物联网系统的综合开发与应用。本书采用案例式的讲解方法，通过贴近社会和生活的物联网系统应用项目，详细地介绍物联网系统的设计和软硬件的开发，并给出了开发验证和总结拓展。本书将理论学习和开发实践紧密结合起来，每个案例均给出了完整的开发代码，读者可以在开发代码的基础上快速地进行二次开发。

本书既可作为高等院校相关专业的教材或教学参考书，也可供相关领域的工程技术人员查阅。对于嵌入式开发和物联网系统开发的爱好者来说，本书也是一本深入浅出、贴近社会应用的技术读物。

本书提供详尽的开发代码和配套 PPT 课件，读者可登录华信教育资源网（www.hxedu.com.cn）免费注册后下载。

图书在版编目（CIP）数据

物联网系统综合开发与应用 / 廖建尚，杨尚森，潘必超编著. —北京：电子工业出版社，2020.4
（物联网开发与应用丛书）
ISBN 978-7-121-38944-3

Ⅰ. ①物… Ⅱ. ①廖… ②杨… ③潘… Ⅲ. ①互联网络－应用②智能技术－应用
Ⅳ. ①TP393.4②TP18

中国版本图书馆 CIP 数据核字（2020）第 060134 号

责任编辑：田宏峰
印　　刷：北京盛通数码印刷有限公司
装　　订：北京盛通数码印刷有限公司
出版发行：电子工业出版社
　　　　　北京市海淀区万寿路 173 信箱　邮编：100036
开　　本：787×1 092　1/16　印张：18.5　字数：470 千字
版　　次：2020 年 4 月第 1 版
印　　次：2024 年 2 月第10次印刷
定　　价：69.00 元

凡所购买电子工业出版社图书有缺损问题，请向购买书店调换。若书店售缺，请与本社发行部联系，联系及邮购电话：（010）88254888，88258888。

质量投诉请发邮件至 zlts@phei.com.cn，盗版侵权举报请发邮件至 dbqq@phei.com.cn。

本书咨询联系方式：tianhf@phei.com.cn。

FOREWORD 前言

近年来，物联网、移动互联网、大数据和云计算的迅猛发展，逐步改变了社会的生产方式，大大提高了生产效率和社会生产力。工业和信息化部发布的《物联网发展规划（2016—2020 年）》总结了“十二五”规划中物联网发展所获得的成就，并分析了“十三五”期间面临的形势，明确了物联网的发展思路和目标，提出了物联网发展的 6 大任务，分别是强化产业生态布局、完善技术创新体系、推动物联网规模应用、构建完善标准体系、完善公共服务体系、提升安全保障能力；提出了 4 大关键技术，分别是传感器技术、体系架构共性技术、操作系统，以及物联网与移动互联网、大数据融合关键技术；提出了 6 大重点领域应用示范工程，分别是智能制造、智慧农业、智能家居、智能交通和车联网、智慧医疗和健康养老，以及智慧节能环保；指出要健全多层次多类型的物联网人才培养和服务体系，支持高校、科研院所加强跨学科交叉整合，加强物联网学科建设，培养物联网复合型专业人才。该发展规划为物联网发展指出了一条鲜明的道路，同时也表明了我国在推动物联网应用方面的坚定决心，相信物联网的规模会越来越大。

物联网系统涉及的技术有很多，从感知层到应用层都有不同的开发技术，包括微处理器的接口驱动开发技术、传感器的驱动开发技术、无线通信技术、客户端（Web 端和 Android 端）开发技术等。本书基于短距离无线通信技术（ZigBee、BLE、Wi-Fi）和长距离无线通信技术（LoRa、NB-IoT、LTE），详细分析物联网系统的综合开发与应用。

全书采用案例式的讲解方法，利用贴近社会和生活的案例，由浅入深地介绍物联网系统的开发技术，每个案例均给出了完整的开发代码。

第 1 章简要介绍物联网基本概念与特征、物联网产业发展现状、物联网的应用前景，以及在物联网中常用的无线通信技术。

第 2 章介绍物联网的开发基础，主要内容包括物联网的开发平台、数据通信协议、智云平台的开发接口，以及 Android 端和 Web 端的开发接口及应用。

第 3 章通过基于 ZigBee 的城市环境信息采集系统、城市景观照明控制系统、智能燃气控制系统和家庭安防监控系统，全面介绍了 ZigBee 物联网系统的架构和应用，对每个开发案例均进行了系统架构分析与设计，设计了基于 ZigBee 的采集类传感器、控制类传感器和安防类传感器的驱动程序，实现了 Android 端和 Web 端的应用开发。

第 4 章通过基于 BLE 的家庭灯光控制系统和智能门禁管理系统，全面介绍了 BLE 物联网系统的架构和应用，对每个开发案例均进行了系统架构分析与设计，设计了基于 BLE 的控制类传感器和识别类传感器的驱动程序，实现了 Android 端和 Web 端的应用开发。

第 5 章通过基于 Wi-Fi 的楼宇消防控制系统和楼宇通风控制系统，全面介绍了 Wi-Fi

物联网系统的架构和应用，对每个开发案例均进行了系统架构分析与设计，设计了基于 Wi-Fi 的采集类传感器、控制类传感器和安防类传感器的驱动程序，实现了 Android 端和 Web 端的应用开发。

第 6 章通过基于 LoRa 的农业土壤调节系统和农业光照度调节系统，全面介绍了 LoRa 物联网系统的架构和应用，对每个开发案例均进行了系统架构分析与设计，设计了基于 LoRa 的采集类传感器、控制类传感器和安防类传感器的驱动程序，实现了 Android 端和 Web 端的应用开发。

第 7 章通过基于 NB-IoT 的停车收费管理系统和智能水表抄表系统，全面介绍了 NB-IoT 物联网系统的架构和应用，对每个开发案例均进行了系统架构分析与设计，设计了基于 NB-IoT 的采集类传感器、控制类传感器、安防类传感器和识别类传感器的驱动程序，实现了 Android 端和 Web 端的应用开发。

第 8 章通过基于 LTE 的仓库环境管理系统和自动化生产线计数系统，全面介绍了 LTE 物联网系统的架构和应用，对每个开发案例均进行了系统架构分析与设计，设计了基于 LTE 的识别类传感器、控制类传感器和安防类传感器的驱动程序，实现了 Android 端和 Web 端的应用开发。

本书在编写过程中，借鉴和参考了国内外专家、学者、技术人员的相关研究成果，我们尽可能按学术规范予以说明，但难免会有疏漏之处，在此谨向有关作者表示深深的敬意和谢意。如有疏漏，请及时通过出版社与我们联系。

本书的出版得到了广东省自然科学基金项目（2018A030313195）、广东省高校省级重大科研项目（2017GKTSCX021）、广东省科技计划项目（2017ZC0358）和广州市科技计划项目（201804010262）的资助。感谢中智讯（武汉）科技有限公司在本书编写过程中提供的帮助，特别感谢电子工业出版社有限公司的编辑在本书出版过程中给予的大力支持。

本书涉及的知识面较广，限于时间仓促，以及作者的水平和经验，疏漏之处在所难免，恳请专家和读者批评指正。

作　者

2020 年 2 月

CONTENTS 目录

第8章 LTE高级应用开发 ……253

参考文献 ……285

第1章 物联网及其无线通信技术

本章主要内容包括物联网概述和物联网无线通信技术。物联网短距离无线通信技术主要包括 ZigBee、低功耗蓝牙（BLE）和 Wi-Fi，物联网长距离无线通信技术主要包括 LoRa、NB-IoT 和 LTE。

1.1 物联网概述

1.1.1 物联网的基本概念与特征

物联网（Internet of Things）是指利用各种信息传感设备，如射频识别（RFID）装置、无线传感器、红外感应器、全球定位系统、激光扫描器等对现有物品信息进行感知、采集，通过网络支撑下的可靠传输技术，将各种物品信息汇入互联网，并进行基于海量信息资源的智能决策、安全保障及管理技术与服务的全球公共的信息综合服务平台。

物联网有两层含义：第一，物联网的核心和基础仍然是互联网，是在互联网基础上进行延伸和扩展而形成的网络；第二，物联网的用户端延伸和扩展到了任何物品，并且可以在物品之间进行信息交换和通信。因此，物联网是指运用传感器、射频识别（RFID）、智能嵌入式等技术，使信息传感设备可以感知任何需要的信息，按照约定的协议，通过可能的网络（如 Wi-Fi、3G、4G）把物体与互联网连接起来进行信息交换，在物与物、物与人的泛在连接基础上，实现对物体的智能识别、定位、跟踪、控制和管理。物联网的架构如图 1.1 所示。

作为新一代信息技术的重要组成部分，物联网技术有三方面的特征：第一，物联网技术具有互联网的特征，接入物联网的物体要由能够实现互连互通的网络来支撑；第二，物联网技术具有识别与通信的特征，接入物联网的物体要具备自动识别和物物通信的功能；第三，物联网技术具有智能化的特征，基于物联网技术构造的网络应该具有自动化、自我反馈和智能控制的功能。

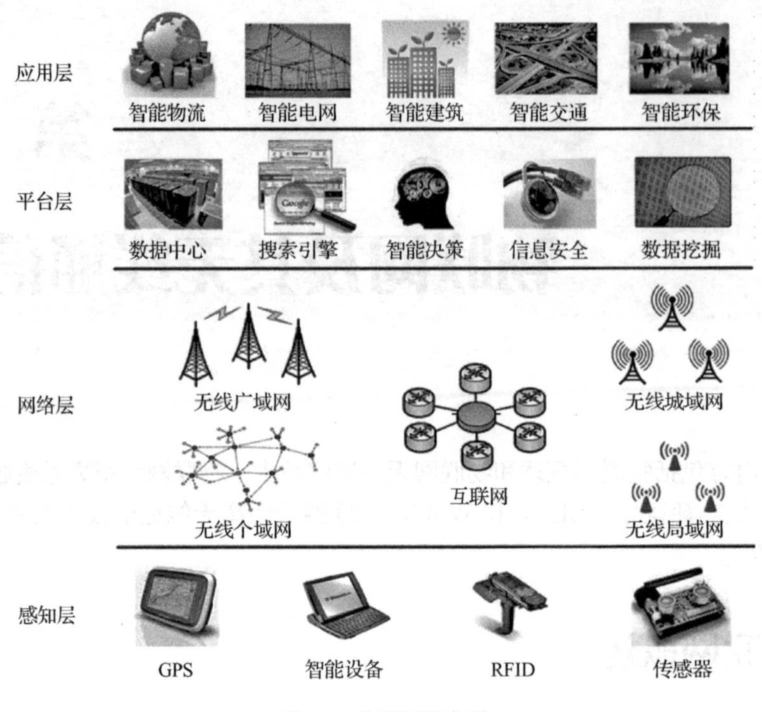

图 1.1 物联网的架构

1.1.2 我国物联网产业发展现状

2010 年 10 月，《国民经济和社会发展第十二个五年规划纲要》出台，指出战略性新兴产业是国家未来重点扶持的对象，下一代通信网络、物联网、三网融合、新型平板显示、高性能集成电路和高端软件等范畴的新一代信息技术产业将是未来扶持的重点。工业和信息化部印发的《物联网“十二五”发展规划》将以下 9 个方面纳入重点发展领域。

（1）智能工业：生产过程控制、生产环境监测、制造供应链跟踪、产品全生命周期监测，促进安全生产和节能减排。

（2）智能农业：农业资源利用、农业生产精细化管理、生产养殖环境监控、农产品质量安全管理与产品溯源。

（3）智能物流：建设库存监控、配送管理、安全追溯等现代流通应用系统，建设跨区域、行业、部门的物流公共服务平台，实现电子商务与物流配送一体化管理。

（4）智能交通：交通状态感知与交换、交通诱导与智能化管控、车辆定位与调度、车辆远程监测与服务、车路协同控制，建设开放的综合智能交通平台。

（5）智能电网：电力设施监测、智能变电站、配网自动化、智能用电、智能调度、远程抄表，建设安全、稳定、可靠的智能电网。

（6）智能环保：污染源监控、水质监测、空气监测、生态监测，建立智能环保信息采集网络和信息平台。

（7）智能安防：社会治安监控，危险化学品运输监控，食品安全监控，重要桥梁、建筑、轨道交通、水利设施、市政管网等基础设施安全监测、预警和应急联动。

（8）智能医疗：药品流通和医院管理，以人体生理和医学参数采集及分析为切入点，面向家庭和社区开展远程医疗服务。

（9）智能家居：家庭网络、家庭安防、家电智能控制、能源智能计量、节能低碳、远程教育等。

物联网应用已进入到实质性的推进阶段。中国信息通信研究院发布的一系列物联网白皮书列出了很多应用领域的例子，涉及工业、农业、交通、M2M、智能电网等多个领域。

1.1.3 物联网的应用前景

工业和信息化部印发的《信息通信行业发展规划（2016—2020 年）》明确提出要大力推动物联网规模应用。

（1）大力发展物联网与制造业的融合应用。围绕重点行业的制造单元、生产线、车间、工厂建设等关键环节进行数字化、网络化、智能化改造，推动生产制造全过程、全产业链、产品全生命周期的深度感知、动态监控、数据汇聚和智能决策。通过对现场级工业数据的实时感知与高级建模分析，形成智能决策与控制。完善工业云与智能服务平台，提升工业大数据开发利用水平，实现工业体系个性化定制、智能化生产、网络化协同和服务化转型，加快智能制造试点示范，开展信息物理系统、工业互联网在离散与流程制造行业的广泛部署应用，初步形成跨界融合的制造业新生态。

（2）加快物联网与行业领域的深度融合。面向农业、物流、能源、环保、医疗等重要领域，组织实施行业重大应用示范工程，推进物联网集成创新和规模化应用，支持物联网与行业深度融合。实施农业物联网区域试验工程，推进农业物联网应用，提高农业智能化和精准化水平。深化物联网在仓储、运输、配送、港口等物流领域的规模应用，支撑多式联运，构建智能高效的物流体系。加大物联网在污染源监控和生态环境监测等方面的推广应用，提高污染治理和环境保护水平。深化物联网在电力、油气、公共建筑节能等能源生产、传输、存储、消费等环节应用，提升能源管理智能化和精细化水平，提高能源利用效率。推动物联网技术在药品流通和使用、病患看护、电子病历管理等领域中的应用，积极推动远程医疗、临床数据应用示范等医疗应用。

（3）推进物联网在消费领域的应用创新。鼓励物联网技术创新、业务创新和模式创新，积极培育新模式新业态，促进车联网、智能家居、健康服务等消费领域应用快速增长。加强车联网技术创新和应用示范，发展车联网自动驾驶、安全节能、地理位置服务等应用。推动家庭安防、家电智能控制、家居环境管理等智能家居应用的规模化发展，打造繁荣的智能家居生态系统。发展社区健康服务物联网应用，开展基于智能可穿戴设备远程健康管理、老人看护等健康服务，推动健康大数据创新应用和服务发展。

（4）深化物联网在智慧城市领域的应用。推进物联网感知设施规划布局，结合市政设施、通信网络设施以及行业设施建设，同步部署视频采集终端、RFID 标签、多类条码、复合传感器节点等多种物联网感知设施，深化物联网在地下管网监测、消防设施管理、城市用电平衡管理、水资源管理、城市交通管理、电子政务、危化品管理和节能环保等重点领域的应用。建立城市级物联网接入管理与数据汇聚平台，推动感知设备统一接入、集中管理和数据共享利用。建立数据开放机制，制定政府数据共享开放目录，推进数据资源向社会开放，鼓励和引导企业、行业协会等开放和交易数据资源，深化政府数据和社会数据融合利用。支持建立数据共享服务平台，提供面向公众、行业和城市管理的智能信息服务。

1.2 物联网无线通信技术

常用的物联网无线通信技术可分为短距离无线通信技术和长距离无线通信技术，分析如下。

1.2.1 短距离无线通信技术

短距离无线通信主要特点是通信距离短，覆盖范围一般在几十米或上百米之内，发射器的发射功率较低，一般小于 100 mW。短距离无线通信技术的特征是低成本、低功耗和对等通信。

（1）ZigBee。ZigBee 是 IEEE 802.15.4 标准的代名词，是符合这个标准的一种短距离、低功耗的无线通信技术。使用 ZigBee 的设备能耗特别低，在进行自组网时无须人工干预，成本低廉，设备复杂度低且网络容量大。ZigBee 本身是针对低数据量、低成本、低功耗、高可靠性的无线数据通信的需求而产生的，在多方面领域有广泛应用，如国防安全、工业应用、交通物流、节能、生产现代化和智能家居等。

（2）低功耗蓝牙（BLE）。BLE 在传统蓝牙的基础上，对传统蓝牙的协议栈做了进一步简化，将数据传输速率和功耗作为主要的技术指标。BLE 可通过单模形式和双模形式来实现，双模形式是指在蓝牙芯片中集成了传统蓝牙协议栈和低功耗蓝牙协议栈，实现了两种蓝牙协议栈的共用；单模形式是指在蓝牙芯片集成了低功耗蓝牙协议栈，它是对传统蓝牙协议栈的简化，从而降低了功耗，提高了数据传输速率。

（3）Wi-Fi。Wi-Fi 是 IEEE 802.11 标准的别名，工作频率为 2.4～2.48 GHz，许多终端设备（如笔记本电脑、视频游戏机、智能手机、数码相机、平板电脑等）都配有 Wi-Fi 模块。通过 Wi-Fi 可以方便快捷地连网，可以使用户摆脱传统的有线连网的束缚。

1.2.2 长距离无线通信技术

（1）LoRa。LoRa 是一种基于 Sub-GHz 技术的无线网络，其特点是传输距离远，易于建设和部署，功耗和成本低，适用于大范围环境的数据采集。

（2）NB-IoT。NB-IoT 是基于蜂窝网络构建的，可直接部署于 GSM、UMTS 或 LTE 网络，由运营商提供连接服务，其特点是覆盖广泛、功耗极低。

（3）LTE。LTE 采用 FDD 和 TDD 技术，其特点是数据传输速率快、容量大、覆盖范围广、移动性好，具有一定的空间定位功能。

第2章 物联网开发基础

本章结合物联网学习平台了解物联网的学习路线、开发环境、应用场景。共分5个模块：

（1）物联网开发平台，学习物联网开发平台的基本构成，详细分析了智能网关和 xLab 未来开发平台。

（2）物联网数据通信协议，学习智云 ZXBee 数据通信协议基本构成和特点，分析数据通信协议的参数定义。

（3）智云平台应用开发接口，学习智云平台的编程接口，分析 Android 端应用开发接口和 Web 端应用开发接口。

（4）Android 端应用开发实例，通过实例学习 Android 端应用开发接口。

（5）Web 端应用开发实例，通过实例学习 Web 端应用开发接口。

2.1 物联网开发平台

本书的物联网开发平台采用智云物联平台（也称为智云平台），一个基本的智云物联项目系统模型如图 2.1 所示，其特点如下：

图 2.1 智云物联项目系统模型

（1）可通过 ZigBee、BLE、Wi-Fi、NB-IoT、LoRa 和 LTE 等无线网络将各种智能设备连接起来，其中的协调器/汇集器节点作为整个网络的汇集中心。

（2）协调器/汇集器节点与智能网关进行交互，通过在智能网关上运行的服务程序，将无线网络与电信网和移动网连接起来，同时将数据推送给智云中心，也可将数据推送到本地局域网。

（3）智云平台提供数据的存储服务、数据推送服务、自动控制服务等深度的项目接口，本地服务仅支持数据的推送服务。

（4）物联网应用项目通过智云 API 进行具体应用的开发，能够实现对无线网络内的节点进行采集、控制、决策等。

2.1.1 物联网中常用开发硬件

智云平台支持各种智能设备的接入，常见的硬件模型如图 2.2 所示。

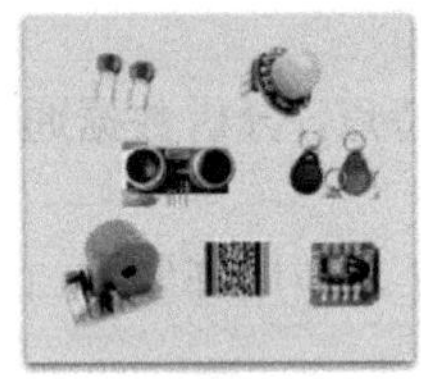

传感器

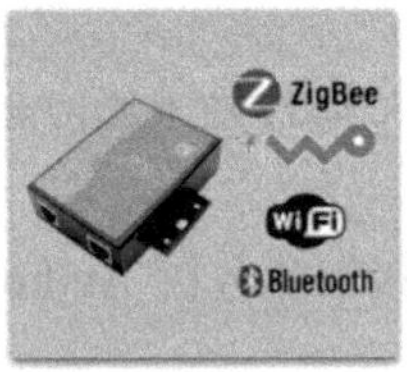

节点

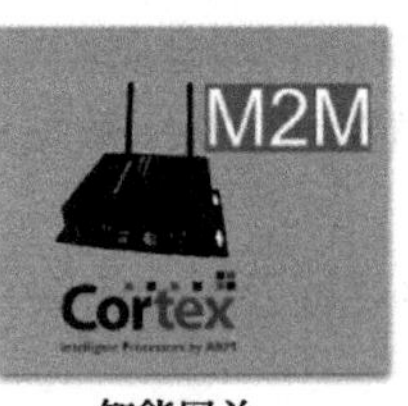

智能网关

智云服务器

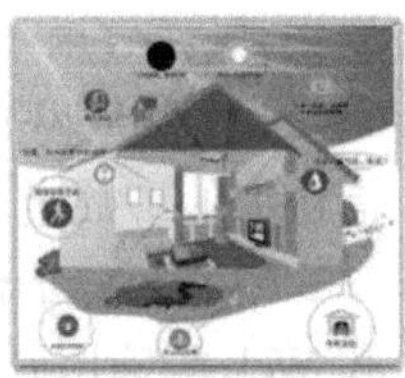

应用终端

图 2.2　常见的硬件模型

（1）传感器：主要用于采集物理世界中信息，包括各类物理量、标识、音频、视频等。

（2）节点：通常基于单片机、ARM 等构建，具有传感器的数据采集、传输、组网等功能，能够构建无线网络。

（3）智能网关：实现无线网络与互联网的数据交互，支持 ZigBee、Wi-Fi、BLE、LoRa、NB-IoT、LTE 等无线网络的数据解析，支持网络路由转发，可实现 M2M 数据交互。

（4）智云服务器：负责对物联网中的海量数据进行处理，采用云计算、大数据技术实现数据的存储、分析、计算、挖掘和推送，并采用统一的开放接口为上层应用提供数据服务。

（5）应用终端：通常是运行物联网应用的移动终端，如智能手机、平板电脑等设备。

2.1.2 智能网关

智能网关（基于 Android 系统构建，也称为 Android 网关）采用基于 ARM Cortex-A9 内核的 S5P4418 四核微处理器，板载 10.1 英寸的电容触摸液晶屏，集成了 Wi-Fi 无线模组、BLE 无线模组、MIPI 高清摄像头模块，可选北斗 GPS 模块、4G 模块。智能网关如图 2.3 所示。

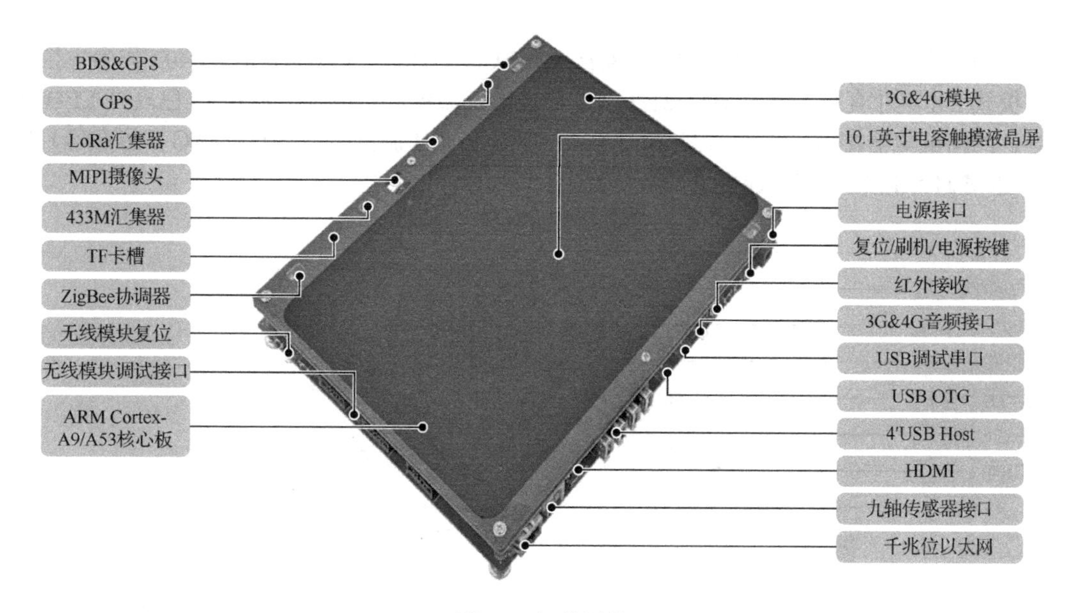

图 2.3　智能网关

2.1.3　xLab 未来开发平台

本书采用的 xLab 未来开发平台提供了经典型无线节点 ZXBeeLiteB 和增强型无线节点 ZXBeePlusB，集成了锂电池供电接口、调试接口、外设控制电路、RJ45 传感器接口等。

经典型无线节点 ZXBeeLiteB 采用 CC2530 作为主控制器，板载信号指示灯包括电源、电池、网络、数据，具有两路功能按键，集成锂电池接口和电源管理芯片，支持电池的充电管理和电量测量，集成 USB 调试串口、TI JTAG 接口和 ARM JTAG 接口，集成两路 RJ45 工业接口，提供主芯片 P0_0～P0_7 输出（包含 I/O、DC 3.3 V、DC 5 V、UART、RS-485），提供两路继电器接口，提供两路 3.3 V、5 V、12 V 电源输出。经典型无线节点 ZXBeeLiteB 如图 2.4 所示。

图 2.4　经典型无线节点 ZXBeeLiteB

增强型无线节点 ZXBeePlusB 采用基于 ARM Cortex-M4 内核的 STM32F407 作为主控制器，板载 2.8 英寸真彩 LCD、HTU21D 型高精度数字温湿度传感器、RGB 三色高亮 LED 指示灯、两路继电器接口、蜂鸣器接口、摄像头接口、USB 调试串口、TI JTAG（仿真器）接口、ARM JTAG 接口、以太网接口等，如图 2.5 所示。

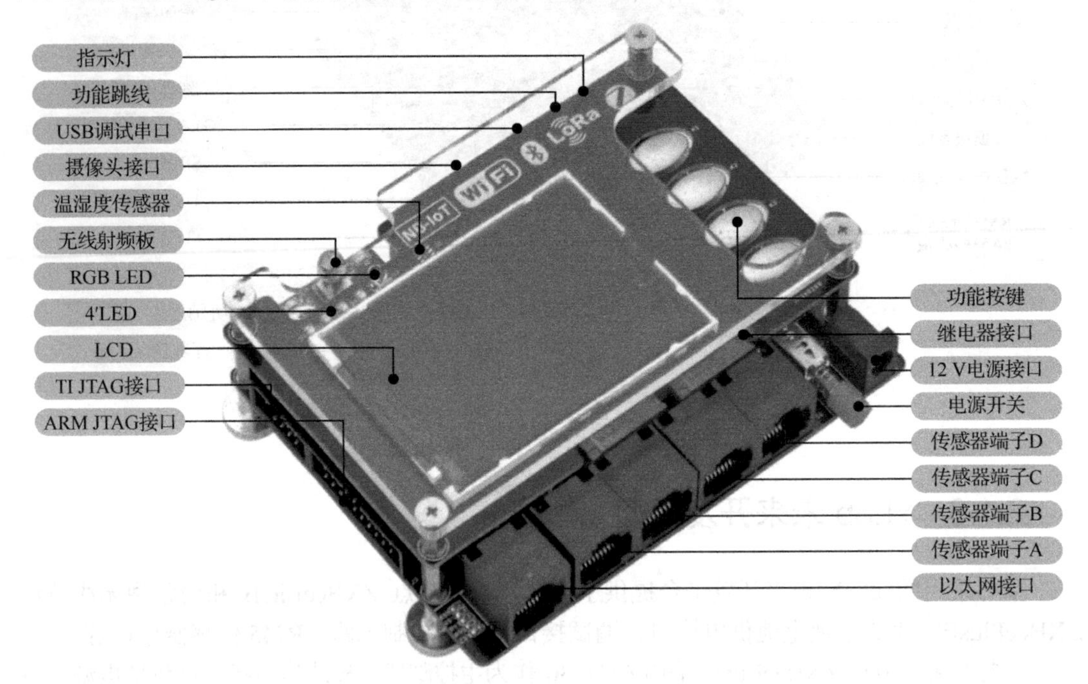

图 2.5　增强型无线节点 ZXBeePlusB

xLab 未来开发平台支持多种无线模组，包括 ZigBee、BLE、Wi-Fi、LoRa、NB-IoT、LTE。无线模组功能如表 2.1 所示。

表 2.1　无线模组功能一览表

无 线 模 组	产 品 图 片	功 能 描 述
ZigBee 无线模组		（1）TI 公司 CC2530 ZigBee 无线芯片，高性能、低功耗的 8051 微处理器内核，适应 2.4 GHz IEEE 802.15.4 的 RF 收发器。 （2）SMA 胶棒天线，数据传输速率达 250 kbps，传输距离可达 200 m
BLE 无线模组		（1）TI 公司 CC2540 BLE 无线芯片，高性能、低功耗的 8051 微处理器内核，适应 2.4 GHz BLE 的 RF 收发器。 （2）SMA 胶棒天线，数据传输速率达 1 Mbps，传输距离可达 100 m
Wi-Fi 无线模组		（1）TI 公司 CC3200 Wi-Fi 无线芯片，内置工业级低功耗 ARM Cortex-M4 微处理器内核，主频为 80 MHz，支持 IEEE 802.11b/g/n 标准，内置强大的加密引擎。 （2）内置 TCP/IP 和 TLS/SSL 协议栈，支持 Http、Server 等多种协议。 （3）板载陶瓷天线，支持主从操作模式，数据传输速率可达 400 kbps

续表

无 线 模 组	产 品 图 片	功 能 描 述
LoRa 无线模组		（1）Semtech 公司 SX1278 LoRa 无线芯片，采用 LoRa 扩频调制技术，工作频率为 410～525 MHz，灵敏度为-148 dBm，输出功率为 20 dBm。 （2）集成 STM32F103 微处理器，采用 Contiki 操作系统。 （3）SMA 胶棒天线，超远通信距离，可达 3 km
NB-IoT 无线模组		（1）BC95 NB-IoT 无线芯片，采用华为 Hi2110 芯片组，支持电信网络，频段为 850 MHz，支持 3GPP Rel-13 以及增强型 AT 指令，数据传输速率可达 100 kbps，灵敏度为-129 dBm，输出功率为 23 dBm。 （2）集成 STM32F103 微处理器，采用 Contiki 操作系统。 （3）SMA 胶棒天线，采用标准 SIM 卡槽
LTE 无线模组		（1）EC20 4G&3G&2G 三合一无线模组，支持 LTE、WCDMA、GPRS 数据传输，支持联通网络，频段为 GSM900/DCS1800、HSUPA、HSDPA 3GPP R5、WCDMA 3GPP R99 EDGE EGPRS Class12、TDD-LTE　Band38/39/40/41、FDD-LTE Band1/3/7、TDS Band34/39、GSM Band2/3/8，支持提供 UART 和 USB 双通道接口，下行数据传输速率为 150 Mbps，上行数据传输速率为 50 Mbps。 （2）集成 STM32F103 微处理器，采用 Contiki 操作系统。 （3）SMA 胶棒天线，采用标准 SIM 卡槽

为了深化在无线网络中对节点的使用，本书的项目实例均用到了传感器和控制设备。xLab 未来开发平台按照传感器类别设计了丰富的传感设备，涉及采集类、控制类、安防类、显示类、识别类、创意类等开发平台。本书实例使用采集类开发平台（Sensor-A）、控制类开发平台（Sensor-B）和安防类开发平台（Sensor-C）。

1. 采集类开发平台（Sensor-A）

采集类开发平台包括：温湿度传感器、光照度传感器、空气质量传感器、气压海拔传感器、三轴加速度传感器、距离传感器、继电器、语音识别传感器等，如图 2.6 所示。

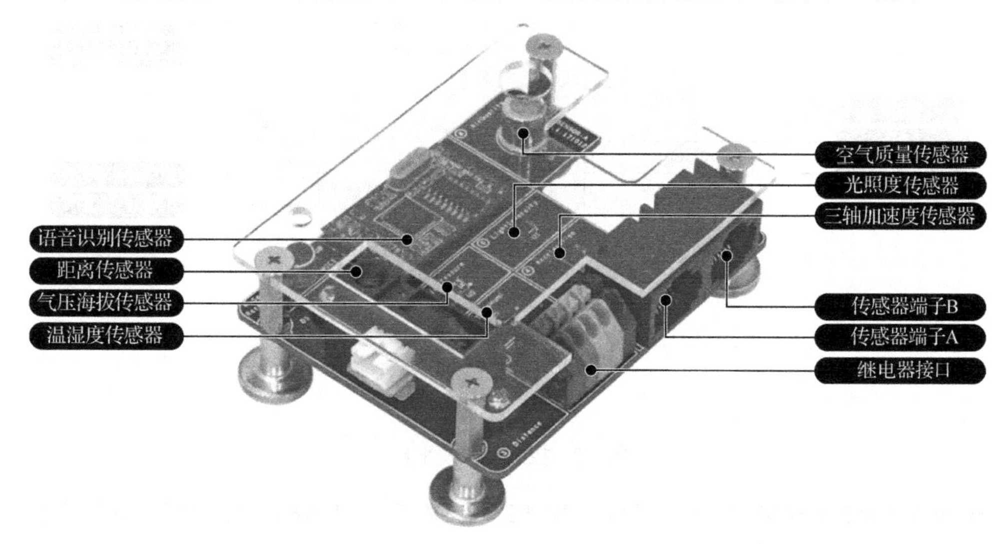

图 2.6　采集类开发平台

- 两路 RJ45 工业接口，包含 I/O、DC 3.3 V、DC 5 V、UART、RS-485、两路继电器输出等功能，提供两路 3.3 V、5 V、12 V 电源输出。
- 采用磁吸附设计，可通过磁力吸附并通过 RJ45 工业接口接入节点进行数据通信。
- 温湿度传感器的型号为 HTU21D，采用数字信号输出和 IIC 总线通信接口，测量范围为-40～125℃，以及 5%～95%RH。
- 光照度传感器的型号为 BH1750FVI-TR，采用数字信号输出和 IIC 总线通信接口，对应广泛的输入光范围，相当于 1～65535 lx。
- 空气质量传感器的型号为 MP503，采用模拟信号输出，可以监测气体酒精、烟雾、异丁烷、甲醛，监测浓度为 10～1000 ppm（酒精）。
- 气压海拔传感器的型号为 FBM320，采用数字信号输出和 IIC 总线通信接口，测量范围为 300～1100 hPa。
- 三轴加速度传感器的型号为 LIS3DH，采用数字信号输出和 IIC 总线通信接口，量程可设置为±2*g*、±4*g*、±8*g*、±16*g*（*g* 为重力加速度），16 位数据输出。
- 距离传感器的型号为 GP2D12，采用模拟信号输出，测量范围为 10～80 cm，更新频率为 40 ms。
- 采用继电器控制，输出节点有两路继电器接口，支持 5 V 电源开关控制。
- 语音识别传感器的型号为 LD3320，支持非特定人识别，具有 50 条识别容量，返回形式丰富，采用串口通信。

2. 控制类开发平台（Sensor-B）

控制类开发平台包括：风扇、步进电机、蜂鸣器、LED、RGB 灯、继电器接口，如图 2.7 所示。

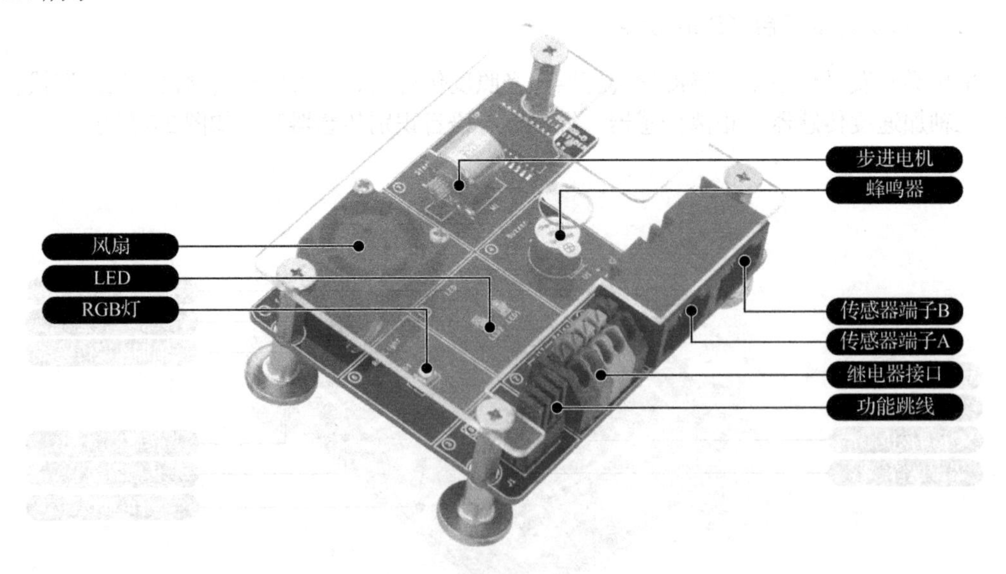

图 2.7　控制类开发平台

- 两路 RJ45 工业接口，包含 I/O、DC 3.3 V、DC 5 V、UART、RS-485、两路继电器输出等功能，提供两路 3.3 V、5 V、12 V 电源输出。
- 采用磁吸附设计，可通过磁力吸附并通过 RJ45 工业接口接入节点进行数据通信。

- 风扇为小型风扇，采用低电平驱动。
- 步进电机为小型42步进电机，驱动芯片为A3967SLB，逻辑电源电压范围为3.0～5.5 V。
- 使用小型蜂鸣器，采用低电平驱动。
- 两路高亮LED，采用低电平驱动。
- RGB灯采用低电平驱动，可组合出任何颜色。
- 采用继电器控制，输出节点有两路继电器接口，支持5 V电源开关控制。

3. 安防类开发平台（Sensor-C）

安防类开发平台包括：火焰传感器、光栅传感器、人体红外传感器、燃气传感器、触摸传感器、振动传感器、霍尔传感器、继电器接口、语音合成传感器等，如图2.8所示。

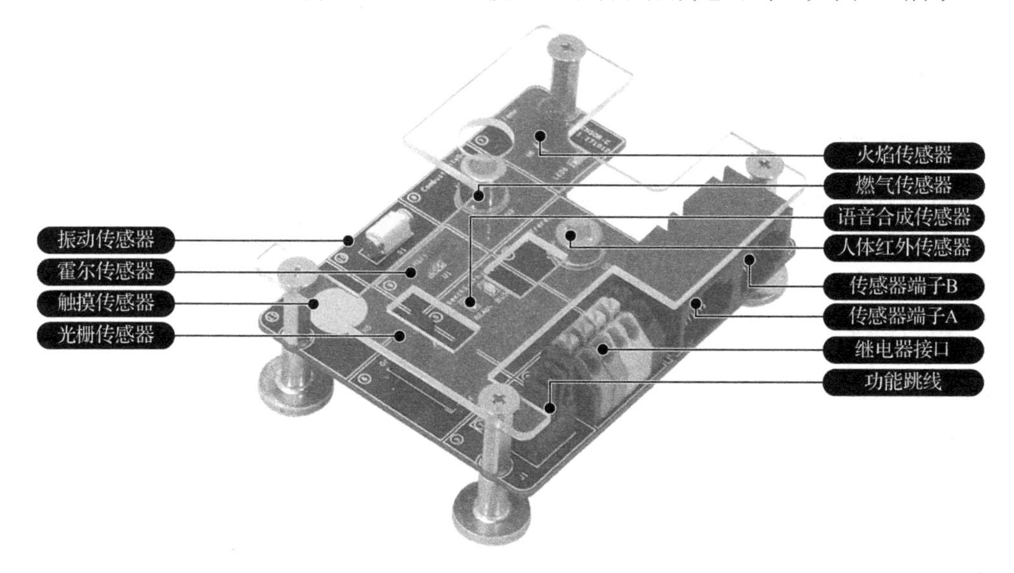

图2.8 安防类开发平台

- 两路RJ45工业接口，包含I/O、DC 3.3 V、DC 5 V、UART、RS-485、两路继电器输出等功能，提供两路3.3 V、5 V、12 V电源输出。
- 采用磁吸附设计，可通过磁力吸附并通过RJ45工业接口接入节点进行数据通信。
- 火焰传感器采用5 mm的探头，可监测火焰或波长为760～1100 nm的光源，探测温度为60℃左右，采用数字开关量输出。
- 光栅传感器的槽式光耦槽宽为10 mm，工作电压为5 V，采用数字开关量信号输出。
- 人体红外传感器的型号为AS312，电源电压为3 V，感应距离为12 m，采用数字开关量信号输出。
- 燃气传感器的型号为MP-4，采用模拟信号输出，传感器加热电压为5 V，供电电压为5 V，可测量天然气、甲烷、瓦斯、沼气等。
- 触摸传感器的型号为SOT23-6，采用数字开关量信号输出，当监测到触摸时，输出电平翻转。
- 振动传感器在低电平时有效，采用数字开关量信号输出。
- 霍尔传感器的型号为AH3144，电源电压为5 V，采用数字开关量输出，工作频率为0～100 kHz。

● 采用继电器控制，输出节点有两路继电器接口，支持 5 V 电源开关控制。
● 语音合成传感器的型号为 SYN6288，采用串口通信，支持 GB2312、GBK、UNICODE 等编码，可设置音量、背景音乐等。

2.2 物联网数据通信协议

一个完整的物联网综合系统，数据贯穿了感知层、网络层、服务层和应用层的各个部分，数据在这四个层之间层层传递，这就需要构建完整的物联网数据通信协议。本书中的物联网数据通信协议采用 ZXBee 数据通信协议。

2.2.1 ZXBee 数据通信协议

1. 数据通信协议的格式

数据通信协议的格式为“{[参数]=[值],[参数]=[值]…}”。

● 每条数据以“{”作为起始字符；
●“{}”内的多个参数以“,”分隔。

例如，{CD0=1,D0=?}。

2. 数据通信协议参数说明

数据通信协议参数说明如下。

（1）参数名称定义如下。

● 变量：A0～A7、D0、D1、V0～V3。
● 指令：CD0、OD0、CD1、OD1。
● 特殊参数：ECHO、TYPE、PN、PANID、CHANNEL。

（2）可以对变量的值进行查询，如“{A0=?}”。

（3）变量 A0～A7 在数据中心中可以保存为历史数据。

（4）指令是对位进行操作的。

具体参数解释如下。

（1）A0～A7：用于传递传感器数据及其携带的信息，只能通过“?”来进行查询当前变量的值，支持上传到物联网云数据中心存储，示例如下。

● 温湿度传感器用 A0 表示温度值，用 A1 表示湿度值，数据类型为浮点型，精度为 0.1。
● 火焰报警传感器用 A3 表示警报状态，数据类型为整型，取值为 0（未监测到火焰）或者 1（监测到火焰）。
● 高频 RFID 模块用 A0 表示卡片 ID，数据类型为字符串型。

ZXBee 数据通信协议的格式为“{参数=值,参数=值…}”，即用一对大括号“{}”包含每条数据，“{ }”内参数如果有多个条目，则用“,”进行分隔，例如，{CD0=1,D0=?}。

（2）D0：D0 中的 Bit0～Bit7 分别对应 A0～A7 的状态（是否主动上传状态），只能通过“?”来查询当前变量的值，0 表示禁止上传，1 表示允许主动上传，示例如下。

● 温湿度传感器用 A0 表示温度值，用 A1 表示湿度值，D0=0 表示不上传温度值和湿度值，D0=1 表示主动上传温度值，D0=2 表示主动上传湿度值，D0=3 表示主动上传温度值和湿度值。

- 火焰报警传感器用 A0 表示警报状态，D0=0 表示不监测火焰，D0=1 表示实时监测火焰。
- 高频 RFID 模块用 A0 表示卡片 ID，D0=0 表示不上报 ID，D0=1 表示上报 ID。

（3）CD0/OD0：对 D0 的位进行操作，CD0 表示位清 0 操作，OD0 表示位置 1 操作，示例如下。

- 温湿度传感器用 A0 表示温度值，用 A1 表示湿度值，CD0=1 表示关闭温度值的主动上报。
- 火焰报警传感器用 A0 表示警报状态，OD0=1 表示开启火焰报警监测，当有火焰报警时，会主动上报 A0 的值。

（4）D1：D1 表示控制编码，只能通过"?"来查询当前变量的数值，用户可根据传感器属性来自定义功能，示例如下。

- 温湿度传感器：D1 的 Bit0 表示电源开关状态。例如，D1=0 表示电源处于关闭状态，D1=1 表示电源处于打开状态。
- 继电器：D1 的位表示各路继电器的状态。例如，D1=0 表示关闭继电器 S1 和 S2，D1=1 表示仅开启继电器 S1，D1=2 表示仅开启继电器 S2，D1=3 表示开启继电器 S1 和 S2。
- 风扇：D1 的 Bit0 表示电源开关状态，Bit1 表示正转或反转。例如，D1=0 或者 D1=2 表示风扇停止转动（电源断开），D1=1 表示风扇处于正转状态，D1=3 表示风扇处于反转状态。
- 红外电器遥控：D1 的 Bit0 表示电源开关状态，Bit1 表示工作模式/学习模式。例如，D1=0 或者 D1=2 表示电源处于关闭状态，D1=1 表示电源处于开启状态且为工作模式，D1=3 表示电源处于开启状态且为学习模式。

（5）CD1/OD1：对 D1 的位进行操作，CD1 表示位清 0 操作，OD1 表示位置 1 操作。

（6）V0～V3：用于表示传感器的参数，用户可根据传感器属性自定义功能，权限为可读写，示例如下。

- 温湿度传感器：V0 表示主动上传数据的时间间隔。
- 风扇：V0 表示风扇转速。
- 红外电器遥控：V0 表示学习的键值。
- 语音合成传感器：V0 表示需要合成的语音字符。

（7）特殊参数：ECHO、TYPE、PN、PANID、CHANNEL。

- ECHO：用于监测节点在线的指令，将发送的值进行回显。例如，发送"{ECHO=test}"，若节点在线则回复"{ECHO=test}"。
- TYPE：表示节点类型，该信息包含了节点类别、节点类型、节点名称，只能通过"?"来查询当前值。TYPE 的值由 5 个字节表示（ASCII 码），例如，1 1 001，第 1 个字节表示节点类别（1 表示 ZigBee、2 表示 RF433、3 表示 Wi-Fi、4 表示 BLE、5 表示 IPv6、9 表示其他）；第 2 个字节表示节点类型（0 表示汇集节点、1 表示路由/中继节点、2 表示终端节点）；第 3～5 个字节合起来表示节点名称（编码由开发者自定义）。
- PN（仅针对 ZigBee、IEEE 802.15.4 IPv6 节点）：表示上行节点地址信息和所有邻居节点地址信息，只能通过"?"来查询当前值。PN 的值为上行节点地址和所有邻居节点地址的组合，其中每 4 个字节表示一个节点地址后 4 位，第 1 个 4 字节上行节点后

4 位，第 2～*n* 个 4 字节表示其所有邻居节点地址后 4 位。

- PANID：表示节点组网的标志 ID，权限为可读写，此处 PANID 的值为十进制，而底层代码定义的 PANID 的值为十六进制，需要自行转换。例如，8200（十进制）= 0x2008（十六进制），通过指令"{PANID=8200}"可将节点的 PANID 修改为 0x2008。PANID 的取值范围为 1～16383。
- CHANNEL：表示节点组网的通信通道，权限为可读写，此处 CHANNEL 的取值范围为 11～26（十进制）。例如，通过指令"{CHANNEL=11}"可将节点的 CHANNEL 修改为 11。

2.2.2 数据通信协议参数定义

xLab 未来开发平台传感器的 ZXBee 数据通信协议参数定义如表 2.2 所示。

表 2.2 ZXBee 数据通信协议参数定义

开发平台	属　性	参　数	权 限	说　明
Sensor-A（601）	温度	A0	R	温度值为浮点型数据，精度为 0.1，范围为-40.0～105.0，单位为℃
	湿度	A1	R	湿度值为浮点型数据，精度为 0.1，范围为 0～100，单位为%
	光照度	A2	R	光照度值为浮点型数据，精度为 0.1，范围为 0～65535，单位为 lx
	空气质量	A3	R	空气质量，表示空气污染程度
	气压	A4	R	气压值，浮点型数据，精度为 0.1，单位为百帕
	三轴（跌倒状态）	A5	—	三轴：通过计算上报跌倒状态，1 表示跌倒（主动上报）
	距离	A6	R	距离（单位为 cm），浮点型数据，精度为 0.1 精度，范围为 20～80 cm
	语音识别返回码	A7	—	语音识别码，整型数据，范围为 1～49（主动上报）
	上报状态	D0(OD0/CD0)	R/W	D0 的 Bit0～Bit7 分别代表 A0～A7 的上报状态，1 表示允许上报
	继电器	D1(OD1/CD1)	R/W	D1 的 Bit6～Bit7 分别代表继电器 K1、K2 的状态，0 表示断开，1 表示吸合
	数据上报时间间隔	V0	R/W	循环上报的时间间隔
Sensor-B（602）	RGB 灯	D1(OD1/CD1)	R/W	D1 的 Bit0～Bit1 代表 RGB 灯的颜色状态，00 表示关闭 RGB 灯，01 表示红色（R），10 表示绿色（G），11 表示蓝色（B）
	步进电机	D1(OD1/CD1)	R/W	D1 的 Bit2 分别代表步进电机的正反转动状态，0 表示正转（5 s 后停止），1 表示反转（5 s 后反转）
	风扇/蜂鸣器	D1(OD1/CD1)	R/W	D1 的 Bit3 代表风扇/蜂鸣器的开关状态，0 表示关闭，1 表示打开
	LED	D1(OD1/CD1)	R/W	D1 的 Bit4、Bit5 代表 LED1、LED2 的开关状态，0 表示关闭，1 表示打开
	继电器	D1(OD1/CD1)	R/W	D1 的 Bit6、Bit7 分别代表继电器 K1、K2 的开关状态，0 表示断开，1 表示吸合
	数据上报时间间隔	V0	R/W	循环上报时间间隔

续表

开发平台	属　性	参　数	权　限	说　明
Sensor-C（603）	人体/触摸状态	A0	R	人体红外状态值，0或1变化；1表示监测到人体/触摸
	振动状态	A1	R	振动状态值，0或1变化；1表示监测到振动
	霍尔状态	A2	R	霍尔状态值，0或1变化；1表示监测到磁场
	火焰状态	A3	R	火焰状态值，0或1变化；1表示监测到明火
	燃气状态	A4	R	燃气泄漏状态值，0或1变化；1表示燃气泄漏
	光栅（红外对射）状态	A5	R	光栅状态值，0或1变化，1表示监测到阻挡
	上报状态	D0(OD0/CD0)	R/W	D0的Bit0～Bit5分别表示A0～A5的上报状态
	继电器	D1(OD1/CD1)	R/W	D1的Bit6～Bit7分别代表继电器K1、K2的开关状态，0表示断开，1表示吸合
	数据上报时间间隔	V0	R/W	循环上报时间间隔
	语音合成数据	V1	W	文字的Unicode编码
Sensor-D（604）	五向开关状态	A0	R	触发上报，状态值为：1（UP）、2（LEFT）、3（DOWN）、4（RIGHT）、5（CENTER）
	电视的开关	D1(OD1/CD1)	R/W	D1的Bit0代表电视开关状态，0表示关闭，1表示打开
	电视频道	V1	R/W	电视频道，范围为0～19
	电视音量	V2	R/W	电视音量，范围为0～99
Sensor-EL（605）	卡号	A0	—	字符串（主动上报，不可查询）
	卡类型	A1	R	整型数据，0表示125K，1表示13.56M
	卡余额	A2	R	整型数据，范围为0～800000，手动查询
	设备余额	A3	R	浮点型数据，设备金额
	设备单次消费金额	A4	R	浮点型数据，本次消费扣款金额
	设备累计消费	A5	R	浮点型数据，设备累计扣款金额
	门锁/设备状态	D1(OD1/CD1)	R/W	D1的Bit0～Bit1表示门锁、设备的开关状态，0表示关闭，1表示打开
	充值金额	V1	R/W	返回充值状态，0或1，1表示操作成功
	扣款金额	V2	R/W	返回扣款状态，0或1，1表示操作成功
	充值金额（设备）	V3	R/W	返回充值状态，0或1，1表示操作成功
	扣款金额（设备）	V4	R/W	返回扣款状态，0或1，1表示操作成功
Sensor-EH（606）	卡号	A0	—	字符串（主动上报，不可查询）
	卡余额	A2	R	整型数据，范围为0～800000，手动查询
	ETC杆开关	D1(OD1/CD1)	R/W	D1的Bit0表示ETC杆开关，0表示关闭，1表示抬起一次3s后自动关闭，同时将Bit0置0
	充值金额	V1	R/W	返回充值状态，0或1，1表示操作成功
	扣款金额	V2	R/W	返回扣款状态，0或1，1表示操作成功
Sensor-F（611）	GPS状态	A0	R	整型数据，0为不在线，1为在线
	GPS经纬度	A1	R	字符串型，形式为*a*&*b*，*a*表示经度，*b*表示维度，精度为0.000001

续表

开发平台	属　性	参　数	权　限	说　明
Sensor-F（611）	九轴计步数	A2	R	整型数据
		A3	R	加速度传感器 x、y、z 数据，格式为 x&y&z
	九轴传感器	A4	R	陀螺仪传感器 x、y、z 数据，格式为 x&y&z
		A5	R	地磁仪传感器 x、y、z 数据，格式为 x&y&z
	数据上报时间间隔	V0	R/W	传感器的循环上报时间间隔

2.3　智云平台应用开发接口

2.3.1　Android 端应用开发接口

智云平台提供了五个应用程序接口供开发者使用，包括实时连接（WSNRTConnect）、历史数据（WSNHistory）、摄像头（WSNCamera）、自动控制（WSNAutoctrl）、用户数据（WSNProperty），其框架如图 2.9 所示。

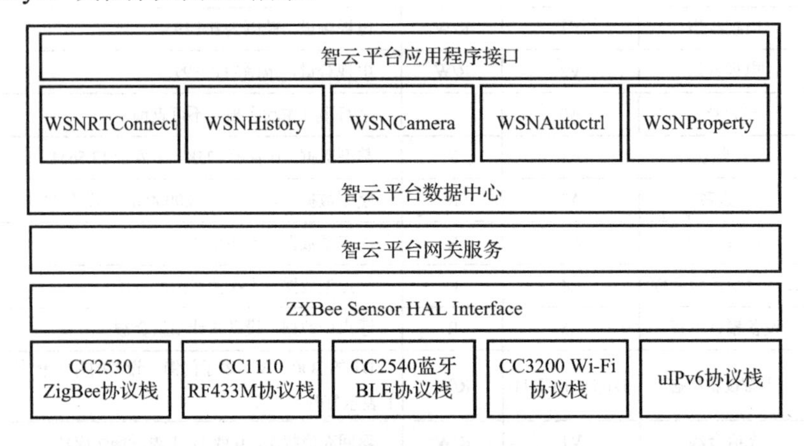

图 2.9　智云平台应用程序接口框架

针对 Android 端应用开发，智云平台提供了应用程序接口库 libwsnDroid2.jar，开发者只需要在编写 Android 端应用程序时，先导入该接口库，然后在代码中调用相应的方法即可。

1. 基于 Android 的实时连接接口

实时连接接口基于智云平台的消息推送服务，该服务是利用云端与客户端之间建立的稳定、可靠的长连接来向客户端应用推送实时消息的。智云平台的消息推送服务针对物联网的特征，支持多种推送类型，如传感器实时数据、执行控制指令、地理位置信息、SMS 等，同时提供关于用户信息及通知消息的统计信息，方便开发者进行后续开发及运营。基于 Android 的实时连接接口如表 2.3 所示。

表 2.3　基于 Android 的实时连接接口

函　　数	参数说明	功　　能
new WSNRTConnect(String myZCloudID, String myZCloudKey);	myZCloudID：智云账号。 myZCloudKey：智云密钥	创建实时数据，并初始化智云账号及密钥
connect()	无	建立实时数据服务连接
disconnect()	无	断开实时数据服务连接

setRTConnectListener(){
onConnect()
onConnectLost(Throwable arg0)
onMessageArrive(String mac, byte[] dat)
}
	mac：传感器的 MAC 地址。
	dat：发送的消息内容
		设置监听，接收实时数据服务推送的消息：
		onConnect：连接成功操作。
		onConnectLost：连接失败操作。
		onMessageArrive：数据接收操作

sendMessage(String mac, byte[] dat)	mac：传感器的 MAC 地址。 dat：发送的消息内容	发送消息
setServerAddr(String sa)	sa：数据中心服务器地址及端口	设置/改变数据中心服务器的地址及端口号
setIdKey(String myZCloudID, String myZCloudKey);	myZCloudID：账号。 myZCloudKey：密钥	设置/改变智云账号及密钥（需要重新断开连接）

2. 基于 Android 的历史数据接口

历史数据接口是基于智云平台数据中心提供的智云数据库接口开发的，智云数据库采用 Hadoop 后端分布式数据库集群，并且支持多机房自动冗余备份，以及自动读写分离，开发者不需要关注后端机器及数据库的稳定性、网络问题、机房灾难、单库压力等各种风险。传感器数据可以在智云数据库中永久保存，通过提供的应用程序接口可以完成与云存储服务器的数据连接、数据访问存储、数据使用等。基于 Android 的历史数据接口如表 2.4 所示。

表 2.4　基于 Android 的历史数据接口

函　　数	参数说明	功　　能
new WSNHistory(String myZCloudID, String myZCloudKey);	myZCloudID：智云账号。 myZCloudKey：智云密钥	初始化历史数据对象，并初始化智云账号及密钥
queryLast1H(String channel);	channel：传感器数据通道	查询最近 1 小时的历史数据
queryLast6H(String channel);	channel：传感器数据通道	查询最近 6 小时的历史数据
queryLast12H(String channel);	channel：传感器数据通道	查询最近 12 小时的历史数据
queryLast1D(String channel);	channel：传感器数据通道	查询最近 1 天的历史数据
queryLast5D(String channel);	channel：传感器数据通道	查询最近 5 天的历史数据
queryLast14D(String channel);	channel：传感器数据通道	查询最近 14 天的历史数据
queryLast1M(String channel);	channel：传感器数据通道	查询最近 1 个月（30 天）的历史数据
queryLast3M(String channel);	channel：传感器数据通道	查询最近 3 个月（90 天）的历史数据
queryLast6M(String channel);	channel：传感器数据通道	查询最近 6 个月（180 天）的历史数据
queryLast1Y(String channel);	channel：传感器数据通道	查询最近 1 年（365 天）的历史数据
query();	无	获取所有通道最后一次数据

续表

函　数	参数说明	功　能
query(String channel);	channel：传感器数据通道	获取该通道中最后一次数据
query(String channel, String start, String end);	channel：传感器数据通道。 start：起始时间。 end：结束时间。 时间为 ISO 8601 格式的日期，例如：2010-05-20T11:00:00Z	通过起止时间查询指定时间段的历史数据（根据时间范围默认选择时间间隔）
query(String channel, String start, String end, String interval);	channel：传感器数据通道。 start：起始时间。 end：结束时间。 interval：采样点的时间间隔。 时间为 ISO 8601 格式的日期，例如：2010-05-20T11:00:00Z	通过起止时间查询指定时间段、指定时间间隔的历史数据
setServerAddr(String sa)	sa：数据中心服务器地址及端口	设置/改变数据中心服务器地址及端口号
setIdKey(String myZCloudID, String myZCloudKey);	myZCloudID：智云账号。 myZCloudKey：智云密钥	设置/改变智云账号及密钥

3．基于 Android 的摄像头接口

智云平台提供了对摄像头进行远程控制的接口，支持远程对视频、图像进行实时采集、图像抓拍、控制云台转动等操作。基于 Android 的摄像头接口如表 2.5 所示。

表 2.5　基于 Android 的摄像头接口

函　数	参数说明	功　能
new WSNCamera(String myZCloudID, String myZCloudKey);	myZCloudID：账号。 myZCloudKey：密钥	初始化摄像头对象，并初始化智云账号及密钥
initCamera(String myCameraIP, String user, String pwd, String type);	myCameraIP：摄像头外网域名和 IP 地址。 user：摄像头用户名。 pwd：摄像头密码。 type：摄像头类型。 以上参数可从摄像头手册获取	设置摄像头域名、用户名、密码、类型等参数
openVideo();	无	打开摄像头
closeVideo();	无	关闭摄像头
control(String cmd);	cmd：云台控制指令，参数如下： UP：向上移动一次。 DOWN：向下移动一次。 LEFT：向左移动一次。 RIGHT：向右移动一次。 HPATROL：水平巡航转动。 VPATROL：垂直巡航转动。 360PATROL：360°巡航转动	发送指令控制云台转动
checkOnline();	无	监测摄像头是否在线

续表

函　　数	参 数 说 明	功　　能
snapshot();	无	抓拍照片

setCameraListener(){
onOnline(String myCameraIP, boolean online)
onSnapshot(String myCameraIP, Bitmap bmp)
onVideoCallBack(String myCameraIP, Bitmap bmp)
}
	myCameraIP：摄像头外网域名和IP地址。
	online：摄像头在线状态（0或1）。
	bmp：图片资源
		监测摄像头返回数据：
		onOnline：摄像头在线状态返回。
		onSnapshot：返回摄像头截图。
		onVideoCallBack：返回实时的摄像头视频图像

freeCamera(String myCameraIP);	myCameraIP：摄像头外网域名和IP地址	释放摄像头资源
setServerAddr(String sa)	sa：数据中心服务器地址及端口	设置/改变数据中心服务器地址及端口号
setIdKey(String myZCloudID, String myZCloudKey);	myZCloudID：智云账号。 myZCloudKey：智云密钥	设置/改变智云账号及密钥

4．基于Android的自动控制接口

智云平台内置了一个操作简单但功能强大的逻辑编辑器，可用于编辑复杂的控制逻辑，可以实现传感器数据更新、传感器状态查询、定时硬件系统控制、定时发送短消息，以及根据各种变量触发某个复杂的控制策略来实现系统复杂控制等功能。实现步骤如下：

（1）为每个传感器、执行器的关键数据和控制量创建一个变量。

（2）新建基本的控制策略，控制策略里可以运用上一步新建的变量。

（3）新建复杂的控制策略，复杂控制策略可以使用运算符，可以组合基本的控制策略。

基于Android的自动控制接口如表2.6所示。

表2.6　基于Android的自动控制接口

函　　数	参 数 说 明	功　　能
new WSNAutoctrl(String myZCloudID, String myZCloudKey);	myZCloudID：智云账号。 myZCloudKey：智云密钥	初始化自动控制对象，并初始化智云账号及密钥
createTrigger(String name, String type, JSONObject param);	name：触发器名称。 type：触发器类型。 param：触发器内容，JSON对象格式，创建成功后返回该触发器ID（JSON格式）	创建触发器
createActuator(String name,String type, JSONObject param);	name：执行器名称。 type：执行器类型。 param：执行器内容，JSON对象格式，创建成功后返回该执行器ID（JSON格式）	创建执行器
createJob(String name, boolean enable, JSONObject param);	name：任务名称。 enable：true（使能任务）、false（禁止任务）。 param：任务内容，JSON对象格式，创建成功后返回该任务ID（JSON格式）	创建任务
deleteTrigger(String id);	id：触发器ID	删除触发器

续表

函　数	参 数 说 明	功　能
deleteActuator(String id);	id：执行器 ID	删除执行器
deleteJob(String id);	id：任务 ID	删除任务
setJob(String id,boolean enable);	id：任务 ID。 enable：true（使能任务）、false（禁止任务）	设置任务使能开关
deleteSchedudler(String id);	id：任务记录 ID	删除任务记录
getTrigger();	无	查询当前智云账号下的所有触发器内容
getTrigger(String id);	id：触发器 ID	查询该触发器 ID
getTrigger(String type);	type：触发器类型	查询当前智云账号下的所有该类型的触发器内容
getActuator();	无	查询当前智云账号下的所有执行器内容
getActuator(String id);	id：执行器 ID	查询该执行器 ID
getActuator(String type);	type：执行器类型	查询当前智云账号下的所有该类型的执行器内容
getJob();	无	查询当前智云账号下的所有任务内容
getJob(String id);	id：任务 ID	查询该任务 ID
getSchedudler();	无	查询当前智云账号下的所有任务记录内容
getSchedudler(String jid,String duration);	id：任务记录 ID。 duration:duration=x<year|month|day|hours|minute>　//默认返回 1 天的记录	查询该任务记录 ID 某个时间段的内容
setServerAddr(String sa)	sa：数据中心服务器地址及端口	设置/改变数据中心服务器地址及端口号
setIdKey(String myZCloudID, String myZCloudKey);	myZCloudID：智云账号。 myZCloudKey：智云密钥	设置/改变智云账号及密钥

5．基于 Android 的用户数据接口

智云平台的用户数据接口提供私有的数据库使用权限，可对多客户端间共享的私有数据进行存储、查询。私有数据存储采用 Key-Value 型数据库服务，编程接口更简单高效。基于 Android 的用户数据接口如表 2.7 所示。

表 2.7　基于 Android 的用户数据接口

函　数	参 数 说 明	功　能
new WSNProperty(String myZCloudID, String myZCloudKey);	myZCloudID：智云账号。 myZCloudKey：智云密钥	初始化用户数据对象，并初始化智云账号及密钥
put(String key,String value);	key：名称。 value：内容	创建用户应用数据
get();	无	获取所有的键值对

续表

函　　数	参 数 说 明	功　　能
get(String key);	key：名称	获取指定 key 的 value 值
setServerAddr(String sa)	sa：数据中心服务器地址及端口	设置/改变数据中心服务器地址及端口号
setIdKey(String myZCloudID, String myZCloudKey);	myZCloudID：智云账号。 myZCloudKey：智云密钥	设置/改变智云账号及密钥

2.3.2　Web 端应用开发接口

针对 Web 端应用开发，智云平台提供了 JavaScript 接口库，开发者调用相应的接口即可完成简单的 Web 端应用开发。

1．基于 Web 的实时连接接口

基于 Web 的实时连接接口如表 2.8 所示。

表 2.8　基于 Web 的实时连接接口

函　　数	参 数 说 明	功　　能
new WSNRTConnect(myZCloudID, myZCloudKey);	myZCloudID：智云账号。 myZCloudKey：智云密钥	创建实时数据，并初始化智云账号及密钥
connect()	无	建立实时数据服务连接
disconnect()	无	断开实时数据服务连接
onConnect()	无	监测连接智云服务器成功
onConnectLost()	无	监测连接智云服务器失败
onMessageArrive(mac, dat)	mac：传感器的 MAC 地址。 dat：发送的消息内容	监测收到的数据
sendMessage(mac, dat)	mac：传感器的 MAC 地址。 dat：发送的消息内容	发送消息
setServerAddr(sa)	sa：数据中心服务器地址及端口	设置/改变数据中心服务器地址及端口号
setIdKey(myZCloudID, myZCloudKey);	myZCloudID：智云账号。 myZCloudKey：智云密钥	设置/改变智云账号及密钥（需要重新断开连接）

2．基于 Web 的历史数据接口

基于 Web 的历史数据接口如表 2.9 所示。

表 2.9　基于 Web 的历史数据接口

函　　数	参 数 说 明	功　　能
new WSNHistory(myZCloudID, myZCloudKey);	myZCloudID：智云账号。 myZCloudKey：智云密钥	初始化历史数据对象，并初始化智云账号及密钥
queryLast1H(channel, cal);	channel：传感器数据通道。 cal：回调函数（处理历史数据）	查询最近 1 小时的历史数据

续表

函　　数	参 数 说 明	功　　能
queryLast6H(channel, cal);	channel：传感器数据通道。 cal：回调函数（处理历史数据）	查询最近 6 小时的历史数据
queryLast12H(channel, cal);	channel：传感器数据通道。 cal：回调函数（处理历史数据）	查询最近 12 小时的历史数据
queryLast1D(channel, cal);	channel：传感器数据通道。 cal：回调函数（处理历史数据）	查询最近 1 天的历史数据
queryLast5D(channel, cal);	channel：传感器数据通道。 cal：回调函数（处理历史数据）	查询最近 5 天的历史数据
queryLast14D(channel, cal);	channel：传感器数据通道。 cal：回调函数（处理历史数据）	查询最近 14 天的历史数据
queryLast1M(channel, cal);	channel：传感器数据通道。 cal：回调函数（处理历史数据）	查询最近 1 个月（30 天）的历史数据
queryLast3M(channel, cal);	channel：传感器数据通道。 cal：回调函数（处理历史数据）	查询最近 3 个月（90 天）的历史数据
queryLast6M(channel, cal);	channel：传感器数据通道。 cal：回调函数（处理历史数据）	查询最近 6 个月（180 天）的历史数据
queryLast1Y(channel, cal);	channel：传感器数据通道。 cal：回调函数（处理历史数据）	查询最近 1 年（365 天）的历史数据
query(cal);	cal：回调函数（处理历史数据）	获取所有通道最后一次数据
query(channel, cal);	channel：传感器数据通道。 cal：回调函数（处理历史数据）	获取该通道下最后一次数据
query(channel, start, end, cal);	channel：传感器数据通道。 cal：回调函数（处理历史数据）。 start：起始时间。 end：结束时间。 时间为 ISO 8601 格式的日期，例如：2010-05-20T11:00:00Z	通过起止时间查询指定时间段的历史数据
query(channel, start, end, interval, cal);	channel：传感器数据通道。 cal：回调函数（处理历史数据）。 start：起始时间。 end：结束时间。 interval：采样点的时间间隔，详细见后续说明。 时间为 ISO 8601 格式的日期，例如：2010-05-20T11:00:00Z	通过起止时间查询指定时间段、指定时间间隔的历史数据
setServerAddr(sa)	sa：数据中心服务器地址及端口	设置/改变数据中心服务器地址及端口号
setIdKey(myZCloudID, myZCloudKey);	myZCloudID：智云账号。 myZCloudKey：智云密钥	设置/改变智云账号及密钥

3．基于 Web 的摄像头接口

基于 Web 的摄像头接口如表 2.10 所示。

表 2.10 基于 Web 的摄像头接口

函 数	参 数 说 明	功 能
new WSNCamera(myZCloudID, myZCloudKey);	myZCloudID：智云账号。 myZCloudKey：智云密钥	初始化摄像头对象，并初始化智云账号及密钥
initCamera(myCameraIP, user, pwd, type);	myCameraIP：摄像头外网域名和 IP 地址。 user：摄像头用户名。 pwd：摄像头密码。 type：摄像头类型。 以上参数可从摄像头手册获取	设置摄像头域名、用户名、密码、类型等参数
openVideo();	无	打开摄像头
closeVideo();	无	关闭摄像头
control(cmd);	cmd：云台控制指令，参数如下： UP：向上移动一次。 DOWN：向下移动一次。 LEFT：向左移动一次。 RIGHT：向右移动一次。 HPATROL：水平巡航转动。 VPATROL：垂直巡航转动。 360PATROL：360° 巡航转动	发送指令控制云台转动
checkOnline(cal);	cal：回调函数（处理检查结果）	监测摄像头是否在线
snapshot();	无	抓拍照片
setDiv(divID);	divID：网页标签	设置展示摄像头视频、图像的标签
freeCamera(myCameraIP);	myCameraIP：摄像头外网域名/IP 地址	释放摄像头资源
setServerAddr(sa)	sa：数据中心服务器地址及端口	设置/改变数据中心服务器地址及端口号
setIdKey(myZCloudID, myZCloudKey);	myZCloudID：智云账号 myZCloudKey：智云密钥	设置/改变智云账号及密钥

4．基于 Web 的自动控制接口

基于 Web 的自动控制接口如表 2.11 所示。

表 2.11 基于 Web 的自动控制接口

函 数	参 数 说 明	功 能
new WSNAutoctrl(myZCloudID, myZCloudKey);	myZCloudID：智云账号。 myZCloudKey：智云密钥	初始化自动控制对象，并初始化智云账号及密钥
createTrigger(name, type, param, cal);	name：触发器名称。 type：触发器类型。 param：触发器内容，JSON 对象格式。 创建成功后返回该触发器 ID（JSON 格式）。 cal：回调函数	创建触发器

续表

<table>
<tr><th>函　数</th><th>参 数 说 明</th><th>功　能</th></tr>
<tr><td>createActuator(name, type, param, cal);</td><td>name：执行器名称。
type：执行器类型。
param：执行器内容，JSON 对象格式。
创建成功后返回该执行器 ID（JSON 格式）。
cal：回调函数</td><td>创建执行器</td></tr>
<tr><td>createJob(name, enable, param, cal);</td><td>name：任务名称。
enable：true（使能任务）、false（禁止任务）。
param：任务内容，JSON 对象格式。
创建成功后返回该任务 ID（JSON 格式）。
cal：回调函数</td><td>创建任务</td></tr>
<tr><td>deleteTrigger(id, cal);</td><td>id：触发器 ID。
cal：回调函数</td><td>删除触发器</td></tr>
<tr><td>deleteActuator(id, cal);</td><td>id：执行器 ID。
cal：回调函数</td><td>删除执行器</td></tr>
<tr><td>deleteJob(id, cal);</td><td>id：任务 ID。
cal：回调函数</td><td>删除任务</td></tr>
<tr><td>setJob(id, enable, cal);</td><td>id：任务 ID。
enable：true（使能任务）、false（禁止任务）。
cal：回调函数</td><td>设置任务使能开关</td></tr>
<tr><td>deleteSchedudler(id, cal);</td><td>id：任务记录 ID。
cal：回调函数</td><td>删除任务记录</td></tr>
<tr><td>getTrigger(cal);</td><td>cal：回调函数</td><td>查询当前智云账号下的所有触发器内容</td></tr>
<tr><td>getTrigger(id, cal);</td><td>id：触发器 ID。
cal：回调函数</td><td>查询该触发器账号内容</td></tr>
<tr><td>getTrigger(type, cal);</td><td>type：触发器类型。
cal：回调函数</td><td>查询当前智云账号下的所有该类型的触发器内容</td></tr>
<tr><td>getActuator(cal);</td><td>cal：回调函数</td><td>查询当前智云账号下的所有执行器内容</td></tr>
<tr><td>getActuator(id, cal);</td><td>id：执行器 ID。
cal：回调函数</td><td>查询该执行器 ID</td></tr>
<tr><td>getActuator(type, cal);</td><td>type：执行器类型。
cal：回调函数</td><td>查询当前智云账号下的所有该类型的执行器内容</td></tr>
<tr><td>getJob(cal);</td><td>cal：回调函数</td><td>查询当前智云账号下的所有任务内容</td></tr>
<tr><td>getJob(id, cal);</td><td>id：任务 ID。
cal：回调函数</td><td>查询该任务 ID</td></tr>
<tr><td>getSchedudler(cal);</td><td>cal：回调函数</td><td>查询当前智云账号下的所有任务记录内容</td></tr>
<tr><td>getSchedudler(jid, duration, cal);</td><td>id：任务记录 ID。
duration:duration=x<year|month|day|hours|minute>
//默认返回 1 天的记录
cal：回调函数</td><td>查询该任务记录账号某个时间段的内容</td></tr>
</table>

续表

函 数	参 数 说 明	功 能
setServerAddr(sa)	sa：数据中心服务器地址及端口	设置/改变数据中心服务器地址及端口号
setIdKey(myZCloudID, myZCloudKey);	myZCloudID：智云账号。 myZCloudKey：智云密钥	设置/改变智云账号及密钥

5. 基于 Web 的用户数据接口

基于 Web 的用户数据接口如表 2.12 所示。

表 2.12 基于 Web 的用户数据接口

函 数	参 数 说 明	功 能
new WSNProperty(myZCloudID, myZCloudKey);	myZCloudID：智云账号。 myZCloudKey：智云密钥	初始化用户数据对象，并初始化智云账号及密钥
put(key, value, cal);	key：名称。 value：内容。 cal：回调函数	创建用户应用数据
get(cal);	cal：回调函数	获取所有的键值对
get(key, cal);	key：名称。 cal：回调函数	获取指定 key 的 value 值
setServerAddr(sa)	sa：数据中心服务器地址及端口	设置/改变数据中心服务器地址及端口号
setIdKey(myZCloudID, myZCloudKey);	myZCloudID：智云账号。 myZCloudKey：智云密钥	设置/改变智云账号及密钥

2.4 Android 端应用开发实例

2.4.1 基于 Android 的实时连接接口的应用

要实现传感器数据的发送，就需要在 SensorActivity.java 文件中调用类 WSNRTConnect 的方法，具体方法是：先导入智云接口的相关文件包，再定义实时连接对象，在通过 new WSNRTConnect 实例化实时连接对象时，智云服务器连接参数 myZCloudID 与 myZCloudKey 需要在主页面定义，然后通过“wRTConnect.setServerAddr("zhiyun360.com");”设置智云服务器地址，最后调用 wRTConnect.connect()方法即可连接到智云服务器。实现代码如下：

```
import com.zhiyun360.wsn.droid.WSNRTConnect;
import com.zhiyun360.wsn.droid.WSNRTConnectListener;
public class SensorActivity extends Activity {
    private Button mBTNOpen,mBTNClose;
    private TextView mTVInfo;
    private WSNRTConnect wRTConnect;
    ……
```

```
    @Override
    public void onCreate(Bundle savedInstanceState) {
    ……
wRTConnect=new WSNRTConnect(DemoActivity.myZCloudID,
DemoActivity.myZCloudKey);
wRTConnect.setServerAddr("zhiyun360.com");                    //设置智云服务器地址
wRTConnect.connect();
```

通过在设置按钮的 setOnClickListener 方法中调用 wRTConnect.sendMessage 接口，应用程序可以向传感器节点发送数据，该方法首先需要设置 MAC 地址与协议指令，实现代码如下：

```
mBTNClose.setOnClickListener(new View.OnClickListener() {
    @Override
    public void onClick(View v) {
        String mac = "00:12:4B:00:10:27:A5:19";
        String dat = "{CD1=64,D1=?}";
        textInfo(mac + " <<< " + dat);
        wRTConnect.sendMessage(mac, dat.getBytes());
    }
});
```

通过调用 WSNRTConnectListener 接口 onMessageArrive(String arg0, byte[] arg1)方法，应用程序可以接收实时数据，实现代码如下：

```
wRTConnect.setRTConnectListener(new WSNRTConnectListener() {
    ……
    public void onMessageArrive(String arg0, byte[] arg1) {
        textInfo(arg0 + " >>> " + new String(arg1));
    }
    ……
}
```

应用程序可以通过下面的代码来关闭与智云服务器的连接。

```
@Override
public void onDestroy() {
    wRTConnect.disconnect();
    super.onDestroy();
}
```

2.4.2 基于 Android 的历史数据接口的应用

历史数据查询的页面如图 2.10 所示。

在进行历史数据查询时，首先通过“new WSNHistory(DemoActivity.myZCloudID, DemoActivity.myZCloudKey);”实例化历史数据对象，然后通过“wHistory.setServerAddr("zhiyun360.com:8080");”设置智云服务器的地址，注意加上后面的端口号 8080。

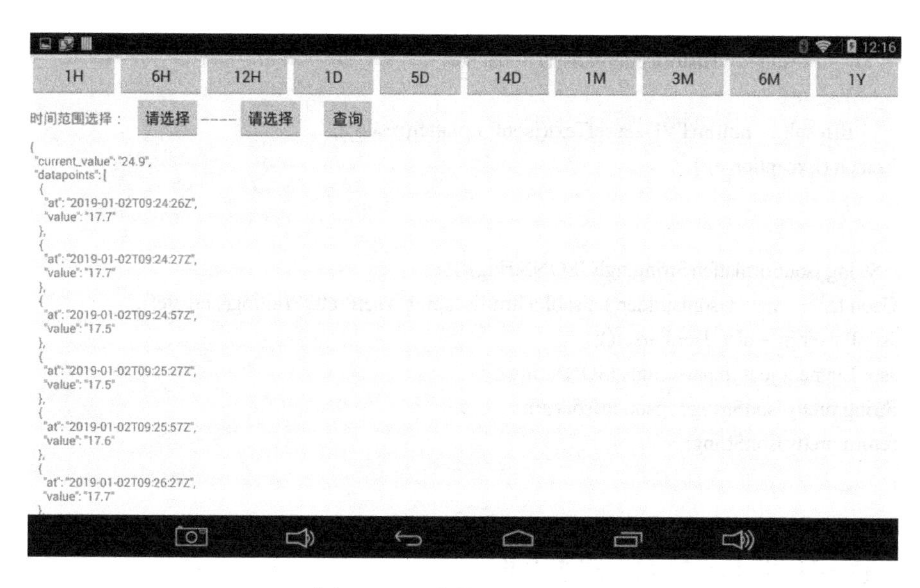

图 2.10　历史数据查询的页面

历史数据的查询还需要设置传感器节点的通道号，如“00:12:4B:00:10:27:A5:19_A0”，其中“00:12:4B:00:10:27:A5:19”是传感器节点的 MAC 地址，“_A0”是温度数据。

```
import com.zhiyun360.wsn.droid.WSNHistory;
public class HistoryActivity extends Activity implements OnClickListener {
    //设置要查询的传感器节点的通道号
    private String channel = "00:12:4B:00:10:27:A5:19_A0";
    private Button mBTN1H, mBTN6H…;
    private TextView mTVData;
    private WSNHistory wHistory;
    @Override
    public void onCreate(Bundle savedInstanceState) {
        mBTN1H = (Button) findViewById(R.id.btn1h);
        mBTN1H.setOnClickListener(this);
        wHistory = new WSNHistory(DemoActivity.myZCloudID, DemoActivity.myZCloudKey);
        wHistory.setServerAddr("zhiyun360.com:8080");
    }
}
```

用户可以查询不同时间段的历史数据，这里以查询最近一小时的历史数据为例进行说明。使用历史数据对象的 wHistory.queryLast1H(channel)方法，参数是要查询的通道号（通道号需要在应用程序中进行初始化），该方法的返回值是 String 类型的字符串，通过“if(result != null)mTVData.setText(jsonFormatter(result));”语句来调用格式转换的方法，最后以 JSON 格式的字符串显示在历史数据查询的页面上。

```
@Override
public void onClick(View arg0) {
    String result = null;
    try {
        if (arg0 == mBTN1H) {
```

```
                result = wHistory.queryLast1H(channel);
            }
            if(result != null)mTVData.setText(jsonFormatter(result));
        } catch (Exception e) {
            }
    }
    public String jsonFormatter(String uglyJSONString) {
        Gson gson = new GsonBuilder().disableHtmlEscaping().setPrettyPrinting().create();
        JsonParser jp = new JsonParser();
        JsonElement je = jp.parse(uglyJSONString);
        String prettyJsonString = gson.toJson(je);
        return prettyJsonString;
    }
```

2.5 Web 端应用开发实例

2.5.1 基于 Web 的实时连接接口的应用

智云平台提供了实时数据推送服务 API，通过提供的 API 可以与底层的传感器节点进行信息交互。理解这些 API 后，用户还可以在底层自定义一些数据通信协议，从而控制底层的传感器节点，实现数据采集的功能。实时数据查询与推送如图 2.11 所示。

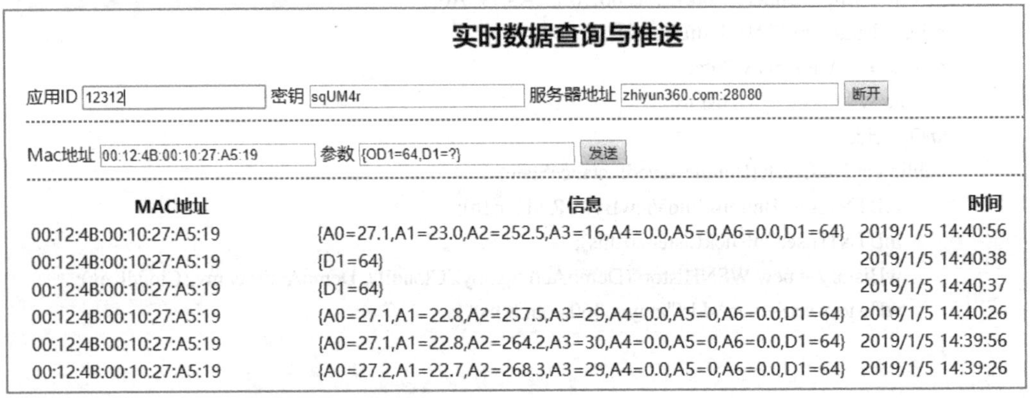

实时数据查询与推送

应用ID 12312 密钥 sqUM4r 服务器地址 zhiyun360.com:28080 断开

Mac地址 00:12:4B:00:10:27:A5:19 参数 {OD1=64,D1=?} 发送

MAC地址	信息	时间
00:12:4B:00:10:27:A5:19	{A0=27.1,A1=23.0,A2=252.5,A3=16,A4=0.0,A5=0,A6=0.0,D1=64}	2019/1/5 14:40:56
00:12:4B:00:10:27:A5:19	{D1=64}	2019/1/5 14:40:38
00:12:4B:00:10:27:A5:19	{D1=64}	2019/1/5 14:40:37
00:12:4B:00:10:27:A5:19	{A0=27.1,A1=22.8,A2=257.5,A3=29,A4=0.0,A5=0,A6=0.0,D1=64}	2019/1/5 14:40:26
00:12:4B:00:10:27:A5:19	{A0=27.1,A1=22.8,A2=264.2,A3=30,A4=0.0,A5=0,A6=0.0,D1=64}	2019/1/5 14:39:56
00:12:4B:00:10:27:A5:19	{A0=27.2,A1=22.7,A2=268.3,A3=29,A4=0.0,A5=0,A6=0.0,D1=64}	2019/1/5 14:39:26

图 2.11　实时数据查询与推送

在 Web 程序中，首先要包含智云 Web 接口的 JS 文件 WSNRTConnect.js（这是因为使用到了 jQuery 库，所以应包含对应的库文件），然后调用 new WSNRTConnect()来初始化实时连接对象。

```
<script src="../../js/jquery-1.11.0.min.js"></script>
<script src="../../js/WSNRTConnect.js"></script>
<script>
var rtc = new WSNRTConnect();
```

在 Web 程序中，连接功能是通过“连接”按钮来实现的，连接智云服务器之后该按钮显示“断开”，没有连接时显示“连接”。单击“连接”按钮时，click 事件代码通过“if (!rtc.isconnect)”判断当前实时连接对象是否连接到了智云服务器。如果没有连接，就先通过“rtc.setIdKey($("#aid").val(), $("#xkey").val());”获得用户输入的 ID 与 KEY，然后通过“rtc.setServerAddr($("#saddr").val());”获得服务器地址，最后通过“rtc.connect();”连接到智云服务器。

```
$(document).ready(function(){
    $("#btn_con").click(function(){
        if (!rtc.isconnect) {
            rtc.setIdKey($("#aid").val(), $("#xkey").val());
            rtc.setServerAddr($("#saddr").val());
            rtc.connect();
        } else {
            rtc.disconnect();
        }
    }
}
```

如果实时连接对象连接到了智云服务器，则上述的按钮显示“断开”；如果没有连接到智云服务器，则显示“连接”。按钮上字符动态切换功能的实现代码如下：

```
function onConnect(){
    rtc.isconnect = true;
    $("#btn_con").val("断开");
    $("#btn_con").attr("class","btn btn-warning");
    console.log("断开");
}
function onConnectLost() {
    rtc.isconnect = false;
    $("#btn_con").val("连接");
    $("#btn_con").attr("class","btn btn-success");
    console.log("连接");
}
```

如果实时连接对象连接到了智云服务器，可通过 onmessageArrive 函数监听接收到的无线数据包并显示出来。

```
function onmessageArrive(mac, msg) {
    var d=new Date();
    var   time=d.toLocaleDateString()+" "+d.getHours()+":"+d.getMinutes()+":"+d.getSeconds();
    var ul_mac = $(".filter").children("ul").attr("mac");
    var html = "<tr mac='"+mac+"'><td>"+mac+"</td><td>"+msg+"</td><td>"+time+"</td></tr>";
    $("table").find("tbody").prepend(html);
}
```

数据发送功能是通过“rtc.sendMessage($("#mac").val(), $("#pa").val());”实现的，需要传感器节点的 MAC 地址与数据通信协议命令作为参数。

```
$(document).ready(function(){
    ......
```

```
        $("#query").click(function(){
            if (!rtc.isconnect) {
                return;
            }
            rtc.sendMessage($("#mac").val(), $("#pa").val());
        });
    });
```

2.5.2　基于 Web 的历史数据接口的应用

历史数据查询是通过“查询”按钮的 click 事件来处理的，可通过“new WSNHistory($("#aid").val(), $("#xkey").val());”来获取的 ID 与 KEY 来初始化历史数据对象，通过“myHisData.setServerAddr($("#saddr").val());”设置智云服务器的地址，通过“time = $("#history_time").val();”获得要查询的时间段，通过“$("#history_channel").val();”获得要查询的通道号，通过“myHisData[time](channel, function(dat){});”实现历史数据的查询与显示。历史数据查询如图 2.12 所示。

历史数据查询

应用ID 34533　密钥 56KIHsqUM4r　服务器地址 zhiyun360.com:8080

查询时间段 最近1小时　通道 00:12:4B:00:10:27:A5:19_A0　查询

{"current_value":"24.7","datapoints":[{"at":"2019-01-05T11:34:29Z","value":"22.8"},{"at":"2019-01-05T11:34:59Z","value":"22.9"},
{"at":"2019-01-05T11:35:29Z","value":"23.0"},{"at":"2019-01-05T11:35:59Z","value":"23.0"},{"at":"2019-01-05T11:36:29Z","value":"23.1"},
{"at":"2019-01-05T11:36:59Z","value":"23.1"},{"at":"2019-01-05T11:37:29Z","value":"23.2"},{"at":"2019-01-05T11:37:59Z","value":"23.2"},
{"at":"2019-01-05T11:38:29Z","value":"23.3"},{"at":"2019-01-05T11:38:59Z","value":"23.3"},{"at":"2019-01-05T11:39:29Z","value":"23.4"},
{"at":"2019-01-05T11:39:59Z","value":"23.4"},{"at":"2019-01-05T11:40:29Z","value":"23.5"},{"at":"2019-01-05T11:40:59Z","value":"23.5"},
{"at":"2019-01-05T11:41:29Z","value":"23.5"},{"at":"2019-01-05T11:41:59Z","value":"23.6"},{"at":"2019-01-05T11:42:29Z","value":"23.6"},
{"at":"2019-01-05T11:42:59Z","value":"23.7"},{"at":"2019-01-05T11:43:29Z","value":"23.7"},{"at":"2019-01-05T11:43:59Z","value":"23.8"},
{"at":"2019-01-05T11:44:29Z","value":"23.9"},{"at":"2019-01-05T11:44:59Z","value":"23.9"},{"at":"2019-01-05T11:45:29Z","value":"24.0"},

图 2.12　历史数据查询

```
$(function(){
    $("#history_query").click(function(){
        //初始化 API
        var myHisData = new WSNHistory($("#aid").val(), $("#xkey").val());
        myHisData.setServerAddr($("#saddr").val());
        $("#data_show").text("");
        var time = $("#history_time").val();
        var channel = $("#history_channel").val();
        myHisData[time](channel, function(dat){
            var data = JSON.stringify(dat);                //JSON 对象变为字符串
            $("#data_show").text(data);
        })
    })
})
```

第3章 ZigBee高级应用开发

ZigBee是一种短距离、低功耗的无线通信技术，符合IEEE 802.15.4标准，其特点是传输距离短、低复杂度、自组织、低功耗、低数据传输速率。ZigBee已广泛应用于物联网产业链中的M2M行业，如智慧农业、智能交通、智能家居、金融、供应链自动化、工业自动化、智能建筑、消防、环境保护、气象、农业、水务等领域。有关更详尽的ZigBee内容请参考《物联网短距离无线通信技术应用与开发》。

本章通过4个贴近生活的开发案例来介绍ZigBee物联网系统的软硬件开发，从而全面了解ZigBee物联网系统的架构和应用。具体开发案例包括基于ZigBee的城市环境信息采集系统、基于ZigBee的城市景观照明控制系统、基于ZigBee的智能燃气控制系统、基于ZigBee的家庭安防监控系统。通过这4个开发案例，本章设计了基于ZigBee的采集类节点、控制类节点和安防类节点的驱动程序，实现了Android端和Web端的系统软件的设计与开发。

3.1 基于ZigBee的城市环境信息采集系统

伴随着城市人口的增加，城市的相关问题也会增加。例如，车辆的增多造成空气质量下降、密集的建筑物使消防安全隐患加重、发生不可抗的破坏时必需的配套防护设施等。城市环境信息采集系统是基于物联网的智慧城市的一部分，部署在城市中的采集类节点可以采集环境的相关信息，然后通过无线通信技术将这些信息上传到物联网信息管理平台。城市环境信息采集系统如图3.1所示。

3.1.1 系统开发目标

（1）熟悉温湿度传感器、光照度传感器、空气质量传感器以及气压海拔传感器的硬件原理和数据通信协议，并实现基于CC2530和ZigBee的传感器驱动开发，传感器将采集到的数据通过汇集节点发送至智云平台，用户通过App可以实时监控城市的环境信息。

（2）实现城市环境信息采集系统的Android端应用开发和Web端应用开发。

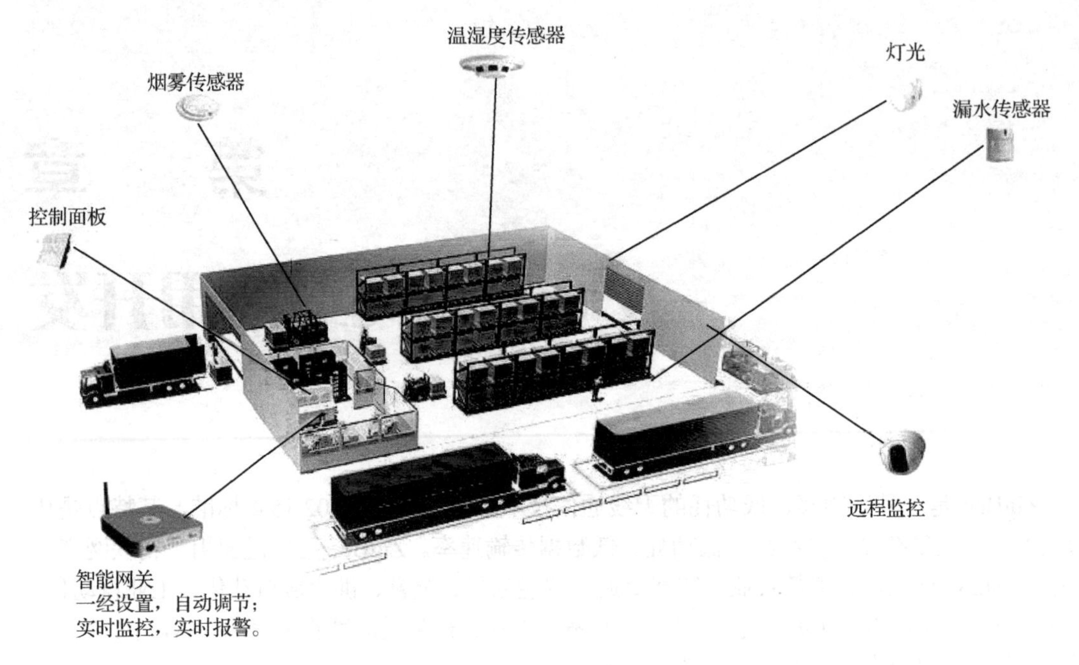

图 3.1 城市环境信息采集系统

3.1.2 系统设计分析

1. 系统的功能设计

城市环境信息采集系统的主要功能是通过传感器实时采集温湿度、光照度、空气质量以及大气压强等数据，并将采集到的数据主动推送到智云平台，从而实现对温湿度、光照度、空气质量以及大气压强等的实时监测。从系统功能的角度来看，城市环境信息采集系统可分为设备采集模块和系统设置模块，如图 3.2 所示。

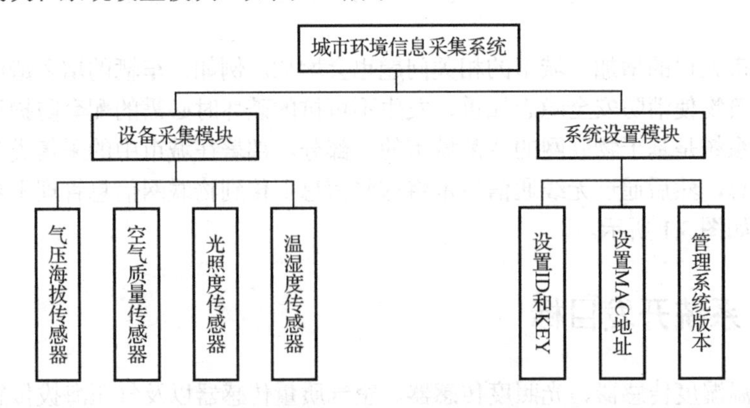

图 3.2 城市环境信息采集系统的组成模块

设备采集模块的主要功能是将传感器采集的数据推送到智云平台，系统设置模块的主要功能是设置智云服务器的 ID 和 KEY、设置 MAC 地址、管理系统版本。

城市环境信息采集系统功能需求分析如表 3.1 所示。

表 3.1 城市环境信息采集系统功能需求分析

功　能	功能说明
采集数据显示	实时更新并显示采集到的温湿度、光照度、空气质量以及大气压强的数据
定时上报功能	定时上报当前温湿度、光照度、空气质量以及大气压强的数据
智云连接设置	服务器的参数设置与连接，MAC 地址的设置

2. 系统的总体架构设计

城市环境信息采集系统是基于物联网四层架构模型来设计的，其总体架构如图 3.3 所示。

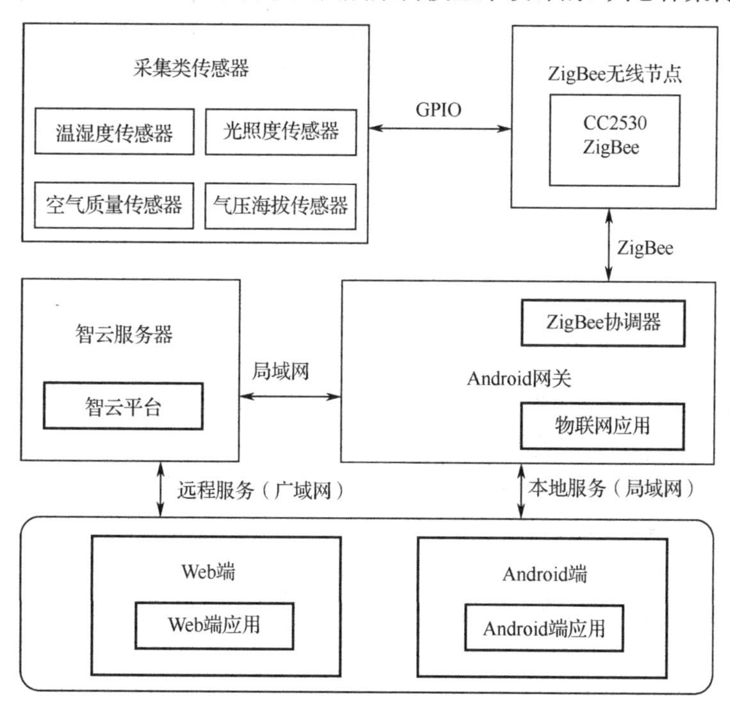

图 3.3 城市环境信息采集系统的总体架构

感知层：主要由采集类传感器构成，如温湿度传感器、光照度传感器、空气质量传感器以及气压海拔传感器，用于采集环境的相关信息。

网络层：感知层的采集类传感器和智能网关（Android 网关）之间是通过 ZigBee 连接的，智能网关和智云服务器、上层应用设备之间是通过局域网（互联网）来传输数据的。

平台层：平台层提供物联网设备之间的基于互联网的存储、访问、控制。

应用层：提供物联网系统的人机交互接口，通过 Web 端、Android 端来提供页面友好、操作交互性强的应用。

3. 系统的数据传输

城市环境信息采集系统的数据传输是在传感器节点、智能网关以及客户端（包括 Web 端和 Android 端）之间进行的，如图 3.4 所示。

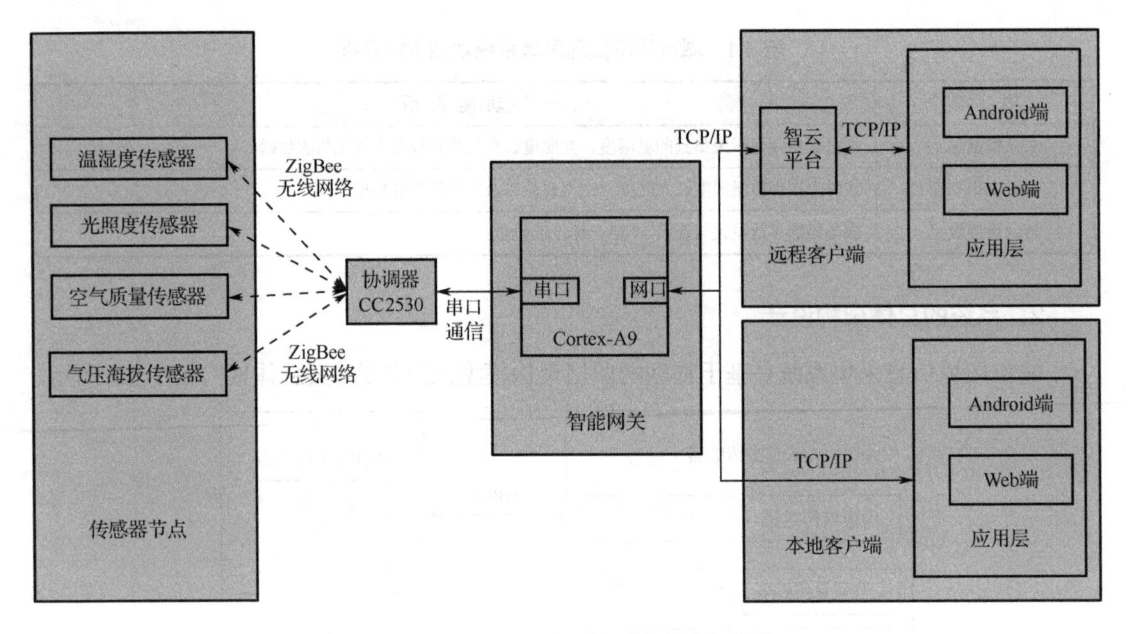

图 3.4 城市环境信息采集系统的数据传输

（1）传感器节点与协调器组建 ZigBee 无线网络，协调器通过串口与智能网关进行数据传输。

（2）智能网关通过局域网（采用 TCP/IP 协议）将传感器采集的数据推送给本地客户端（包括 Web 端和 Android 端）并接收本地客户端发送的信息。

（3）智能网关通过智云平台将传感器采集的数据推送给远程客户端（包括 Web 端和 Android 端）并接收远程客户端发送的信息。

3.1.3 系统的软硬件开发：城市环境信息采集系统

1. 系统底层软硬件设计

1）感知层硬件设计

城市环境信息采集系统的感知层包括 xLab 未来开发平台的智能网关、经典型无线节点 ZXBeeLiteB、采集类开发平台 Sensor-A 三类设备。本系统使用的传感器包括温湿度传感器、光照度传感器、空气质量传感器和气压海拔传感器，各传感器的硬件接口电路如图 3.5 到图 3.8 所示。

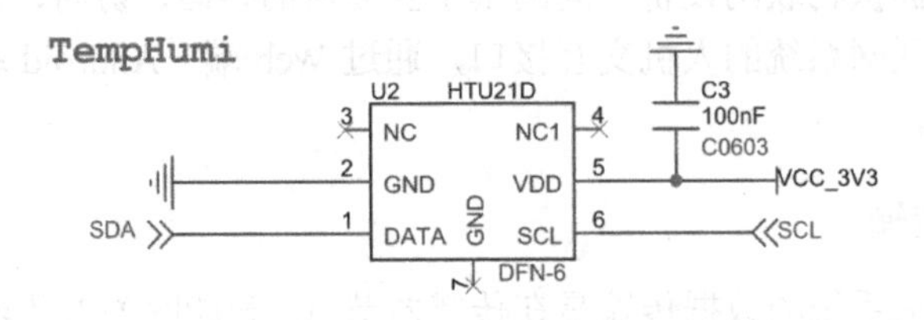

图 3.5 温湿度传感器的硬件接口电路

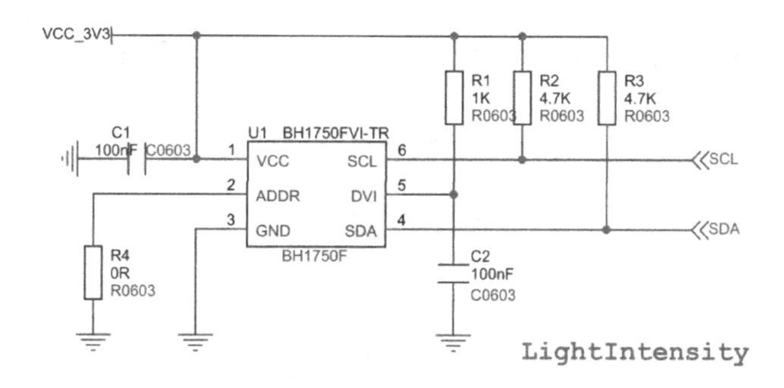

图 3.6 光照度传感器的硬件接口电路

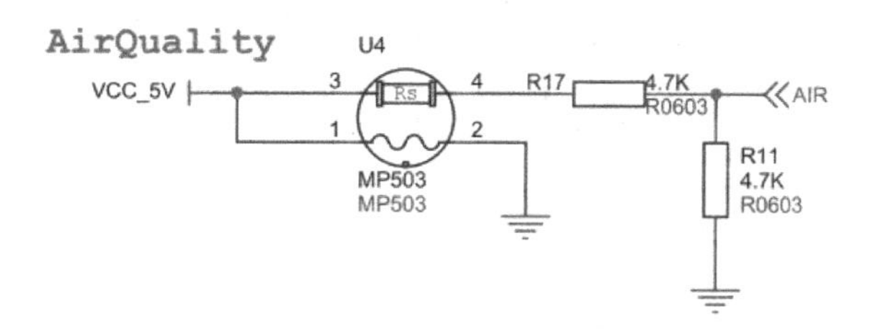

图 3.7 空气质量传感器的硬件接口电路

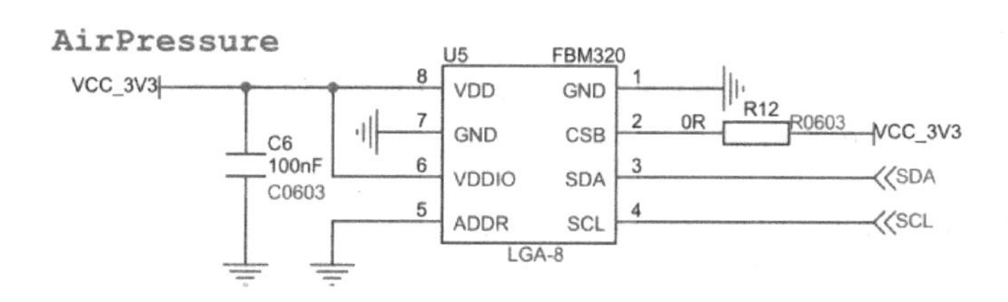

图 3.8 气压海拔传感器的硬件接口电路

2）系统底层开发

系统底层开发是基于智云框架进行的，智云框架是在传感器应用程序接口和 SAPI 框架的基础上搭建起来的。

（1）传感器应用程序接口。通过合理调用传感器应用程序接口，可以形成一套系统的 ZigBee 开发逻辑。传感器应用程序接口是在 sensor.c 文件中实现的，具体的接口函数如表 3.2 所示。

表 3.2 传感器应用程序接口函数

函 数 名 称	函 数 说 明
sensorInit()	传感器初始化
sensorLinkOn()	传感器节点入网成功调用的函数
sensorUpdate()	传感器数据定时上报
sensorControl()	传感器控制函数
sensorCheck()	传感器预警监测及处理函数
ZXBeeInfRecv()	处理节点接收到的无线数据包
MyEventProcess()	自定义事件处理函数，启动定时器触发上报事件 MY_REPORT_EVT

（2）智云框架下传感器程序执行流程。智云框架下传感器程序执行流程如图 3.9 所示。

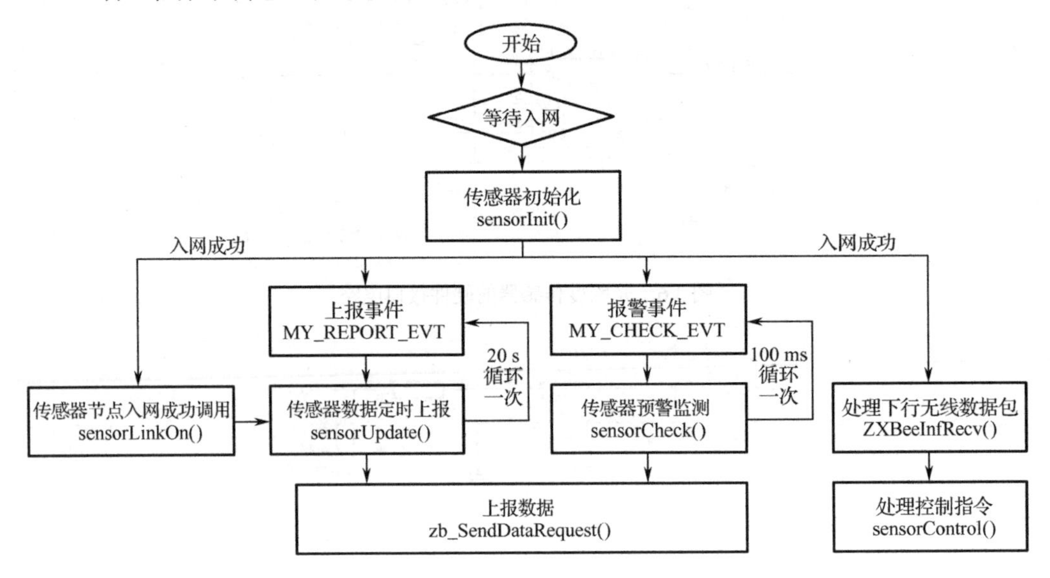

图 3.9　智云框架下传感器程序执行流程

智云框架为 ZigBee 协议栈（ZStack）的上层应用提供分层的软件设计结构，将传感器的私有操作部分封装在 sensor.c 文件中，用户事件和节点类型选择在 sensor.h 文件中定义。sensor.h 文件中用户事件和节点类型的宏定义如下：

```
#define MY_REPORT_EVT      0x0001
#define MY_CHECK_EVT       0x0002
#define NODE_NAME          "601"            //节点名称
#define NODE_CATEGORY      1                //节点类型
#define NODE_TYPE NODE_ENDDEVICE   //NODE_ROUTER 表示路由节点，NODE_ENDDEVICE 表示终端节点
```

用户事件中定义的内容分别是上报事件（MY_REPORT_EVT）和报警事件（MY_CHECK_EVT），上报事件用于上报传感器采集到的数据，报警事件用于对传感器监测到的危险信息进行响应。在 sensor.h 文件中，通过节点类型的宏定义可以选择将节点设置为路由节点（NODE_ROUTER）或终端节点（NODE_ENDDEVICE），同时还声明了智云框架下的传感器应用文件 sensor.c 中的函数。

sensorInit()函数用于进行传感器的初始化，以及触发上报事件 MY_REPORT_EVT 和报警事件 MY_CHECK_EVT，相关代码如下：

```
/*********************************************************************************************
*名称：sensorInit()
*功能：传感器初始化
*********************************************************************************************/
void sensorInit(void)
{
    printf("sensor->sensorInit(): Sensor init!\r\n");
    //传感器初始化
```

```
    ……
    //启动定时器，触发上报事件 MY_REPORT_EVT 和报警事件 MY_CHECK_EVT
    osal_start_timerEx(sapi_TaskID, MY_REPORT_EVT, (uint16)((osal_rand()%10) *1000));
    osal_start_timerEx(sapi_TaskID, MY_CHECK_EVT, 100));
}
```

传感器节点入网成功后会调用 sensorLinkOn()函数来执行相关的操作，相关代码如下：

```
/*****************************************************************************************
*名称：sensorLinkOn()
*功能：传感器节点入网成功调用函数
*****************************************************************************************/
void sensorLinkOn(void)
{
    printf("sensor->sensorLinkOn(): Sensor Link on!\r\n");
    sensorUpdate();                                    //入网成功后上报一次传感器数据
}
```

sensorUpdate()函数用于更新传感器的数据，并将更新后的数据打包上报，相关代码如下：

```
/*****************************************************************************************
*名称：sensorUpdate()
*功能：处理主动上报的数据
*****************************************************************************************/
void sensorUpdate(void)
{
    char pData[32];
    char *p = pData;
    //光照度采集（采用 0～1000 之间的随机数来模拟光照度）
    lightIntensity = (uint16)(osal_rand()%1000);
    //更新采集数值
    sprintf(p, "lightIntensity=%.1f", lightIntensity);
    zb_SendDataRequest( 0, 0, strlen(p), (uint8*)p, 0, 0, AF_DEFAULT_RADIUS );
    HalLedSet( HAL_LED_1, HAL_LED_MODE_OFF );
    HalLedSet( HAL_LED_1, HAL_LED_MODE_BLINK );

    printf("sensor->sensorUpdate(): lightIntensity=%.1f\r\n", lightIntensity);
}
```

MyEventProcess()函数用于启动和处理用户自定义事件，相关代码如下：

```
/*****************************************************************************************
*名称：MyEventProcess()
*功能：自定义事件处理
*参数：event—事件编号
*****************************************************************************************/
void MyEventProcess( uint16 event )
{
    if (event & MY_REPORT_EVT) {
```

```
        sensorUpdate();                                  //传感器数据定时上报
        //启动定时器，触发上报事件 MY_REPORT_EVT
        osal_start_timerEx(sapi_TaskID, MY_REPORT_EVT, 20*1000);
    }
    if (event & MY_CHECK_EVT) {
        sensorCheck();                                   //传感器状态实时监测
        //启动定时器，触发报警事件 MY_CHECK_EVT
        osal_start_timerEx(sapi_TaskID, MY_CHECK_EVT, 100);
    }
}
```

ZXBeeInfRecv()函数用于处理节点接收到的无线数据包，相关代码如下：

```
/*********************************************************************************************
*名称：ZXBeeInfRecv()
*功能：处理节点接收到的无线数据包
*参数：*pkg—收到的无线数据包
*********************************************************************************************/
void ZXBeeInfRecv(char *pkg, int len)
{
    ……
    printf("sensor->ZXBeeInfRecv(): Receive ZigBee Data!\r\n");
    ptag = pkg;
    p = strchr(pkg, '=');
    if (p != NULL) {
        *p++ = 0;
        pval = p;
    }
    val = atoi(pval);
    //控制命令解析
    if (0 == strcmp("cmd", ptag)){                  //对 D0 的位进行操作，CD0 表示位清零操作
        sensorControl(val);
    }
}
```

sensorControl()函数用于控制传感器，相关代码如下：

```
/*********************************************************************************************
*名称：sensorControl()
*功能：传感器控制
*参数：cmd—控制命令
*********************************************************************************************/
void sensorControl(uint8 cmd)
{
    //根据参数 cmd 执行相应的控制程序
}
```

通过实现 sensor.c 文件中具体接口函数即可快速地完成 ZigBee 项目的开发。

3）传感器驱动设计

城市环境信息采集系统的底层硬件主要是采集类传感器，主要关注采集类传感器的数据上报是否有效。采集类传感器的逻辑事件可分为以下四种。

① 采集类传感器能够进行数据的采集，并且能够根据设定的参数循环进行数据的上报更新。在实际的应用中，通常需要根据应用的需求和能耗设定一个合适的数据上报时间间隔。采集类传感器进行数据采集越频繁，传感器的耗电量就越大，从而增加系统的功耗。如果在无线网络中有多个采集类传感器频繁地发送数据，则会对无线网络的通信造成压力，严重时还会造成网络阻塞、丢包等后果。因此，采集类传感器在定时进行数据上报时需要注意数据上报时间间隔和发送的数据量。

② 在实际应用中可根据需求关闭采集类传感器的数据上报，以便降低功耗。例如，一个无线网络可以采集 CO_2、温度、湿度、光照度等信息，在夜晚时可以关闭光照度传感器的数据上报。

③ 能够远程设定采集类传感器的数据的更新时间。这种功能通常用于物联网的自动场景，将数据上报时间间隔变短，可以达到快速更新数据的目的。

④ 采集类传感器接收到查询指令后应立刻响应并反馈实时数据。这种功能通常用于物联网的人为场景，接收到查询指令后立即响应并反馈实时数据是采集类传感器必需的功能。

传感器的驱动设计主要包括数据通信协议的定义以及驱动程序的开发。

（1）数据通信协议的定义。本系统主要使用的是采集类开发平台 Sensor-A，其 ZXBee 数据通信协议如表 3.3 所示。

表 3.3 采集类开发平台的 ZXBee 数据通信协议

开发平台	属性	参数	权限	说明
Sensor-A（601）	温度	A0	R	温度值，浮点型数据，精度为 0.1，范围为-40～105
	湿度	A1	R	湿度值，浮点型数据，精度为 0.1，范围为 0～100
	光照度	A2	R	光照度值，浮点型数据，精度为 0.1，范围为 0～65535
	空气质量	A3	R	空气质量值
	气压	A4	R	气压值，浮点型数据，精度为 0.1
	三轴（跌倒状态）	A5	R	三轴：通过计算上报跌倒状态
	距离	A6	R	距离值（cm），浮点型数据，精度为 0.1，范围为 10～80
	语音识别返回码	A7	—	语音识别码，整型：1～49（主动上报，不可查询）
	上报状态	D0（OD0/CD0）	R/W	D0 的 Bit0～Bit7 分别代表 A0～A7 传感器数据的上报
	继电器状态	D1（OD1/CD1）	R/W	D1 的 Bit6～Bit7 分别代表开关 K1、K2 的状态
	数据上报时间间隔	V0	R/W	A0～A7 传感器数据的循环上报时间间隔

（2）驱动程序的开发。在智云框架下不仅可以很容易地实现传感器驱动程序的开发，还可以省略无线传感器节点的组网和用户任务的创建等烦琐过程。例如，调用 sensorInit()函数可以实现传感器的初始化；调用 ZXBeeInfRecv()函数可以处理节点接收到的无线数据包；设备状态的定时上报使用 MyEventProcess()作为 sensorUpdate()函数的定时进入接口来反馈设备状态信息。

在 sensor.c 中，需要在 sensorInit()函数中添加传感器初始化的内容，并通过定义上报事件和报警事件来实现设备工作状态的定时反馈。部分代码如下：

```
void sensorInit(void)
{
    //初始化传感器代码
    htu21d_init();                                    //温湿度传感器的初始化
    bh1750_init();                                    //光照度传感器的初始化
    airgas_init();                                    //空气质量传感器的初始化
    fbm320_init();                                    //气压海拔传感器的初始化
    //启动定时器，触发上报事件 MY_REPORT_EVT
    osal_start_timerEx(sapi_TaskID, MY_REPORT_EVT, (uint16)((osal_rand()%10) *1000));
    //启动定时器，触发报警事件 MY_CHECK_EVT
    osal_start_timerEx(sapi_TaskID, MY_CHECK_EVT, 100);
}
```

温湿度传感器的初始化函数是 htu21d_init()，该函数是通过 IIC 总线写寄存器地址来初始化温湿度传感器的。htu21d_read_reg()函数实现读寄存器的操作，htu21d_get_data()函数实现温湿度的测量。IIC 总线的驱动代码请查看项目源码，这里不做讲述。htu21d.c 程序文件部分代码如下：

```
/*********************************************************************************
*名称：htu21d_init()
*功能：HTU21D 型温湿度传感器的初始化
*********************************************************************************/
void htu21d_init(void)
{
    iic_init();                                       //IIC 总线初始化
    iic_start();                                      //开启 IIC 总线
    iic_write_byte(HTU21DADDR&0xfe);                  //写 HTU21D 型温湿度传感器的 IIC 总线地址
    iic_write_byte(0xfe);
    iic_stop();                                       //停止 IIC 总线
}
/*********************************************************************************
*名称：htu21d_read_reg()
*功能：读取寄存器
*参数：cmd—寄存器地址
*返回：data—寄存器数据
*********************************************************************************/
unsigned char htu21d_read_reg(unsigned char cmd)
{
    unsigned char data = 0;
    iic_start();
    if(iic_write_byte(HTU21DADDR & 0xfe) == 0){       //写 HTU21D 型温湿度传感器的 IIC 总线地址
        if(iic_write_byte(cmd) == 0){                 //写寄存器地址
            do{
                delay(30);                            //延时 30 ms
                iic_start();                          //开启 IIC 总线
```

```
                }
                while(iic_write_byte(HTU21DADDR | 0x01) == 1);       //发送读信号
                data = iic_read_byte(0);                              //读取一个字节数据
                iic_stop();                                           //停止 IIC 总线
            }
        }
        return data;
}
/*********************************************************************************************
*名称：htu21d_get_data()
*功能：HTU21D 型温湿度传感器采集的数据
*参数：order—指令
*返回：temperature—温度值；humidity—湿度值
*********************************************************************************************/
int htu21d_get_data(unsigned char order)
{
        float temp = 0,TH = 0;
        unsigned char MSB,LSB;
        unsigned int humidity,temperature;
        iic_start();    //IIC 总线启动
        if(iic_write_byte(HTU21DADDR & 0xfe) == 0){      //写 HTU21D 型温湿度传感器的 IIC 总线地址
            if(iic_write_byte(order) == 0){              //写寄存器地址
                do{
                    delay(30);
                    iic_start();
                }
                while(iic_write_byte(HTU21DADDR | 0x01) == 1);      //发送读信号
                MSB = iic_read_byte(0);                              //读取数据高 8 位
                delay(30);                                           //延时 30 ms
                LSB = iic_read_byte(0);                              //读取数据低 8 位
                iic_read_byte(1);
                iic_stop();                                          //停止 IIC 总线
                LSB &= 0xfc;                                         //取出数据有效位
                temp = MSB*256+LSB;                                  //数据合并
                if (order == 0xf3){                                  //触发开启温度监测
                    TH=(175.72)*temp/65536-46.85;                    //温度为-46.85 + 175.72 *ST/2^16
                    temperature =(unsigned int)(fabs(TH)*100);
                    if(TH >= 0)
                        flag = 0;
                    else
                        flag = 1;
                    return temperature;
                }else{
                    TH = (temp*125)/65536-6;
                    humidity = (unsigned int)(fabs(TH)*100);         //湿度为 RH%= -6 + 125 *SRH/2^16
                    return humidity;
                }
```

```
            }
        }
        iic_stop();
        return 0;
}
```

在 bh1750.c 程序文件中，光照度传感器的初始化函数是 bh1750_init()，该函数是通过初始化 IIC 总线引脚来初始化光照度传感器的。bh1750.c 程序文件部分代码如下：

```
/*********************************************************************************
*名称：bh1750_send_byte()
*功能：光照度传感器发送数据
*********************************************************************************/
uchar bh1750_send_byte(uchar sla, uchar c)
{
    iic_start();                                    //开启 IIC 总线
    if(iic_write_byte(sla) == 0){                   //发送器件地址
        if(iic_write_byte(c) == 0){                 //发送数据
        }
    }
    iic_stop();                                     //停止 IIC 总线
    return(1);
}
/*********************************************************************************
*名称：bh1750_read_nbyte()
*功能：连续读出光照度传感器的内部数据
*返回：应答或非应答信号
*********************************************************************************/
uchar bh1750_read_nbyte(uchar sla,uchar *s,uchar no)
{
    uchar i;
    iic_start();                                    //起始信号
    if(iic_write_byte(sla+1) == 0){                 //发送设备地址+读信号
        for (i=0; i<no-1; i++){                     //连续读取 6 个地址数据，存储在 buf 中
            *s=iic_read_byte(0);
            s++;
        }
        *s=iic_read_byte(1);
    }
    iic_stop();
    return(1);
}
/*********************************************************************************
*名称：bh1750_init()
*功能：初始化光照度传感器
*********************************************************************************/
void bh1750_init()
```

```
{
    iic_init();
}
/*********************************************************************************************
*名称：bh1750_get_data()
*功能：光照度传感器数据的处理函数
*********************************************************************************************/
float bh1750_get_data(void)
{
    uchar *p=buf;
    bh1750_init();                              //初始化光照度传感器
    bh1750_send_byte(0x46,0x01);                //上电
    bh1750_send_byte(0x46,0X20);                //高分辨率模式
    delay_ms(180);                              //延时 180 ms
    bh1750_read_nbyte(0x46,p,2);                //连续读出数据，存储在 buf 中
    unsigned short x = buf[0]<<8 | buf[1];
    return x/1.2;
}
```

空气质量传感器的初始化函数是 airgas_init()，代码如下：

```
/*********************************************************************************************
*名称：airgas_init()
*功能：空气质量传感器初始化
*********************************************************************************************/
void airgas_init(void)
{
    APCFG |= 0x20;                              //模拟 I/O 使能
    P0SEL |= 0x20;                              //端口 P0_5 功能选择外设功能
    P0DIR &= ~0x20;                             //设置输入模式
    ADCCON3  = 0xB5;                            //选择 AVDD5 为参考电压；12 位分辨率；P0_5 连接 ADC
    ADCCON1 |= 0x30;                            //选择 ADC 的启动模式为手动
}
/*********************************************************************************************
*名称：unsigned int get_airgas_data(void)
*功能：获取空气质量传感器状态
*********************************************************************************************/
unsigned int get_airgas_data(void)
{
    unsigned int  value;
    ADCCON3  = 0xB5;                            //选择 AVDD5 为参考电压；12 位分辨率；P0_5 连接 ADC
    ADCCON1 |= 0x30;                            //选择 ADC 的启动模式为手动
    ADCCON1 |= 0x40;                            //启动 A/D 转化
    while(!(ADCCON1 & 0x80));                   //等待 A/D 转化结束
    value =  ADCL >> 2;
    value |= (ADCH << 6)>> 2;                   //取得最终转化结果，存入 value 中
    return value;                               //返回有效值
}
```

fbm320.c程序文件中定义的函数较多，这里主要列出初始化与数据获取的代码。气压海拔传感器的初始化函数是fbm320_init()，该函数通过IIC总线读取ID来判断气压海拔传感器的初始化是否成功，代码如下：

```
//fbm320.h
/*********************************************************************************************
*fbm320 函数原型声明
*********************************************************************************************/
unsigned char fbm320_read_id(void);
unsigned char fbm320_read_reg(unsigned char reg);
void fbm320_write_reg(unsigned char reg,unsigned char data);
long fbm320_read_data(void);
void Coefficient(void);
void Calculate(long UP, long UT);
unsigned char fbm320_init(void);
int fbm320_data_get(float *temperature,long *pressure);

//fbm320.c
/*********************************************************************************************
*名称：fbm320_init()
*功能：气压海拔传感器初始化
*********************************************************************************************/
unsigned char fbm320_init(void)
{
    iic_init();                                                    //IIC 总线初始化
    if(fbm320_read_id() == 0)                                      //判读初始化是否成功
    return 0;
    Coefficient();
    return 1;
}
/*********************************************************************************************
*名称：fbm320_data_get()
*功能：传感器数据读取函数
*********************************************************************************************/
int fbm320_data_get(float *temperature,long *pressure)
{
    //Coefficient();                                               //系数换算
    fbm320_write_reg(FBM320_CONFIG,TEMPERATURE);                   //发送识别信息
    delay_ms(5);                                                   //延时 5 ms
    UT_I = fbm320_read_data();                                     //读取传感器数据
    fbm320_write_reg(FBM320_CONFIG,OSR8192);                       //发送识别信息
    delay_ms(20);                                                  //延时 20 ms
    UP_I = fbm320_read_data();                                     //读取传感器数据
    if (UT_I == -1 || UP_I == -1){
        return -1;
    }
    Calculate( UP_I, UT_I);                                        //传感器数值换算
```

```
    *temperature = RT_I *0.01f;                                    //温度计算
    *pressure = RP_I;                                              //压力计算
    return 0;
}
```

项目的自定义事件处理代码中 event 的传递参数为系统参数，此处不用处理。执行 sensorUpdate()函数的条件就是 event 与 MY_REPORT_EVT 相与为真。虽然 MY_REPORT_EVT 可以自由设置，但一定要与 sensor.h 文件下配置任务名称一致。当 if 条件判断语句为真时执行 sensorUpdate()函数，执行完成后更新用户事件。用户事件更新函数 osal_start_timerEx()中参数 MY_REPORT_EVT 要与用户定义的事件参数一致。sensor.c 程序文件部分代码如下：

```
/*********************************************************************************************
*名称：MyEventProcess()
*功能：自定义事件处理
*参数：event—事件编号
*********************************************************************************************/
void MyEventProcess( uint16 event )
{
    if (event & MY_REPORT_EVT) {
        sensorUpdate();
        //启动定时器，触发事件 MY_REPORT_EVT
        osal_start_timerEx(sapi_TaskID, MY_REPORT_EVT, V0*1000);
    }
    if (event & MY_CHECK_EVT) {
        sensorCheck();
        //启动定时器，触发事件 MY_CHECK_EVT
        osal_start_timerEx(sapi_TaskID, MY_CHECK_EVT, 100);
    }
}
```

设备传感器定时反馈的相关代码如下。

```
/*********************************************************************************************
*名称：sensorUpdate()
*功能：处理主动上报的数据
*********************************************************************************************/
void sensorUpdate(void)
{
    char pData[16];
    char *p = pData;
    ZXBeeBegin();                         //智云数据帧格式的包头
    //根据 D0 的位状态判定需要主动上报的数值
    if ((D0 & 0x01) == 0x01){             //若温度上报允许，则在 pData 数据包中添加温度数据
        updateA0();
        sprintf(p, "%.1f", A0);
        ZXBeeAdd("A0", p);
    }
    if ((D0 & 0x02) == 0x02){             //若湿度上报允许，则在 pData 数据包中添加湿度数据
```

```
            updateA1();
            sprintf(p, "%.1f", A1);
            ZXBeeAdd("A1", p);
        }
        if ((D0 & 0x04) == 0x04){           //若光照度上报允许，则在 pData 数据包中添加光照度数据
            updateA2();
            sprintf(p, "%.1f", A2);
            ZXBeeAdd("A2", p);
    }
        if ((D0 & 0x08) == 0x08){           //若空气质量上报允许，则在 pData 数据包中添加空气质量数据
            updateA3();
            sprintf(p, "%u", A3);
            ZXBeeAdd("A3", p);
        }
        if ((D0 & 0x10) == 0x10){           //若大气压强上报允许，则在 pData 数据包中添加大气压强数据
            updateA4();
            sprintf(p, "%.1f", A4);
            ZXBeeAdd("A4", p);
        }
        sprintf(p, "%u", D1);                //上报控制编码
        ZXBeeAdd("D1", p);

        p = ZXBeeEnd();                      //智云数据帧格式的包尾
        if (p != NULL) {
            ZXBeeInfSend(p, strlen(p)); //将需要上传的数据打包，并通过 zb_SendDataRequest()发送到协调器
        }
}
```

2. Android 端应用设计

1）Android 工程设计框架

打开 Android Studio 开发环境，可以看到本系统的工程目录，如图 3.10 所示。系统的工程框架如表 3.4 所示。

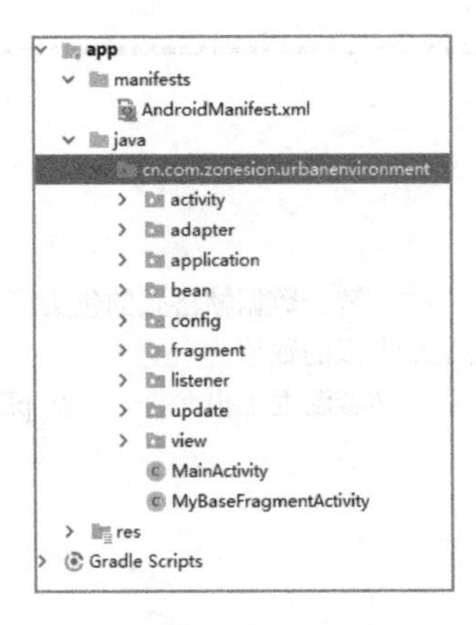

图 3.10　城市环境信息采集系统的工程目录

表 3.4 城市环境信息采集系统的工程框架

类 名	说 明
activity	
IdKeyShareActivity.java	在 IDKey 页面单击“分享”按钮时，可弹出 activity，用于分享二维码图片
adapter	
HdArrayAdapter.java	历史数据显示适配器
application	
LCApplication.java	LCApplication 继承 application 类，使用单例模式（Singleton Pattern）创建 WSNRTConnect 对象
bean	
HistoricalData.java	历史数据的 bean 类，用于将从智云服务器获得的历史数据记录（JSON 形式）转换成该类的对象
IdKeyBean.java	IdKeyBean 用来描述用户设备的 ID、KEY，以及智云服务器的地址 SERVER
config	
Config.java	config 用于修改用户的 ID、KEY，以及智云服务器的地址和 MAC 地址
fragment	
BaseFragment.java	页面基础 Fragment 定义类
HDFragment.java	历史数据页面
HistoricalDataFragment.java	历史数据显示页面
HomepageFragment.java	展示首页页面的 Fragment
IDKeyFragment.java	IDKey 选项的页面
MacSettingFragment.java	当用户设置被监测项的 MAC 地址时显示的页面
MoreInformationFragment.java	更多信息显示页面
RunHomePageFragment.java	运营首页显示页面
VersionInformationFragment.java	显示版本等相关信息的页面
listener	
IOnWSNDataListener.java	传感器数据监听器接口
update	
UpdateService.java	应用下载服务类
view	
APKVersionCodeUtils.java	获取当前本地 apk 的版本
CustomRadioButton.java	自定义按钮类
PagerSlidingTabStrip.java	自定义滑动控件类
MainActivity.java：主页面类	
MyBaseFragmentActivity.java：系统 Fragment 通信类	

2）软件设计

根据智云 Android 端应用程序接口的定义，城市环境信息采集系统的应用设计主要采用实时数据 API 接口，其流程如图 3.11 所示。

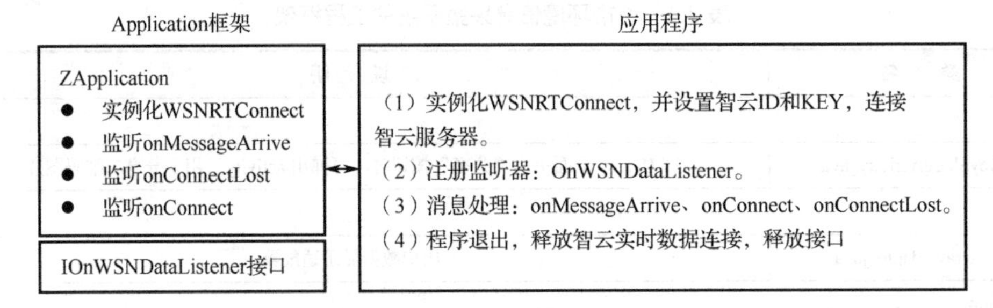

图 3.11　实时数据 API 接口的流程

（1）LCApplication.java 程序代码剖析。实时连接对象应用程序的实现过程是：从 LCApplication 获取 WSNRTConnect，建立智云实时数据连接，注册数据监听器 registerOnWSNDataListener，通过复写 onMessageArrive 方法来处理节点接收到的无线数据包，通过按钮发送传感器控制命令。LCApplication 的关键代码如下：

```
public class LCApplication extends Application implements WSNRTConnectListener{
    private WSNRTConnect wsnrtConnect;    //创建 WSNRTConnect 实例
    private ArrayList<IOnWSNDataListener> mIOnWSNDataListeners = new ArrayList<>();
    //传感器数据监听器数组
    private boolean isDisconnected = true;    //当前是否断开连接，true 表示连接断开，false 表示已经连接
    public boolean getIsDisconnected() {    //属性 isDisconnected 的 getter 方法
        return isDisconnected;
    }
    public void setDisconnected(boolean disconnected) {    //属性 isDisconnected 的 setter 方法
        isDisconnected = disconnected;
    }
    public WSNRTConnect getWSNRConnect() {
        if (wsnrtConnect == null) {
            wsnrtConnect = new WSNRTConnect();    //初始化 WSNRTConnect 实例
        }
        return wsnrtConnect;
    }
    //注册传感器数据监听器
    public void registerOnWSNDataListener(IOnWSNDataListener li) {
        mIOnWSNDataListeners.add(li);
    }
    //取消注册传感器数据监听器
    public void unregisterOnWSNDataListener(IOnWSNDataListener li) {
        mIOnWSNDataListeners.remove(li);
    }
    @Override
    public void onCreate() {
        super.onCreate();
        wsnrtConnect = getWSNRConnect();
        wsnrtConnect.setRTConnectListener(this);
    }
```

```
        @Override
        public void onConnectLost(Throwable throwable) {
            Toast.makeText(this, "数据服务断开连接！", Toast.LENGTH_SHORT).show();
            for (IOnWSNDataListener li : mIOnWSNDataListeners) {
                li.onConnectLost();
            }
        }
        @Override
        public void onConnect() {
            Toast.makeText(this, "数据服务连接成功！", Toast.LENGTH_SHORT).show();
            for (IOnWSNDataListener li : mIOnWSNDataListeners) {
                li.onConnect();
            }
        }
        //消息到达时会自动调用该方法
        @Override
        public void onMessageArrive(String mac, byte[] data) {
            if (data[0] == '{' && data[data.length - 1] == '}') {
                String sData = new String(data, 1, data.length - 2);
                String[] pDatas = sData.split(",");
                for (String pData : pDatas) {
                    String[] tagVal = pData.split("=");
                    if (tagVal.length == 2) {
                        for (IOnWSNDataListener li : mIOnWSNDataListeners) {
                            li.onMessageArrive(mac, tagVal[0], tagVal[1]);
                        }
                    }
                }
            }
        }
    }
```

（2）HomepageFragment.java 程序代码剖析。下面的代码通过(LCApplication) getActivity().getApplication()获取 LCApplication 类中的 WSNRTConnect 对象。

```
    private void initViewAndBindEvent() {
        preferences = getActivity().getSharedPreferences("user_info", Context.MODE_PRIVATE);
        lcApplication = (LCApplication) getActivity().getApplication();
        wsnrtConnect = lcApplication.getWSNRConnect();
        lcApplication.registerOnWSNDataListener(this);
        editor = preferences.edit();
    }
```

下面的代码通过复写 onMessageArrive 方法来处理节点接收到的无线数据包，实现了节点设备 MAC 地址的获取，并在当前的页面显示节点设备的状态。

```
    public void onMessageArrive(String mac, String tag, String val) {
        if (sensorAMAC == null) {
```

```
            wsnrtConnect.sendMessage(mac, "{TYPE=?}".getBytes());
        }
        if ("TYPE".equals(tag) && i.equals(val.substring(2, val.length()))) {
            sensorAMAC = mac;
        }
        if (mac.equals(sensorAMAC) && "A0".equals(tag)) {
            textTemperatureState.setText("在线");
            textTemperatureState.setTextColor(getResources().getColor(R.color.line_text_color));
            temperatureText.setText(val+"°C");
        }
        if (mac.equals(sensorAMAC) && "A1".equals(tag)) {
            textHumidityState.setText("在线");
            textHumidityState.setTextColor(getResources().getColor(R.color.line_text_color));
            humidityText.setText(val+"µg/m3");
        }
        if (mac.equals(sensorAMAC) && "A2".equals(tag)) {
            textIlluminationState.setText("在线");
            textIlluminationState.setTextColor(getResources().getColor(R.color.line_text_color));
            float v = Float.parseFloat(val);
            illuminationCircleView.setCurrentStatus(v);
            illuminationCircleView.invalidate();
        }
        if (mac.equals(sensorAMAC) && "A3".equals(tag)) {
            textQualityState.setText("在线");
            textQualityState.setTextColor(getResources().getColor(R.color.line_text_color));
            float v = Float.parseFloat(val);
            qualityCircleView.setCurrentStatus(v);
            qualityCircleView.invalidate();
        }
        if (mac.equals(sensorAMAC) && "A4".equals(tag)) {
            textPressureState.setText("在线");
            textQualityState.setTextColor(getResources().getColor(R.color.line_text_color));
            float v = Float.parseFloat(val);
            pressureCircleView.setCurrentStatus(v);
            pressureCircleView.invalidate();
        }
}
```

3. Web端应用设计

1）页面功能结构分析

城市环境信息采集系统Web端应用默认显示的是“运营首页”，“运营首页”上有5个模块，即温度数据显示模块、湿度数据显示模块、光照度数据显示模块、空气质量显示模块、大气压强显示模块，如图3.12所示。

“环境数据”页面通过Web端的地图插件实时显示不同城市的环境数据，如图3.13所示。

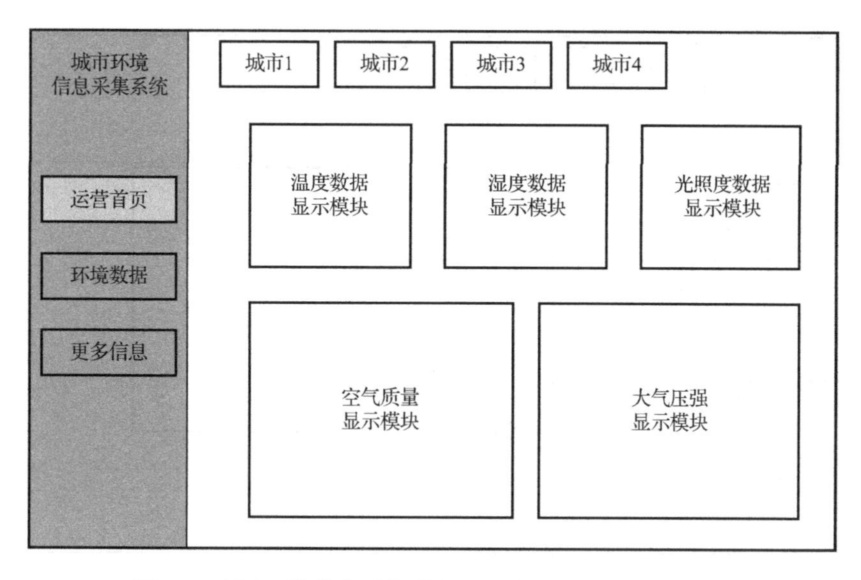

图 3.12 城市环境信息采集系统 Web 端的“运营首页”页面

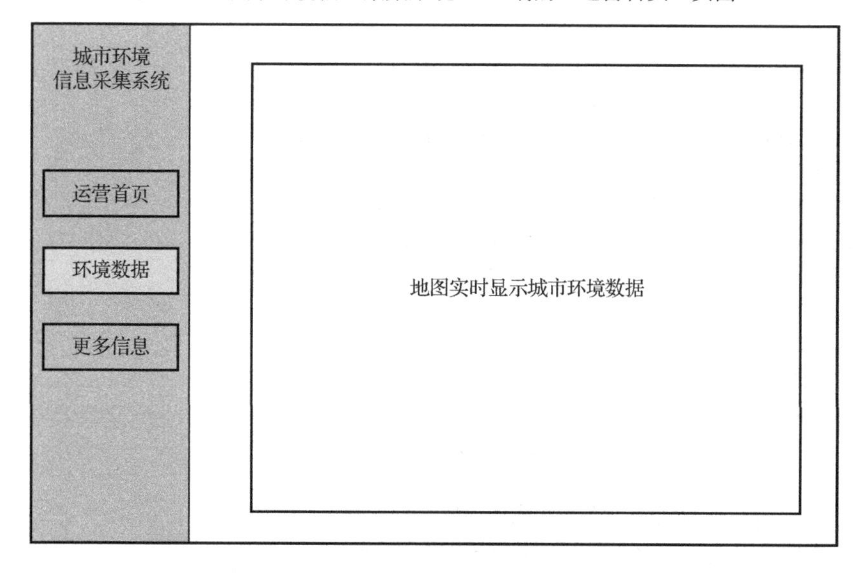

图 3.13 城市环境信息采集系统 Web 端的“环境数据”页面

“更多信息”页面的功能主要是配置智云服务器的连接，本页面分为三个标签，通过标签进行切换显示，“ID Key”标签通过设置智云服务器 ID 与 KEY 调用智云 Web 端应用接口并连接到智云服务器。“MAC 设置”标签显示设备 MAC 地址，“版本信息”标签显示版本信息与升级，如图 3.14 所示。

2）软件设计

城市环境信息采集系统 Web 端的 JS 开发逻辑与 Android 端的开发逻辑相似，首先通过配置 ID 和 KEY 与智云服务器进行连接，再通过实时监听数据的方法来获取相关传感器的数据并进行处理。JS 开发的部分代码如下。

在 getConnect()函数中定义了实时连接对象 rtc，连接成功回调函数是 rtc.onConnect，数据服务掉线回调函数是 rtc.onConnectLost，消息处理回调函数是 rtc.onmessageArrive。

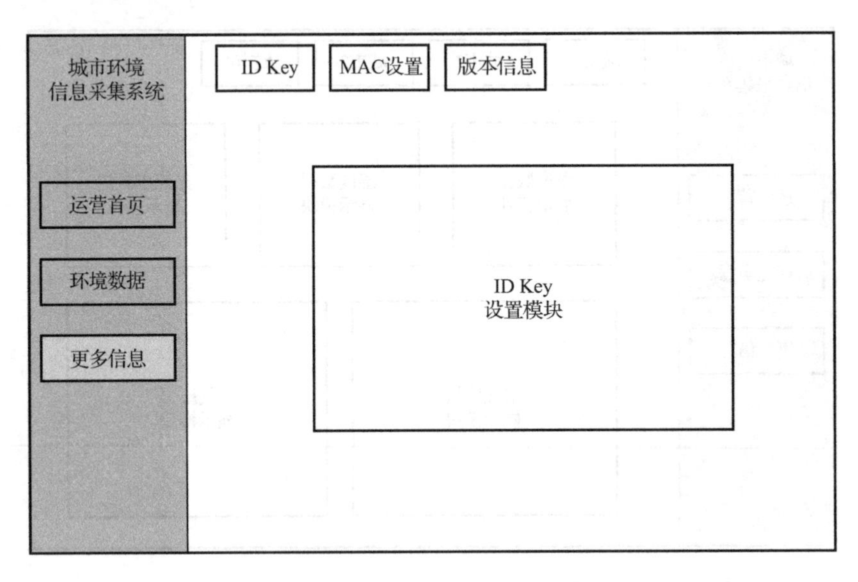

图 3.14 城市环境信息采集系统 Web 端的“更多信息”页面

```
function getConnect() {
    config["id"] = config["id"] ? config["id"] : $("#ID").val();
    config["key"] = config["key"] ? config["key"] : $("#KEY").val();
    config["server"] = config["server"] ? config["server"] : $("#server").val();
    //创建数据连接服务对象
    rtc = new WSNRTConnect(config["id"], config["key"]);
    rtc.setServerAddr(config["server"] + ":28080");
    rtc.connect();
    rtc._connect = false;
    //连接成功回调函数
    rtc.onConnect = function() {
        $("#ConnectState").text("数据服务连接成功！");
        rtc._connect = 1;
        message_show("数据服务连接成功！");
        $("#idkeyInput").text("断开").addClass("btn-danger");
        $("#id,#key,#server").attr('disabled',true)
    };
    //数据服务掉线回调函数
    rtc.onConnectLost = function() {
        rtc._connect = 0;
        $("#ConnectState").text("数据服务连接掉线！");
        $("#idkeyInput").text("连接").removeClass("btn-danger");
        message_show("数据服务连接失败，检查网络或 IDKEY");
        $(".online_601").text("离线").css("color", "#e75d59");
        $(".online_602").text("离线").css("color", "#e75d59");
        $("#id,#key,#server").removeAttr('disabled');
    };
    //消息处理回调函数
    rtc.onmessageArrive = function (mac, dat) {
```

```
            //console.log(mac+" >>> "+dat);
            if (dat[0]=='{' && dat[dat.length-1]=='}') {
                dat = dat.substr(1, dat.length-2);
                var its = dat.split(',');
                for (var i=0; i<its.length; i++) {
                    var it = its[i].split('=');
                    if (it.length == 2) {
                        process_tag(mac, it[0], it[1]);
                    }
                }
                if (!mac2type[mac]) { //如果没有获取到 TYPE 值，主动去查询
                    rtc.sendMessage(mac, "{TYPE=?,A0=?,A1=?,A2=?,A3=?,A4=?,A5=?,A6=?,A7=?,D1=?}");
                }
            }
        }
    }
```

下述 JS 开发代码的功能是根据设备连接情况，在页面更新设备的状态，若设备在线，则可以采集城市环境信息。

```
var wsn_config = {
    "601" : {
        "online" : function() {
            $(".online_601").text("在线").css("color", "#96ba5c");
        },
        "pro" : function (tag, val) {
            if(tag=="A0"){
                thermometer('WH_temp','℃','#ff7850', -20, 80, val);
                thermometer('BJ_temp','℃','#ff7850', -20, 80, val);
                thermometer('SH_temp','℃','#ff7850', -20, 80, val);
                thermometer('SZ_temp','℃','#ff7850', -20, 80, val);
            }
            else if(tag=="A1"){
                thermometer('WH_humi','%','#27A9E3', 0, 100, val);
                thermometer('BJ_humi','%','#27A9E3', 0, 100, val);
                thermometer('SH_humi','%','#27A9E3', 0, 100, val);
                thermometer('SZ_humi','%','#27A9E3', 0, 100, val);
            }
            else if(tag=="A2"){
                dial2('WH_illum','Lx',val);
                dial2('BJ_illum','Lx',val);
                dial2('SH_illum','Lx',val);
                dial2('SZ_illum','Lx',val);
            }
            else if(tag=="A3"){
                dial('WH_PM2.5','μg/m3',val);
                dial('BJ_PM2.5','μg/m3',val);
```

```
                dial('SH_PM2.5','μg/m3',val);
                dial('SZ_PM2.5','μg/m3',val);
                WH_data = val;
                map("PM_Map","#5cadba",BJ_data,SH_data,SZ_data,WH_data);
            }
            else if(tag == "A4") {
                dial3('WH_pressure','kpa',val);
                dial3('BJ_pressure','kpa',val);
                dial3('SH_pressure','kpa',val);
                dial3('SZ_pressure','kpa',val);
            }
        }
    },
};
```

3.1.4 开发验证

1. Web 端应用测试

在 Web 端打开城市环境信息采集系统后，其“运营首页”页面如图 3.15 所示。

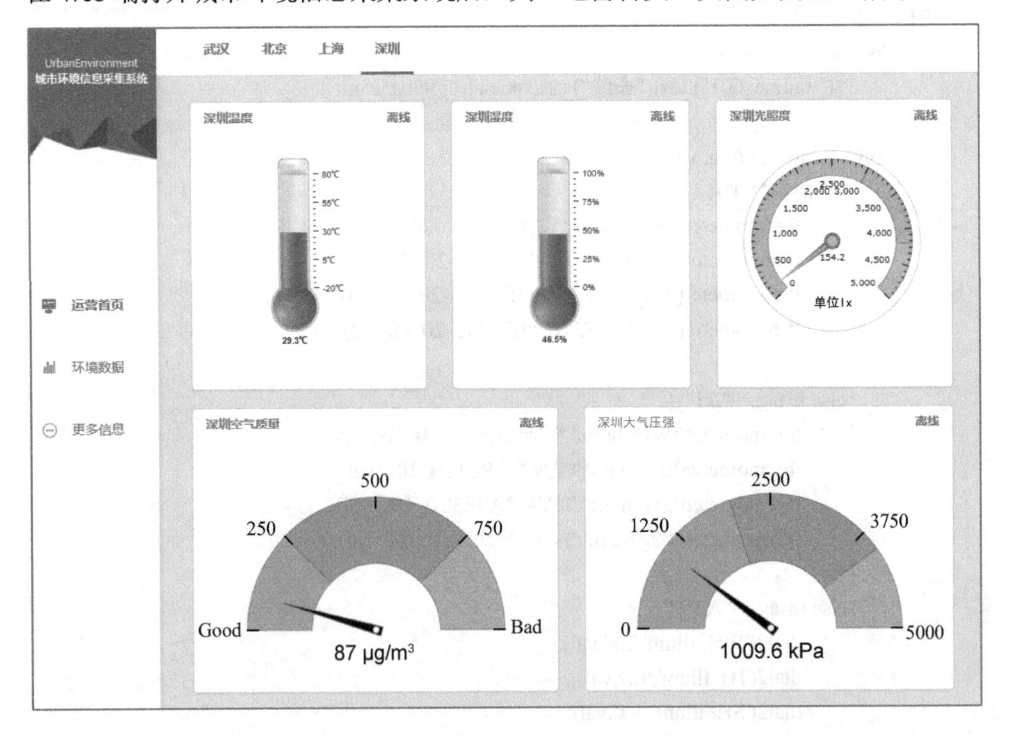

图 3.15 城市环境信息采集系统 Web 端的“运营首页”页面

此时设备的右上角状态显示为“离线”，在“更多信息”页面设置智云服务器的 ID 与 KEY 后可以连接到智云服务器。城市环境信息采集系统的“更多信息”页面如图 3.16 所示。

选择“MAC 设置”可以查看传感器节点的 MAC 地址，如图 3.17 所示。

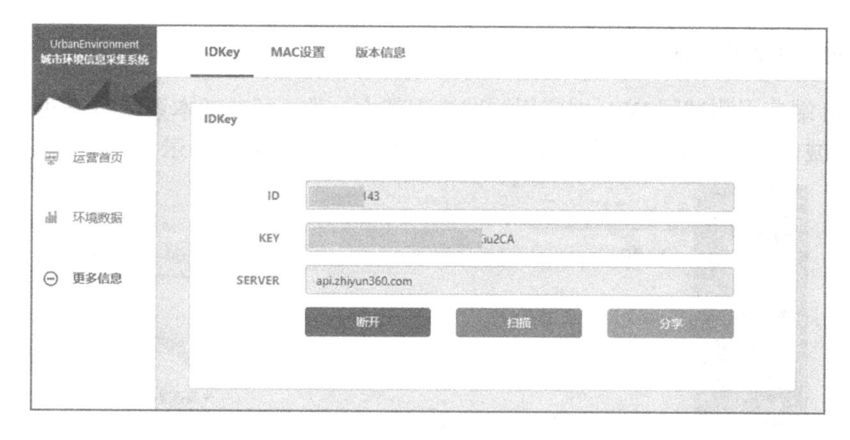

图 3.16 城市环境信息采集系统 Web 端的“更多信息”页面

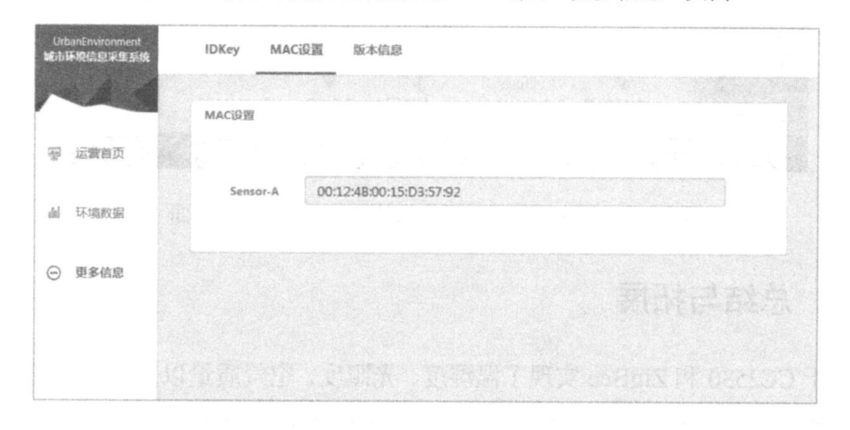

图 3.17 城市环境信息采集系统 Web 端“更多信息”页面的“MAC 设置”

成功连接智云服务器后切换到“运营首页”页面，可以看到设备状态更新为“在线”，并且已更新了传感器采集的数据，如图 3.18 所示。

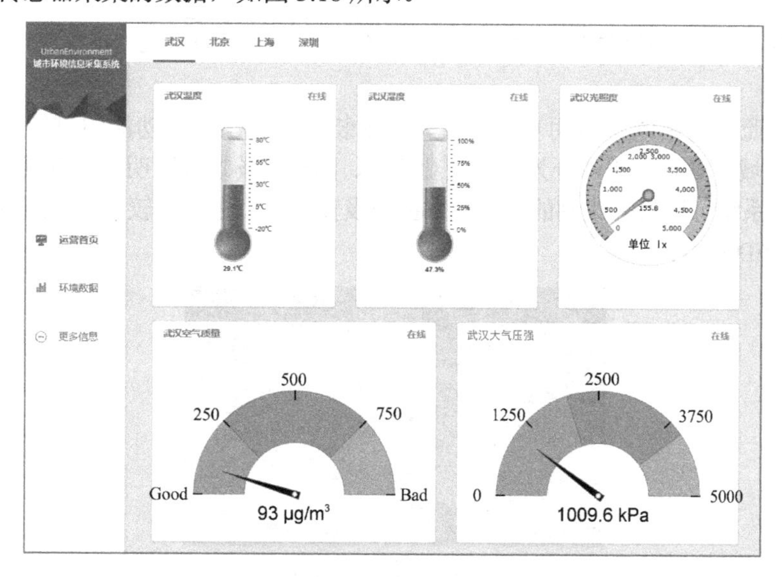

图 3.18 城市环境信息采集系统 Web 端的设备在线状态

2. Android 端应用测试

Android 端应用测试同 Web 端应用测试流程基本一致，可参考本系统的 Web 端应用测试进行操作。城市环境信息采集系统 Android 端的主页面如图 3.19 所示。

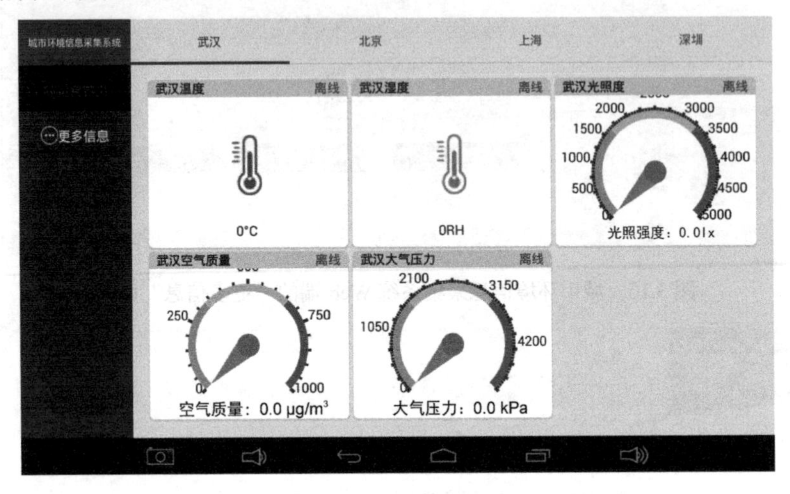

图 3.19　城市环境信息采集系统 Android 端的主页面

3.1.5　总结与拓展

本节基于 CC2530 和 ZigBee 实现了温湿度、光照度、空气质量以及气压海拔等传感器数据的采集，通过 Android 和 HTML5 技术实现了 Android 端和 Web 端的应用设计，可以在页面中显示温湿度、光照度、空气质量以及大气压强等数据，实现了基于 ZigBee 的城市环境信息采集系统。

3.2　基于 ZigBee 的城市景观照明控制系统

城市景观照明是指既有照明功能，又有艺术装饰和美化环境功能的户外照明工程，如图 3.20 所示。城市景观照明可分为道路景观照明、园林广场景观照明、建筑景观照明，城市景观照明控制系统可以将采集到的光照度数据与设置的阈值进行比较，当光照度数据低于阈值时就启动 LED 进行照明。

图 3.20　城市景观照明

3.2.1 系统开发目标

（1）熟悉光照度传感器和 LED 等硬件的原理以及数据通信协议，基于 CC2530 和 ZigBee 实现光照度传感器和 LED 的驱动开发，将实时采集到的光照度数据发送到汇集节点，然后发送至智云平台，通过 App 可以获得这些数据。当光照度数据低于设置的阈值时，系统既可以自动开启 LED，也可以由用户手动开启 LED。

（2）实现城市景观照明控制系统的 Android 端应用开发和 Web 端应用开发。

3.2.2 系统设计分析

1．系统的功能设计

城市景观照明控制系统的主要功能是通过光照度传感器实时地采集光照度数据，并将采集到的光照度数据主动推送到智云平台，在 Android 端和 Web 端获得这些数据后，用户就能够随时随地控制 LED 的开关。从系统功能的角度来看，城市景观照明控制系统可以分为设备采集和控制模块以及系统设置模块，如图 3.21 所示。

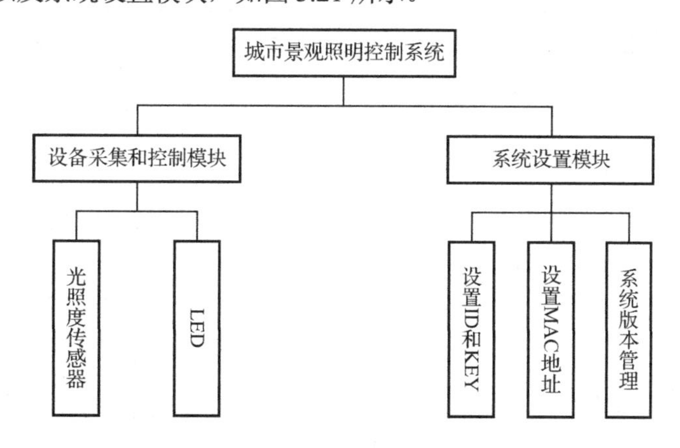

图 3.21 城市景观照明控制系统的组成模块

城市景观照明控制系统的功能需求分析如表 3.5 所示。

表 3.5 城市景观照明控制系统的功能需求分析

功 能	功 能 说 明
数据的更新和显示	在上层应用页面中实时更新和显示采集到的光照度数据
LED 控制	通过上层应用程序，对 LED 进行控制
模式设置	自动模式：定时或者根据光照度数据来自动控制照明设备（LED）。手动模式：通过页面手动开关照明设备
智云连接设置	智云服务器的参数设置与连接

2．系统的总体架构设计

城市景观照明控制系统采用物联网项目架构进行设计，由感知层、网络层、平台层和应用层组成。城市景观照明控制系统的总体架构如图 3.22 所示。

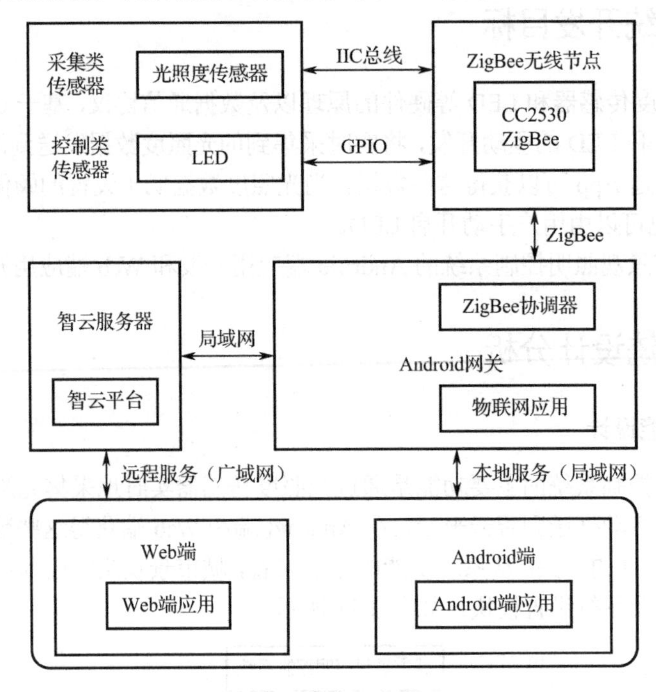

图 3.22　城市景观照明控制系统的总体架构

3．系统的数据传输

城市景观照明控制系统的数据传输是在传感器节点、智能网关以及客户端（包括 Web 端和 Android 端）之间进行的，如图 3.23 所示。

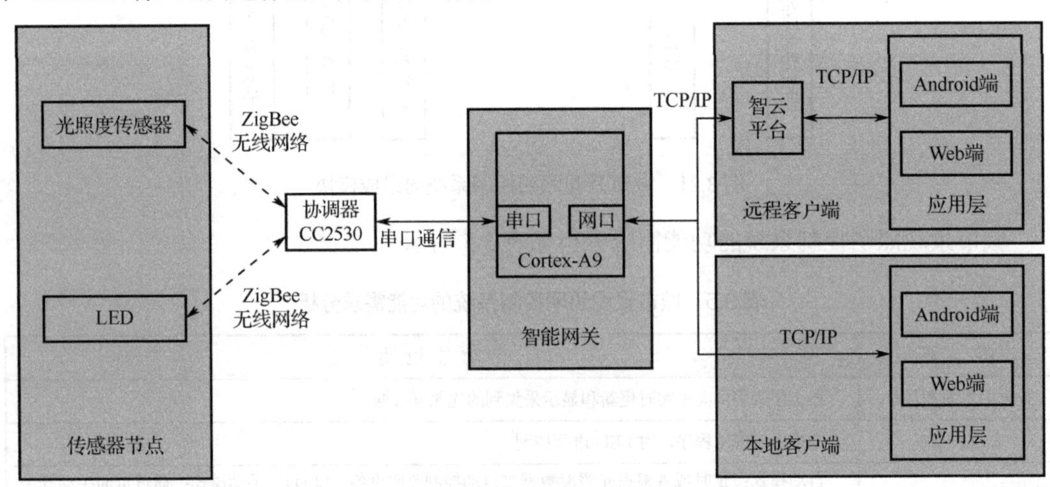

图 3.23　城市景观照明控制系统的数据传输

3.2.3　系统的软硬件开发：城市景观照明控制系统

1. 系统底层软硬件设计

1）感知层硬件设计

城市景观照明控制系统的感知层硬件主要包括 xLab 未来开发平台的智能网关、经典型无线节点 ZXBeeLiteB、采集类开发平台 Sensor-A、控制类开发平台 Sensor-B。其中，智能网关负责汇集传感器采集的数据；ZigBee 无线节点（由经典型无线节点 ZXBeeLiteB 实现）通过无线通信的方式向智能网关发送传感器采集的数据，接收智能网关的控制命令；采集类开发平台 Sensor-A 和控制类开发平台 Sensor-B 连接到 ZigBee 无线节点，由其中的 CC2530 微处理器对相关设备进行控制。本系统主要使用光照度传感器、两路高亮 LED，光照度传感器的硬件接口电路如图 3.6 所示（见 3.1.3 节），LED 的硬件接口电路如图 3.24 所示（图中也包含 RGB 灯的硬件接口电路）。

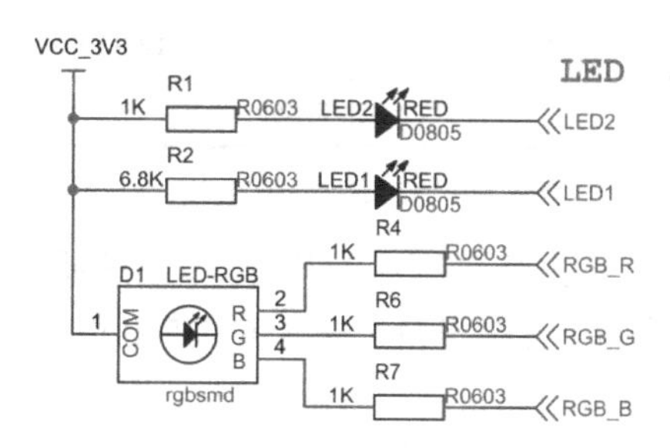

图 3.24　LED 和 RGB 灯的硬件接口电路

2）系统底层开发

基于 ZigBee 的城市景观照明控制系统的底层开发和基于 ZigBee 的城市环境信息采集系统的底层开发相同，详见 3.1.3 节。

3）传感器驱动设计

城市景观照明控制系统的底层硬件主要包括采集类传感器和控制类传感器，传感器驱动设计主要是针对这两类传感器进行的。采集类传感器的逻辑事件有 4 种，详见 3.1.3 节的相关内容。对于控制类传感器，主要关注的是它对远程设备的控制是否有效，以及控制的结果。控制类传感器逻辑事件可分为以下 3 种。

① 远程设备向控制类传感器发送控制指令，控制类传感器能够实时响应并执行相应的操作。

② 远程设备发送查询指令后，控制类传感器能够实时响应并反馈设备的状态。当远程设备向控制类传感器发出控制指令后，远程设备并不了解控制类传感器是否完成了对设备的控制，这种不确定性对于一个调节系统而言是非常危险的，所以需要通过发送查询指令来了解控制类传感器对设备的控制结果，以确保控制指令执行的有效性。

上述的两种逻辑事件在实际的操作中其实是同时发生的，即：发送一条控制指令后紧跟着发送一条查询指令，当控制类传感器完成控制操作后立即返回控制结果。

③ 控制类传感器工作状态的实时上报。当控制类传感器受到外界干扰（如雷击）或人为干扰而造成设备重启时，重启后的状态通常为默认状态。如果控制类节点的状态与远程设备需要的状态不符，远程设备就可以重新发送控制指令使控制类传感器回到正常的工作状态。

（1）数据通信协议的定义。本系统主要使用的是采集类开发平台 Sensor-A 和控制类开发平台 Sensor-B，其 ZXBee 数据通信协议如表 3.6 所示。

表 3.6　采集类开发平台和控制类开发平台的 ZXBee 数据通信协议

开发平台	属　性	参　数	权　限	说　明
Sensor-A（601）	光照度	A2	R	光照度值，浮点型数据，精度为 0.1，范围为 0～65535
	上报状态	D0(OD0/CD0)	R/W	D0 的 Bit0~Bit7 分别代表参数 A0～A7 传感器数据
	数据上报时间间隔	V0	R/W	A0～A7 传感器数据的循环上报时间间隔
Sensor-B（602）	LED	D1(OD1/CD1)	R/W	D1 的 Bit4、Bit5 代表 LED1 和 LED2 状态，0 表示关闭，1 表示开启
	上报状态	D0(OD0/CD0)	R/W	D0 的 Bit0~Bit7 分别是参数 A0～A7 传感器数据
	数据上报时间间隔	V0	R/W	A0～A7 传感器数据的循环上报时间间隔

注：参数 A0～A7 代表不同的数据，这些数据是由对应的传感器采集的，详见 2.2 节。

（2）驱动程序的开发。本系统使用采集类开发平台 Sensor-A 和控制类开发平台 Sensor-B，这两个开发平台是基于 CC2530 微处理器构建的。sensor.c 中 sensorInit()函数负责采集类传感器和控制类传感器的初始化，采集类传感器的初始化详见 3.1.3 节，控制类传感器的初始化过程与采集类传感器的初始化过程相同，部分代码如下：

```
void sensorInit(void)
{
    //初始化传感器代码
    bh1750_init();                                          //光照度传感器初始化
    led_init();                                             //LED 初始化
    //启动定时器，触发上报事件 MY_REPORT_EVT
    osal_start_timerEx(sapi_TaskID, MY_REPORT_EVT, (uint16)((osal_rand()%10) *1000));
    //启动定时器，触发报警事件 MY_CHECK_EVT
    osal_start_timerEx(sapi_TaskID, MY_CHECK_EVT, 100);
}
```

在本系统中，对光照度传感器的操作主要是向光照度传感器写入控制指令，以及读取光照度传感器内部的数据，这些操作是通过 IIC 总线来进行的，在进行操作之前还需要先对传感器进行初始化，其初始化函数是 bh1750_init()，光照度传感器的初始化主要是 IIC 总线的初始化。

```
/*********************************************************************************************
*名　称：bh1750_send_byte()
*功　能：向光照度传感器写入控制指令
*参　数：无
*返回值：如果返回 1 表示操作成功，否则操作有误
*********************************************************************************************/
uchar bh1750_send_byte(uchar sla,uchar c)
{
    iic_start();                                    //开启 IIC 总线
    if(iic_write_byte(sla) == 0){                   //发送器件地址
        if(iic_write_byte(c) == 0){                 //写入控制指令
        }
    }
```

```
        iic_stop();                                     //停止 IIC 总线
        return(1);
    }
    /*****************************************************************************
    *名称：bh1750_read_nbyte()
    *功能：连续读出光照度传感器内部的数据
    *返回：应答或非应答信号
    *****************************************************************************/
    uchar bh1750_read_nbyte(uchar sla,uchar *s,uchar no)
    {
        uchar i;
        iic_start();                                    //起始信号
        if(iic_write_byte(sla+1) == 0){                 //发送设备地址+读信号
            for (i=0; i<no-1; i++){                     //连续读取 6 个地址的数据，存储在 buf 中
                *s=iic_read_byte(0);
                s++;
            }
            *s=iic_read_byte(1);
        }
        iic_stop();
        return(1);
    }
    /*****************************************************************************
    *名称：bh1750_init()
    *功能：初始化光照度传感器
    *****************************************************************************/
    void bh1750_init()
    {
        iic_init();
    }
    /*****************************************************************************
    *名称：bh1750_get_data()
    *功能：光照度传感器数据处理函数
    *****************************************************************************/
    float bh1750_get_data(void)
    {
        uchar *p=buf;
        bh1750_init();                                  //初始化光照度传感器
        bh1750_send_byte(0x46,0x01);                    //power on
        bh1750_send_byte(0x46,0X20);                    //高分辨率模式
        delay_ms(180);                                  //延时 180 ms
        bh1750_read_nbyte(0x46,p,2);                    //读出数据，存储在 buf 中
        unsigned short x = buf[0]<<8 | buf[1];
        return x/1.2;
    }
```

LED 可以看成控制类传感器，对 LED 的初始化同样也是在 sensorInit()函数中实现的。LED 的初始化函数是 led_init()，该函数主要对 LED 控制引脚进行初始化，代码如下：

```
/*********************************************************************************************
*名称：led_init()
*功能：LED 控制引脚初始化
*********************************************************************************************/
void led_init(void)
{
    P0SEL &= ~0x30;                         //配置控制引脚（P0_4 和 P0_5）为 GPIO 模式
    P0DIR |= 0x30;                          //配置控制引脚（P0_4 和 P0_5）为输出模式
    LED1 = OFF;                             //初始状态为关闭
    LED2 = OFF;                             //初始状态为关闭
}
/*********************************************************************************************
*名称：led_on(unsigned char led)
*功能：开启 LED
*参数：led—LED 的编号，在 led.h 中宏定义为 LED1 和 LED2
*********************************************************************************************/
signed char led_on(unsigned char led)
{
    if(led == LED1){                        //开启 LED1
        LED1 = ON;
        return 0;
    }
    if(led == LED2){                        //开启 LED2
        LED2 = ON;
        return 0;
    }
    return -1;                              //参数错误，返回-1
}
/*********************************************************************************************
*名称：led_off()
*功能：关闭 LED
*参数：led—LED 的编号，在 led.h 中宏定义为 LED1 和 LED2
*********************************************************************************************/
signed char led_off(unsigned char led)
{
    if(led == LED1){                        //关闭 LED1
        LED1 = OFF;
        return 0;
    }
    if(led == LED2){                        //关闭 LED2
        LED2 = OFF;
        return 0;
    }
    return -1;                              //参数错误，返回-1
}
```

对 LED 的控制是通过 sensorControl()函数来实现的，相关代码如下：

```
/*********************************************************************************************
*名称：sensorControl()
*功能：传感器控制
*参数：cmd—控制命令
*********************************************************************************************/
void sensorControl(uint8 cmd)
{
    if(cmd & 0x20){                                   //LED2 的控制位：Bit5
        LED2 = ON;                                    //开启 LED2
    } else{
        LED2 = OFF;                                   //关闭 LED2
    }
    if(cmd & 0x10){                                   //LED1 的控制位：Bit4
        LED1 = ON;                                    //开启 LED1
    } else{
        LED1 = OFF;                                   //关闭 LED1
    }
}
```

2. Android 端应用设计

1）Android 工程设计框架

打开 Android Studio 开发环境，可以看到城市景观照明控制系统的工程目录，如图 3.25 所示。该系统的工程框架如表 3.7 所示。

图 3.25 城市景观照明控制系统的工程目录

表 3.7　城市景观照明控制系统的工程框架

类　名	说　明
activity	
IdKeyShareActivity.java	在 IDKey 页面单击“分享”按钮时，可弹出 activity，用于分享二维码图片
TimePickerActivity.java	自动控制中的时间选择器
adapter	
HdArrayAdapter.java	历史数据显示适配器
application	
LCApplication.java	LCApplication 继承 application 类，使用单例模式（Singleton Pattern）创建 WSNRTConnect 对象
bean	
HistoricalData.java	历史数据的 bean 类，用于将从智云服务器获得的历史数据记录（JSON 形式）转换成该类对象
IdKeyBean.java	IdKeyBean 用来描述用户设备的 ID、KEY，以及智云服务器的地址 SERVER
config	
Config.java	config 用于修改用户的 ID、KEY，以及智云服务器的地址和 MAC 地址
fragment	
BaseFragment.java	页面基础 Fragment 定义类
HDFragment.java	历史数据页面
HistoricalDataFragment.java	历史数据显示页面
HomepageFragment.java	展示首页页面的 Fragment
IDKeyFragment.java	IDKey 选项的页面
MacSettingFragment.java	当用户设置被监测项的 MAC 地址时显示的页面
MoreInformationFragment.java	更多信息显示页面
RunHomePageFragment.java	运营首页显示页面
VersionInformationFragment.java	显示版本等相关信息的页面
listener	
IOnWSNDataListener.java	传感器数据监听器接口
update	
UpdateService.java	应用下载服务类
view	
APKVersionCodeUtils.java	获取当前本地 apk 的版本
CustomRadioButton.java	自定义按钮类
PagerSlidingTabStrip.java	自定义滑动控件类
MainActivity.java：主页面类	
MyBaseFragmentActivity.java：系统 Fragment 通信类	

2）软件设计

根据智云 Android 端应用程序接口的定义，城市景观照明控制系统的应用设计主要采用实时数据 API 接口（和城市环境信息采集系统相同），其流程见图 3.11。

（1）LCApplication.java 程序代码剖析。城市景观照明控制系统中的 LCApplication.java 程序代码和城市环境信息采集系统的 LCApplication.java 程序代码相同，详见 3.1.3 节的相关内容。

（2）HomepageFragment.java 程序代码剖析。下面的代码通过(LCApplication) getActivity().getApplication()获取 LCApplication 类中的 WSNRTConnect 对象。

```
private void initViewAndBindEvent() {
    preferences = getActivity().getSharedPreferences("user_info", Context.MODE_PRIVATE);
    lcApplication = (LCApplication) getActivity().getApplication();
    wsnrtConnect = lcApplication.getWSNRConnect();
    lcApplication.registerOnWSNDataListener(this);
    editor = preferences.edit();
}
```

下面的代码通过复写 onMessageArrive 方法来处理接收到的无线数据包，实现了设备 MAC 地址的获取，并在当前页面显示设备的状态。

```
@Override
    public void onMessageArrive(String mac, String tag, String val) {
        if (sensorAMAC == null && sensorBMAC == null) {
            wsnrtConnect.sendMessage(mac, "{TYPE=?}".getBytes());
        }
        if ("TYPE".equals(tag) && "601".equals(val.substring(2, val.length()))) {
            sensorAMAC = mac;
        }
        if ("TYPE".equals(tag) && "602".equals(val.substring(2, val.length()))) {
            sensorBMAC = mac;
        }
        if (mac.equals(sensorAMAC) && "A2".equals(tag)) {
            textIlluminationeState.setText("在线");
            textIlluminationeState.setTextColor(getResources().getColor(R.color.line_text_color));
            illuminationText.setText(val+"Lx");
            lightIntensity = Float.parseFloat(val);
            if(seekbarThreshold.getProgress() != 0) {
                limitofilluminationTooHigh();
            }
        }
        if (mac.equals(sensorBMAC) && "D1".equals(tag)) {
            textEquipmentState.setText("在线");
            textEquipmentState.setTextColor(getResources().getColor(R.color.line_text_color));
            int numResult = Integer.parseInt(val);
            if ((numResult & 0X10) == 0x10) {
                imageEquipmentState.setImageDrawable(getResources().getDrawable(R.drawable.open_
```

```
lamp));
                    openOrCloseLamp.setText("关闭");
                    openOrCloseLamp.setBackground(getResources().getDrawable(R.drawable.close));
                }else {
                    imageEquipmentState.setImageDrawable(getResources().getDrawable(R.drawable.close_
lamp));
                    openOrCloseLamp.setText("开启");
                    openOrCloseLamp.setBackground(getResources().getDrawable(R.drawable.open));
                }
            }
        }
```

3. Web端应用设计

1）页面功能结构分析

城市景观照明控制系统的 Web 端默认显示的是“运营首页”页面，在“运营首页”页面中设计了 5 个模块，分别是光照度数据显示模块、照明设备控制模块、模式切换模块、定时器设置模块、阈值设置模块。城市景观照明控制系统 Web 端的“运营首页”页面如图 3.26 所示。

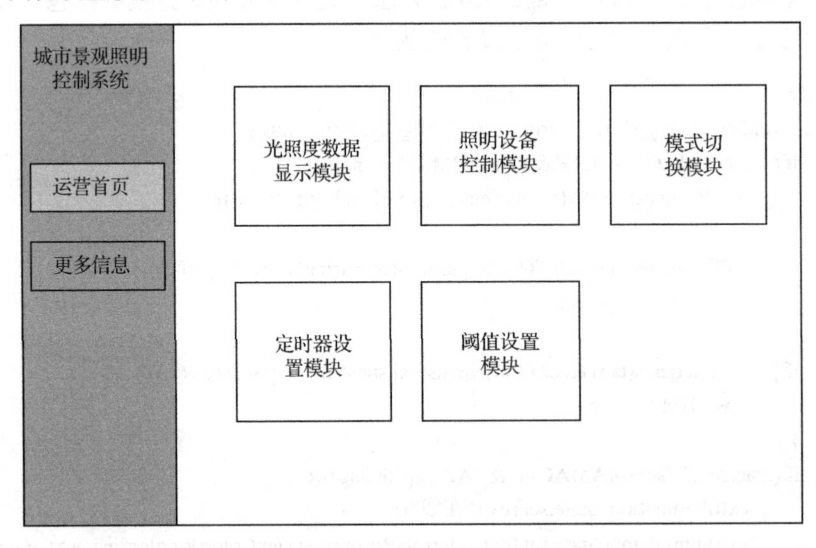

图 3.26　城市景观照明控制系统 Web 端的“运营首页”页面

在城市景观照明控制系统的 Web 端中，“更多信息”页面的主要功能是配置智云服务器的连接，该页面分为三个子功能，通过标签进行切换显示，“ID Key”标签通过设置智云服务器的 ID 与 KEY 来调用 Web 端应用接口，从而连接到智云服务器；“MAC 设置”标签用来显示设备的 MAC 地址；“版本信息”标签用来显示版本的信息并进行升级。城市景观照明控制系统 Web 端的“更多信息”页面如图 3.27 所示。

2）软件设计

城市景观照明控制系统 Web 端的 JS 开发逻辑与 Android 端的开发逻辑相似，首先通过配置 ID 和 KEY 与智云服务器进行连接，再通过实时监听数据的方法来获取相关传感器的数据并进行处理。JS 开发的部分代码如下。

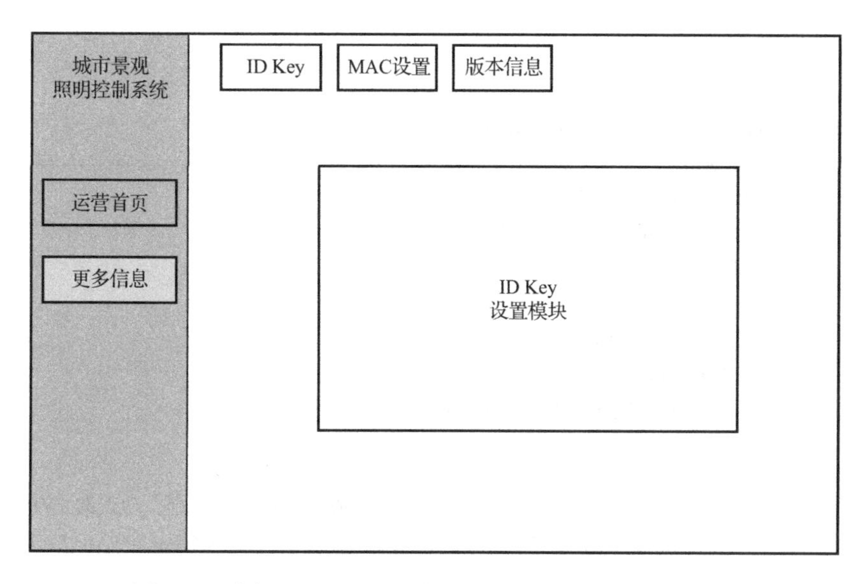

图 3.27　城市景观照明控制系统 Web 端的“更多信息”页面

在 getConnect()函数中定义了实时连接对象 rtc，连接成功回调函数是 rtc.onConnect，数据服务掉线回调函数是 rtc.onConnectLost，消息处理回调函数是 rtc.onmessageArrive。

```
function getConnect() {
    config["id"] = config["id"] ? config["id"] : $("#ID").val();
    config["key"] = config["key"] ? config["key"] : $("#KEY").val();
    config["server"] = config["server"] ? config["server"] : $("#server").val();
    //创建数据连接服务对象
    rtc = new WSNRTConnect(config["id"], config["key"]);
    rtc.setServerAddr(config["server"] + ":28080");
    rtc.connect();
    rtc._connect = false;
    //连接成功回调函数
    rtc.onConnect = function() {
        $("#ConnectState").text("数据服务连接成功！");
        rtc._connect = 1;
        message_show("数据服务连接成功！");
        $("#idkeyInput").text("断开").addClass("btn-danger");
        $("#id,#key,#server").attr('disabled',true)
    };
    //数据服务掉线回调函数
    rtc.onConnectLost = function() {
        rtc._connect = 0;
        $("#ConnectState").text("数据服务连接掉线！");
        $("#idkeyInput").text("连接").removeClass("btn-danger");
        message_show("数据服务连接失败，检查网络或 ID、KEY");
        $(".online_601").text("离线").css("color", "#e75d59");
        $(".online_602").text("离线").css("color", "#e75d59");
        $("#id,#key,#server").removeAttr('disabled');
```

```
        };
        //消息处理回调函数
        rtc.onmessageArrive = function (mac, dat) {
            //console.log(mac+" >>> "+dat);
            if (dat[0]=='{' && dat[dat.length-1]=='}') {
                dat = dat.substr(1, dat.length-2);
                var its = dat.split(',');
                for (var i=0; i<its.length; i++) {
                    var it = its[i].split('=');
                    if (it.length == 2) {
                        process_tag(mac, it[0], it[1]);
                    }
                }
                if (!mac2type[mac]) {            //如果没有获取到 TYPE 值，则主动查询 TYPE 值
                    rtc.sendMessage(mac, "{TYPE=?,A0=?,A1=?,A2=?,A3=?,A4=?,A5=?,A6=?,A7=?,D1=?}");
                }
            }
        }
}
```

下述 JS 开发代码的功能是根据设备连接的情况，在页面中更新设备是否在线的状态，显示照明设备（本系统为 LED）的状态（如打开或关闭），并根据当前光照度数据和设定的光照度阈值的比较结果来控制照明设备。

```
var wsn_config = {
    "601" : {
        "online" : function() {
            $(".online_601").text("在线").css("color", "#96ba5c");
        },
        "pro" : function (tag, val) {
            if(tag=="A2"){
                soilTemper(val);
                if(config["curMode"]=="auto-mode" && val<config["threshold"] && !state.light){
                rtc.sendMessage(config["mac_602"], "{OD1=16,D1=?}");
                message_show("低于光照度阈值，将打开照明设备！");
            } else {
                rtc.sendMessage(config["mac_602"], "{CD1=16,D1=?}");
                message_show("高于光照度阈值，将关闭！");
            }
        }
    },
    "602" : {
        "online" : function() {
            $(".online_602").text("在线").css("color", "#96ba5c");
        },
        "pro" : function (tag, val) {
            console.log('当前灯的状态'+val);
```

```
                if(tag=="D1"){
                    if(val & 16){
                        $("#lightStatus").text("关闭");
                        $("#lightImg").attr("src", "img/LED-on.png");
                        state.light = true;
                    }else{
                        $("#lightStatus").text("打开");
                        $("#lightImg").attr("src", "img/LED-off.png");
                        state.light = false;
                    }
                }
            }
        }
    }
```

实现照明设备控制功能的代码如下：

```
    //照明设备控制开关
    $("#lightStatus").on("click", function() {
        if (page.checkOnline() && page.checkMac("mac_602")){
            var state = $(this).text()=="关闭", cmd;
            if(state){
                cmd = "{CD1=16,D1=?}";
            }else{
                cmd = "{OD1=16,D1=?}";
            }
            console.log(cmd)
            rtc.sendMessage(config["mac_602"], cmd);
        }
    });
```

实现定时器控制功能的代码如下：

```
    function getTime(){
        var nowdate = new Date();
        //获取年、月、日、时、分、秒
        var hours = nowdate.getHours(),
        minutes = nowdate.getMinutes(),
        date = nowdate.getDate();
        var hour_info = hours >=10 ? hours : "0"+hours;
        var minute_info = minutes >=10 ? minutes : "0"+minutes;
        var cur_time = hour_info + ":" + minute_info;
        //保存一个全局变量来缓存上一次的时间字符串，当最新的时间字符串与保存的时间字符串不相同
时，则更新保存的时间字符串，并根据更新后的时间字符串来执行相应的动作
        //console.log("每秒更新："+cur_time);
        if(last_time != cur_time){
            last_time = cur_time;
            console.log("每分钟更新：当前时间："+cur_time);
```

```
            if(config["curMode"]=="auto-mode"  &&  (cur_time==config["open_time"]  ||  cur_time==config
["close_time"])){
                if(page.checkOnline() && page.checkMac("mac_602")){
                    var cmd;
                    if(cur_time==config["open_time"]){
                        message_show("定时器控制照明设备开启！");
                        cmd = "{OD1=16,D1=?}";
                    }
                    else if(cur_time==config["close_time"]){
                        message_show("定时器控制照明设备关闭！");
                        cmd = "{CD1=16,D1=?}";
                    }
                    rtc.sendMessage(config["mac_602"], cmd);
                }else{
                    setTimeout(function() {
                        message_show("定时器控制失败！");
                    },4000)
                }
            }
        }
        setTimeout(getTime, 1000);
    }
```

3.2.4 开发验证

1. Web 端应用测试

在 Web 端打开城市景观照明控制系统，成功连接智云服务器后切换到系统的主页，此时可以看到设备的状态为"在线"，如图 3.28 所示，可以在"模式切换"中选择"手动模式"或"自动模式"。

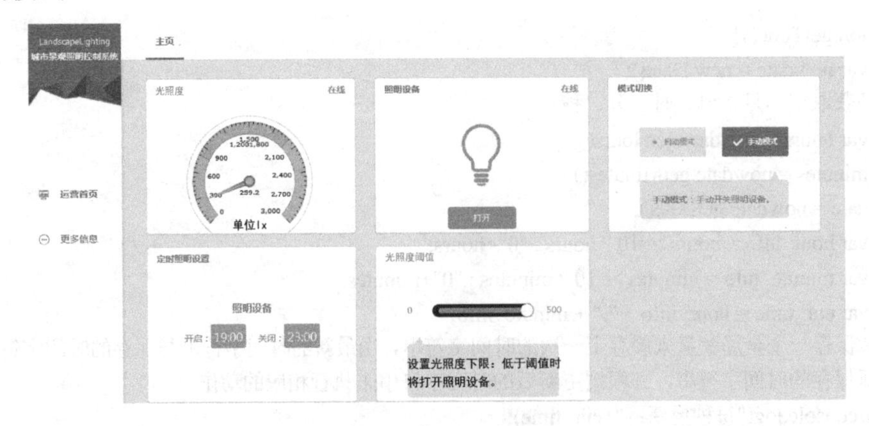

图 3.28　城市景观照明控制系统 Web 端主页

在自动模式下，可以通过"定时照明设备"或"光照度阈值"来控制照明设备；在手动模式下，可以通过"照明设备"中的"打开"或"关闭"按钮来控制照明设备。

2. Android 端应用测试

Android 端应用测试同 Web 端应用测试基本一致，可参考 Web 端应用测试进行操作。城市景观照明控制系统 Android 端运营首页如图 3.29 所示。

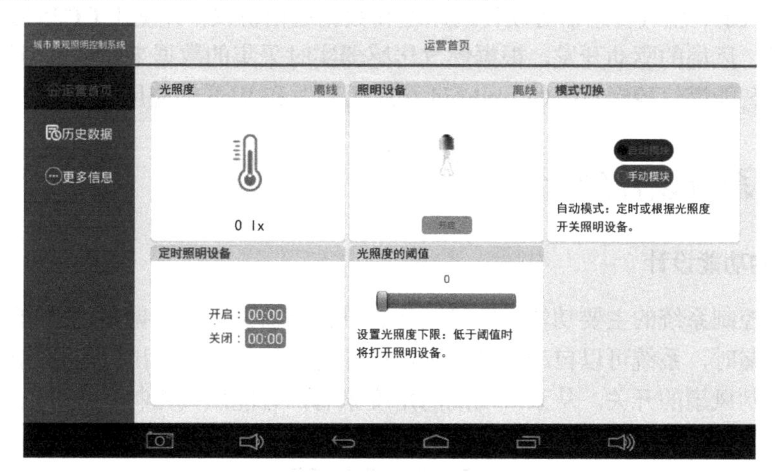

图 3.29　城市景观照明控制系统 Android 端运营首页

3.2.5　总结与拓展

本节基于 ZigBee 实现了光照度传感器的数据采集和 LED 的控制，通过 Android 和 HTML5 技术实现了 Android 端和 Web 端的应用设计，可根据实时获取的光照度数据来控制 LED 的开关，实现了基于 ZigBee 的城市景观照明控制系统。

3.3　基于 ZigBee 的智能燃气控制系统

智能燃气控制系统可以全天候地监控燃气管网中的闸井、调压箱、调压站等关键部位，实时掌握这些部位的压力、流量、燃气浓度、热值、温度等数据，实现智能报警。智能燃气控制系统如图 3.30 所示。

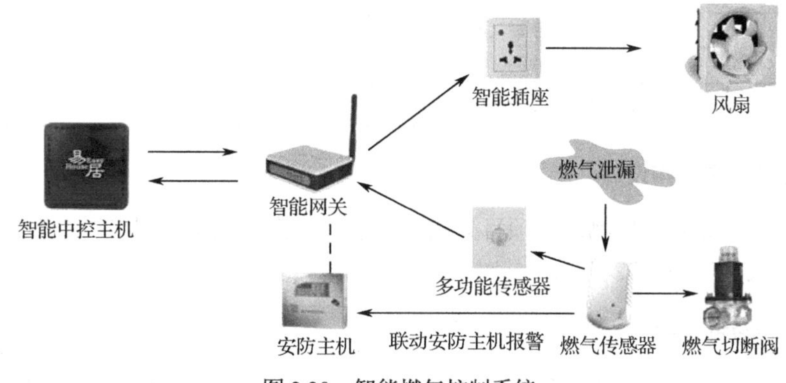

图 3.30　智能燃气控制系统

3.3.1 系统开发目标

（1）熟悉风扇、燃气传感器等硬件的原理和数据通信协议，并基于CC2530和ZigBee实现燃气传感器、风扇的驱动开发，根据燃气传感器实时采集的数据来控制风扇的开关。

（2）实现智能燃气控制系统的Android端应用开发和Web端应用开发。

3.3.2 系统设计分析

1. 系统的功能设计

智能燃气控制系统的主要功能是实时地将燃气传感器采集的数据主动推送到智云平台，当发现燃气泄漏时，系统可以自动打开风扇，以便将泄漏的燃气排出；用户也可以通过系统随时随地地控制风扇的开关。从系统功能的角度来看，智能燃气控制系统可以分为控制类和安防类传感器模块以及系统设置模块，如图3.31所示。

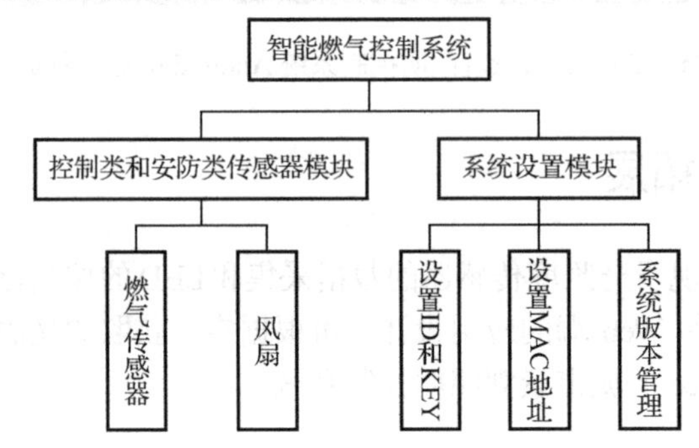

图3.31 智能燃气控制系统的组成模块

智能燃气控制系统的功能需求分析如表3.8所示。

表3.8 智能燃气控制系统的功能需求分析

功能	功能说明
采集类传感器状态	在上层应用页面中实时更新并显示燃气传感器的状态
模式设置	自动模式：当监测到燃气泄漏时，自动关闭燃气开关并启动风扇。手动模式：手动关闭燃气开关，并启动风扇
智云连接设置	设置智云服务器的参数和设备的MAC地址

2. 系统的总体架构设计

智能燃气控制系统采用物联网项目架构进行设计，由感知层、网络层、平台层和应用层组成，其总体架构如图3.32所示。

3. 系统的数据传输

智能燃气控制系统的数据传输是在传感器节点、智能网关以及客户端（包括Web端和Android端）之间进行的，如图3.33所示。

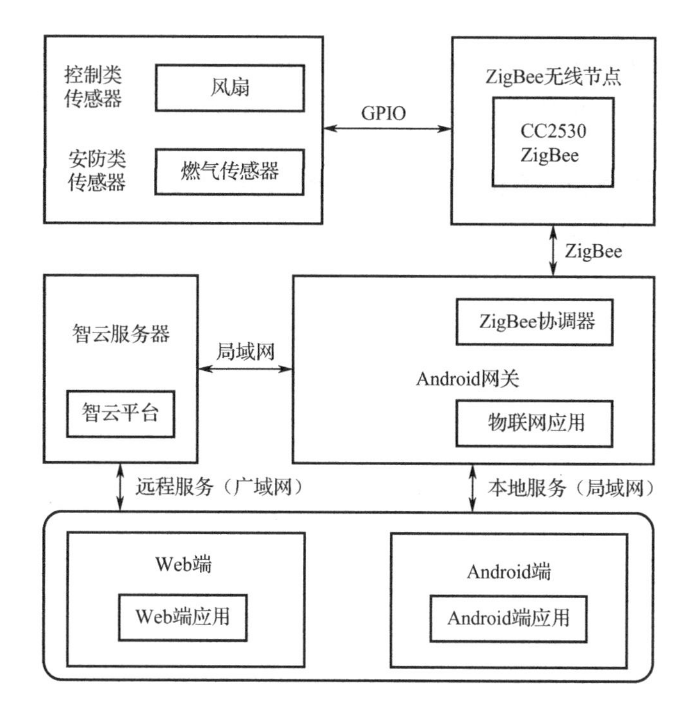

图 3.32　智能燃气控制系统的总体架构

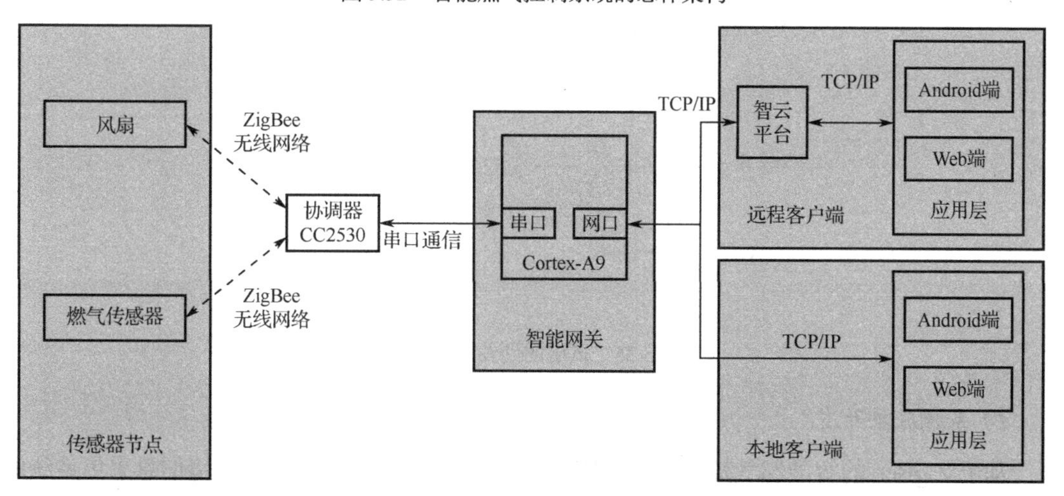

图 3.33　智能燃气控制系统的数据传输

3.3.3　系统的软硬件开发：智能燃气控制系统

1. 系统底层软硬件设计

1）感知层硬件设计

智能燃气控制系统的底层硬件主要包括 xLab 未来开发平台的智能网关、经典型无线节点 ZXBeeLiteB、控制类开发平台 Sensor-B、安防类开发平台 Sensor-C。其中，智能网关负责汇集传感器采集的数据；ZigBee 无线节点（由经典型无线节点 ZXBeeLiteB 实现）通过无线通信的方式向智能网关发送传感器数据，接收智能网关发送的命令；控制类开发平台 Sensor-B 和安防类开发平台 Sensor-C 连接到 ZigBee 无线节点，由其中的 CC2530 微处理器对相关设备进行控制。

本系统主要使用燃气传感器和风扇，燃气传感器的硬件接口电路如图 3.34 所示，风扇的硬件接口电路如图 3.35 所示。

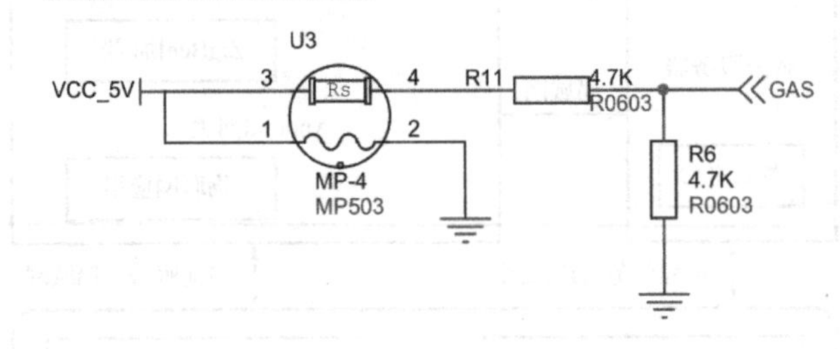

图 3.34　燃气传感器的硬件接口电路

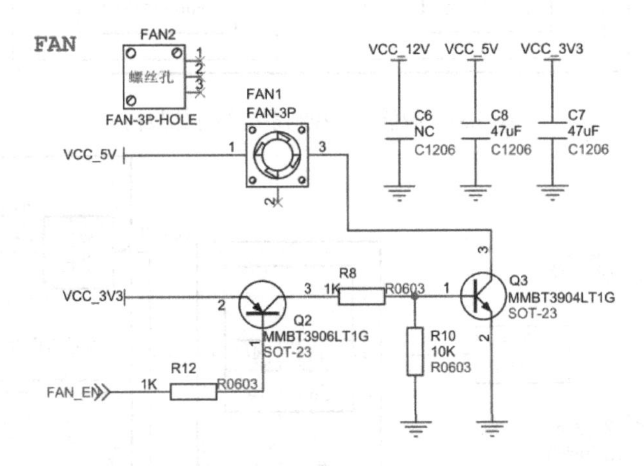

图 3.35　风扇的硬件接口电路

2）系统底层开发

基于 ZigBee 的智能燃气控制系统的底层开发和基于 ZigBee 的城市环境信息采集系统的底层开发相同，详见 3.1.3 节。

3）传感器驱动设计

智能燃气控制系统的底层硬件主要包括控制类传感器和安防类传感器，传感器的驱动设计主要是针对这两类传感器进行的。控制类传感器的逻辑事件有 3 种，详见 3.2.3 节。安防类传感器的逻辑事件有 4 种，如下所述。

① 定时获取并上报安防类传感器的安全信息。在一个监控系统中，远程设备需要不断了解安防类传感器采集的安全信息，只有不断地更新安全信息，监控系统的持续安全性才能得到保障。如果安全信息不能够持续更新，那么在设备出现故障或遭到人为破坏时将造成危险后果，因此安全信息的持续上报可以降低监控系统安全的不确定性。

② 当安防类传感器监测到危险信息时，系统能够迅速上报危险信息。如果安防类传感器

不能够及时上报危险信息，则该传感器的报警功能将是失效的。例如，在消防系统中，如果出现火情，此时火灾的危险信息上报就变得尤为重要，如果危险信息不能够及时上报，火灾将会造成巨大的经济损失。因此，危险信息的及时上报是安防类传感器的关键功能。

③ 当危险信息解除时，系统能够恢复正常。在物联网系统中，设备往往都不是一次性的，很多设备都要重复利用，所以当危险信息解除后系统必须回到安全状态，这就需要安防类传感器能够发出安全信息让系统从危险警戒状态中退出。

另外，安防类传感器发送安全信息与危险信息的实时性是不同的，安全信息可以在一段时间内更新一次，如半分钟或一分钟；而危险信息则相对比较紧急，危险信息的发送要保持每秒一次，要实时监控危险信息的变化，以确保对危险信息的实时掌握。

④ 当接收到查询指令时，安防类传感器能够响应指令并反馈安全信息。当管理员需要对设备进行调试或者主动查询当前的安全状态时，就需要通过远程设备向安防类传感器发送查询指令，用以查询当前设备的安全状态。

（1）数据通信协议的定义。本系统主要使用控制类开发平台Sensor-B和安防类开发平台Sensor-C，其ZXBee数据通信协议如表3.9所示。

表3.9 控制类开发平台和安防类开发平台的ZXBee数据通信协议

开发平台	属　性	参　数	权限	说　明
Sensor-B（602）	风扇	D1(OD1/CD1)	R/W	D1的Bit3代表风扇的开关状态，0表示关闭，1表示打开
	数据上报时间间隔	V0	R/W	数据上报的时间间隔（循环上报）
Sensor-C（603）	燃气状态	A4	R	燃气泄漏的状态，0表示燃气未泄漏，1表示燃气泄漏
	数据上报时间间隔	V0	R/W	数据上报的时间间隔（循环上报）

（2）驱动程序的开发。本系统使用控制类开发平台Sensor-B和安防类开发平台Sensor-C，这两个开发平台是基于CC2530微处理器构建的。sensor.c中sensorInit()函数负责采集类传感器和控制类传感器的初始化，部分代码如下：

```
/*********************************************************************************
*名称：sensorInit()
*功能：传感器初始化
*********************************************************************************/
void sensorInit(void)
{    //初始化传感器代码
     combustiblegas_init();                                         //燃气传感器初始化
     //启动定时器，触发上报事件MY_REPORT_EVT
     osal_start_timerEx(sapi_TaskID, MY_REPORT_EVT, (uint16)((osal_rand()%10) *1000));
     //启动定时器，触发报警事件MY_CHECK_EVT
     osal_start_timerEx(sapi_TaskID, MY_CHECK_EVT, 100);
}
```

燃气传感器的初始化函数是combustiblegas_init()，部分代码如下：

```
/*********************************************************************************
*名称：combustiblegas _init()
*功能：燃气传感器初始化
```

```
**************************************************************************************/
void combustiblegas_init(void)
{
    APCFG |= 0x20;                                    //模拟 I/O 使能
    P0SEL |= 0x20;                                    //将端口 P0_5 设置为外设功能
    P0DIR &= ~0x20;                                   //设置输入模式
}
/**************************************************************************************
*名称：unsigned int get_combustiblegas_data(void)
*功能：获取燃气传感器的数据
**************************************************************************************/
unsigned int get_combustiblegas_data(void)
{
    unsigned int   value;
    value = HalAdcRead (HAL_ADC_CHANNEL_5,HAL_ADC_RESOLUTION_12);

    if(value <= 1500)
        return 0;
    else
        return 1;
}
```

风扇初始化的函数是 fan_init()，是通过将对应的引脚配置为输出模式来实现初始化的，其驱动程序文件为 fan.c，部分代码如下：

```
/**************************************************************************************
*名称：fan_init()
*功能：风扇初始化
**************************************************************************************/
void fan_init(void)
{
    P0SEL &= ~0x08;                                   //配置引脚为 GPIO 模式
    P0DIR |= 0x08;                                    //配置控制引脚为输出模式

    FANIO = FAN_OFF;
}
```

传感器的实时监测函数是 sensorCheck()，相关代码如下：

```
/**************************************************************************************
*名称：sensorCheck()
*功能：传感器的实时监测
**************************************************************************************/
void sensorCheck(void)
{
    static char lastA0 = 0,lastA1=0,lastA2=0,lastA3=0,lastA4=0,lastA5=0;
    static uint32 ct0=0, ct1=0, ct2=0, ct3=0, ct4=0, ct5=0;
    char pData[16];
```

```
    char *p = pData;
    updateA0();  updateA1();  updateA2();  updateA3();  updateA4();  updateA5();
    ZXBeeBegin();
    if (lastA4 != A4 || (ct4 != 0 && osal_GetSystemClock() > (ct4+3000))) {  //燃气传感器的实时监测
        if (D0 & 0x10) {
            sprintf(p, "%u", A4);
            ZXBeeAdd("A4", p);
            ct4 = osal_GetSystemClock();
        }
        if (A4 == 0) {
            ct4 = 0;
        }
        lastA4 = A4;
    }
    p = ZXBeeEnd();
    if (p != NULL) {
        int len = strlen(p);
        ZXBeeInfSend(p, len);
    }
}
```

控制传感器的函数是 sensorControl()，相关代码如下：

```
/*********************************************************************************************
*名称：sensorControl()
*功能：传感器的控制
*参数：cmd—控制命令
*********************************************************************************************/
void sensorControl(uint8 cmd)
{
    //根据 cmd 参数执行对应的控制程序
    if(cmd & 0x08) {                                        //风扇控制位：Bit3
        FANIO = FAN_ON;                                     //开启风扇
    } else {
        FANIO = FAN_OFF;                                    //关闭风扇
    }
}
```

2．Android 端应用设计

1）Android 工程设计框架

打开 Android Studio 开发环境，可以看到本系统的工程目录，如图 3.36 所示，本系统的工程框架如表 3.10 所示。

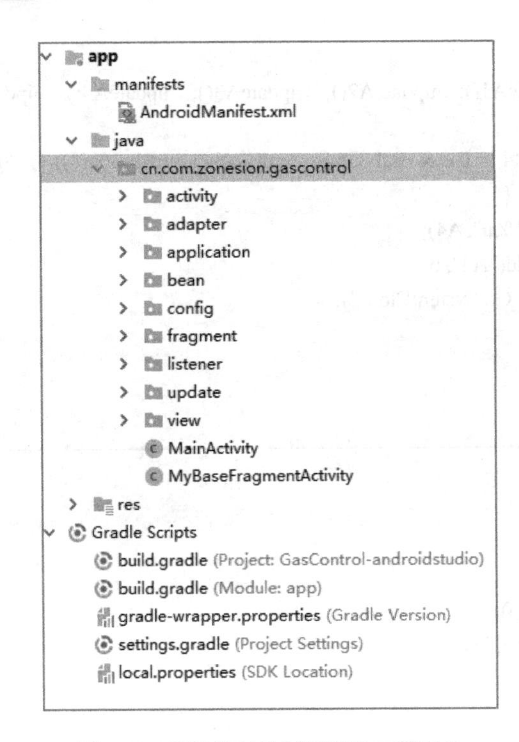

图 3.36　智能燃气控制系统的工程目录

表 3.10　智能燃气控制系统的工程框架

类　名	说　明
activity	
IdKeyShareActivity.java	在 IDKey 页面单击“分享”按钮时，可弹出 activity，用于分享二维码图片
adapter	
HdArrayAdapter.java	历史数据显示适配器
ViewPagerAdapter.java	对 ViewPager 进行适配，可以处理多个 Fragment 的横向滑动
application	
LCApplication.java	LCApplication 继承 application 类，使用单例模式（Singleton Pattern）创建 WSNRTConnect 对象
bean	
HistoricalData.java	历史数据的 bean 类，用于将从智云服务器获得的历史数据记录（JSON 形式）转换成该类对象
IdKeyBean.java	IdKeyBean 用来描述用户设备的 ID、KEY，以及智云服务器的地址 SERVER
config	
Config.java	config 用于修改用户的 ID、KEY，以及智云服务器的地址和 MAC 地址
fragment	
BaseFragment.java	页面基础 Fragment 定义类
BasicsFragment.java：下面一些 Fragment 的基类，定义共有的属性以及 getter 和 setter 方法	
HomepageFragment.java	展示首页页面的 Fragment
IDKeyFragment.java	IDKey 选项的页面
MacSettingFragment.java	当用户设置被监测项的 MAC 地址时显示的页面

续表

类　名	说　明
MoreInformationFragment.java	更多信息显示页面
RunHomePageFragment.java	运营首页显示页面
VersionInformationFragment.java	显示版本等相关信息的页面
listener	
IOnWSNDataListener.java	传感器数据监听器接口
update	
UpdateService.java	应用下载服务类
view	
APKVersionCodeUtils.java	获取当前本地 apk 的版本
CustomRadioButton.java	自定义按钮类
PagerSlidingTabStrip.java	自定义滑动控件类
MainActivity.java：主页面类	
MyBaseFragmentActivity.java：系统 Fragment 通信类	

2）软件设计

根据智云 Android 端应用程序接口的定义，智能燃气控制系统的应用设计主要采用实时数据 API 接口（和城市环境信息采集系统相同），其流程见图 3.11。

（1）LCApplication.java 程序代码剖析。智能燃气控制系统中的 LCApplication.java 程序代码和城市环境信息采集系统的 LCApplication.java 程序代码相同，详见 3.1.3 节的相关内容。

（2）HomepageFragment.java 程序代码剖析。下面的代码通过(LCApplication) getActivity().getApplication()获取 LCApplication 类中的 WSNRTConnect 对象。

```
private void initViewAndBindEvent() {
    preferences = getActivity().getSharedPreferences("user_info", Context.MODE_PRIVATE);
    lcApplication = (LCApplication) getActivity().getApplication();
    wsnrtConnect = lcApplication.getWSNRConnect();
    lcApplication.registerOnWSNDataListener(this);
    editor = preferences.edit();
}
```

下面的代码通过复写 onMessageArrive 方法来处理节点接收到的无线数据包，获取设备的 MAC 地址，并在当前页面显示设备的状态。

```
@Override
    public void onMessageArrive(String mac, String tag, String val) {
        if (seneorBMAC == null && seneorCMAC == null) {
            wsnrtConnect.sendMessage(mac, "{TYPE=?}".getBytes());
        }
        if ("TYPE".equals(tag) && "602".equals(val.substring(2, val.length()))) {
            seneorBMAC = mac;
        }
```

```
        if ("TYPE".equals(tag) && "603".equals(val.substring(2, val.length()))) {
            seneorCMAC = mac;
        }
        if (mac.equals(seneorBMAC) && "D1".equals(tag)) {
            textFanState.setText("在线");
            textFanState.setTextColor(getResources().getColor(R.color.line_text_color));
            int numResult = Integer.parseInt(val);
             if ((numResult & 0X8) == 0x8) {
                imageFanState.setImageDrawable(getResources().getDrawable(R.drawable.fan_on));
                openFanLamp.setText("关闭");
                openFanLamp.setBackground(getResources().getDrawable(R.drawable.close));
            }else {
                imageFanState.setImageDrawable(getResources().getDrawable(R.drawable.fan));
                openFanLamp.setText("开启");
                openFanLamp.setBackground(getResources().getDrawable(R.drawable.open));
            }
        }
        if (mac.equals(seneorCMAC) && "A4".equals(tag)) {
            textGasState.setText("在线");
            textGasState.setTextColor(getResources().getColor(R.color.line_text_color));
            int numResult = Integer.parseInt(val);
            if (isSecurityMode == true) {
                if ((numResult & 0X1) == 0x1) {
                    imageGasState.setImageDrawable(getResources().getDrawable(R.drawable.gas_on));
                    wsnrtConnect.sendMessage(seneorCMAC, "{OD1=8，D1=?}".getBytes());
                }else {
                    imageGasState.setImageDrawable(getResources().getDrawable(R.drawable.gas));
                    wsnrtConnect.sendMessage(seneorCMAC, "{CD1=8，D1=?}".getBytes());
                }
            }
            if (isSecurityMode == false) {
                if ((numResult & 0X1) == 0x1) {
                imageGasState.setImageDrawable(getResources().getDrawable(R.drawable.gas_on));
                }else {
                    imageGasState.setImageDrawable(getResources().getDrawable(R.drawable.gas));
                }
            }
        }
    }
```

3. Web端应用设计

1）页面功能结构分析

智能燃气控制系统的 Web 端默认显示的是“运营首页”页面，在“运营首页”页面上设计了三个模块，分别是燃气监测控制模块、风扇控制模块、模式切换设置模块。智能燃气控制系统 Web 端的“运营首页”页面如图 3.37 所示。

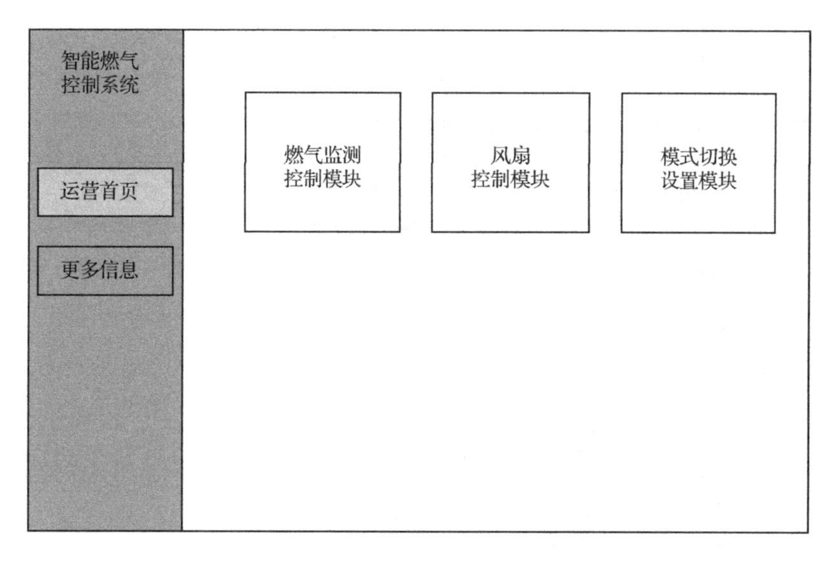

图 3.37　智能燃气控制系统 Web 端的“运营首页”页面

“更多信息”页面的主要功能是进行智云服务器的连接配置，和城市环境信息采集系统中的 Web 端“更多信息”页面类似，参见图 3.14。

2）软件设计

智能燃气控制系统 Web 端的 JS 开发逻辑与 Android 端的开发逻辑相似，首先通过配置 ID 和 KEY 与智云服务器进行连接，再通过实时监听数据的方法来获取相关传感器的数据并进行处理。JS 开发的部分代码如下。

在 getConnect()函数中定义了实时连接对象 rtc，连接成功回调函数是 rtc.onConnect，数据服务掉线回调函数是 rtc.onConnectLost，消息处理回调函数是 rtc.onmessageArrive。

```
function getConnect() {
    config["id"] = config["id"] ? config["id"] : $("#ID").val();
    config["key"] = config["key"] ? config["key"] : $("#KEY").val();
    config["server"] = config["server"] ? config["server"] : $("#server").val();
    //创建数据连接服务对象
    rtc = new WSNRTConnect(config["id"], config["key"]);
    rtc.setServerAddr(config["server"] + ":28080");
    rtc.connect();
    rtc._connect = false;
    //连接成功回调函数
    rtc.onConnect = function() {
        $("#ConnectState").text("数据服务连接成功!");
        rtc._connect = 1;
        message_show("数据服务连接成功!");
        $("#idkeyInput").text("断开").addClass("btn-danger");
        $("#id,#key,#server").attr('disabled',true)
    };
    //数据服务掉线回调函数
    rtc.onConnectLost = function() {
```

```
            rtc._connect = 0;
            $("#ConnectState").text("数据服务连接掉线!");
            $("#idkeyInput").text("连接").removeClass("btn-danger");
            message_show("数据服务连接失败，检查网络或 ID、KEY");
            $(".online_601").text("离线").css("color", "#e75d59");
            $(".online_602").text("离线").css("color", "#e75d59");
            $("#id,#key,#server").removeAttr('disabled');
        };
        //消息处理回调函数
        rtc.onmessageArrive = function (mac, dat) {
            //console.log(mac+" >>> "+dat);
            if (dat[0]=='{' && dat[dat.length-1]=='}') {
                dat = dat.substr(1, dat.length-2);
                var its = dat.split(',');
                for (var i=0; i<its.length; i++) {
                    var it = its[i].split('=');
                    if (it.length == 2) {
                        process_tag(mac, it[0], it[1]);
                    }
                }
                if (!mac2type[mac]) {                    //如果没有获取到 TYPE 值，则主动去查询
                    rtc.sendMessage(mac, "{TYPE=?,A0=?,A1=?,A2=?,A3=?,A4=?,A5=?,A6=?,A7=?,D1=?}");
                }
            }
        }
    }
```

下述 JS 开发代码的功能是根据设备连接的情况，在页面中显示设备是否在线的状态，显示排风设备（本系统为风扇）的状态（如打开或关闭），并根据采集到的燃气数据和设定阈值的比较结果来控制排风设备。

```
var wsn_config = {
    "602" : {
        "online" : function() {
            $(".online_602").text("在线").css("color", "#96ba5c");
        },
        "pro" : function (tag, val) {
            if(tag=="D1"){
                if(val & 8){
                    $("#fanStatus").text("关闭");
                    $("#fanImg").attr("src", "img/fan-on.png").addClass("fan-active");
                }else{
                    $("#fanStatus").text("打开");
                    $("#fanImg").attr("src", "img/fan.png").removeClass("fan-active");
                }
            }
        }
```

```
        },
        "603" : {
            "online" : function() {
                $(".online_603").text("在线").css("color", "#96ba5c");
            },
            "pro" : function (tag, val) {
                if(tag=="A4"){
                    console.log(val);
                    if(val == 1){
                        $("#gasStatus").addClass("gas-on");
                        //$("#gasControl").text("关闭");
                        if(config["curMode"]=="auto-mode"){
                            message_show("自动模式下监测到燃气，将自动打开风扇，关闭燃气开关");
                            rtc.sendMessage(config["mac_602"], "{OD1=8,D1=?}");
                            rtc.sendMessage(config["mac_603"], "{CD1=64,D1=?}");
                        }
                    }else{
                        $("#gasStatus").removeClass("gas-on");
                        rtc.sendMessage(config["mac_603"], "{CD1=64,D1=?}");
                        rtc.sendMessage(config["mac_602"], "{CD1=8,D1=?}");
                    }
                }
                if(tag=="D1"){
                    if(val & 64){
                        $("#gasControl").text("关闭");
                        $("#gasStatus").addClass("gas-on");
                        state.gas = true;
                    } else{
                        $("#gasControl").text("打开");
                        $("#gasStatus").removeClass("gas-on");
                        state.gas = false;
                    }
                }
            }
        }
};
```

设备的控制有自动模式和手动模式，两种模式切换的代码如下：

```
//模式切换
$("#mode-switch input").on("click", function() {
    config["curMode"] = $(this).val();
    console.log("切换到："+config["curMode"]);
    var isManualMode = config["curMode"]=="manual-mode";
    console.log(isManualMode);
    if(isManualMode){
        $("#mode-txt-2").removeClass("hidden").siblings("span").addClass("hidden");
        $("#mode-text").addClass("mode-right");
```

```
        }
        //自动模式下禁用开关按钮
        else{
            $("#mode-txt-1").removeClass("hidden").siblings("span").addClass("hidden");
            $("#mode-text").removeClass("mode-right");
        }
        $(".btn-switch").prop("disabled", !isManualMode);
        storeStorage();
});
```

实现燃气设备与排风设备开关控制的代码如下。

```
$("#gasControl").on("click", function() {
    if (page.checkOnline() && page.checkMac("mac_603")){
        var state = $(this).text()=="打开", cmd;
        if(state){
            cmd = "{OD1=64,D1=?}";
        }else{
            cmd = "{CD1=64,D1=?}";
        }
        console.log(cmd);
        rtc.sendMessage(config["mac_603"], cmd);
    }
});
$("#fanStatus").on("click", function() {
    if (page.checkOnline() && page.checkMac("mac_602")){
        var state = $(this).text()=="打开", cmd;
        if(state){
            cmd = "{OD1=8,D1=?}";
        }else{
            cmd = "{CD1=8,D1=?}";
        }
        rtc.sendMessage(config.mac_602, cmd);
    }
});
```

3.3.4 开发验证

1. Web 端应用测试

在 Web 端打开智能燃气控制系统，成功连接智云服务器后切换到系统主页，此时可以看到设备的状态为“在线”，如图 3.38 所示。

在自动模式下，当监测到燃气泄漏时，系统将自动关闭燃气开关并启动风扇排风，如图 3.39 所示。

在手动模式下，可以控制燃气设备和风扇的开关，如图 3.40 所示。

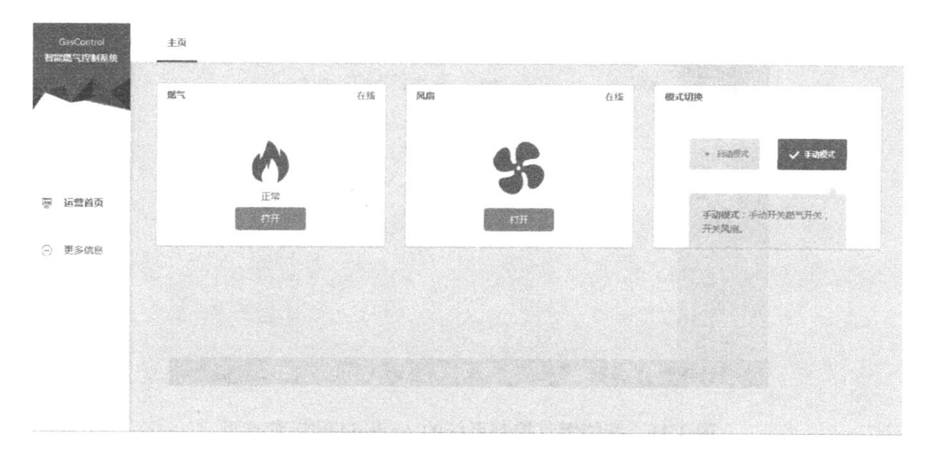

图 3.38　智能燃气控制系统的 Web 端主页

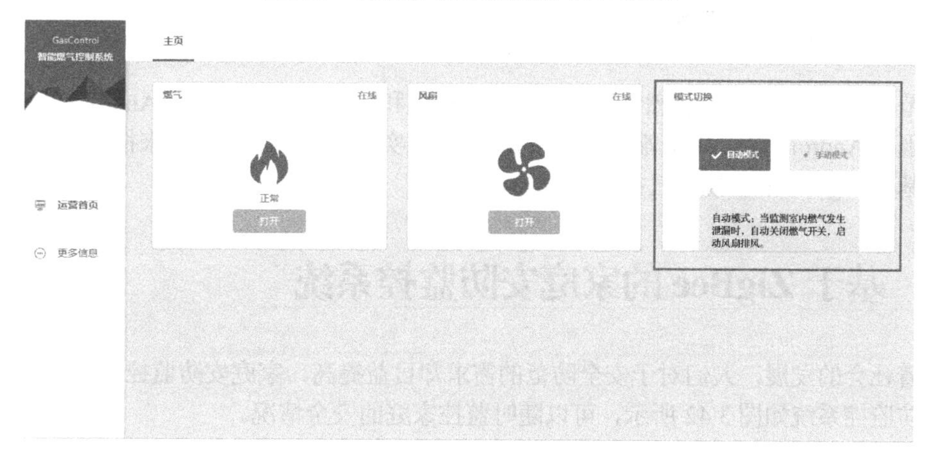

图 3.39　设备控制的自动模式

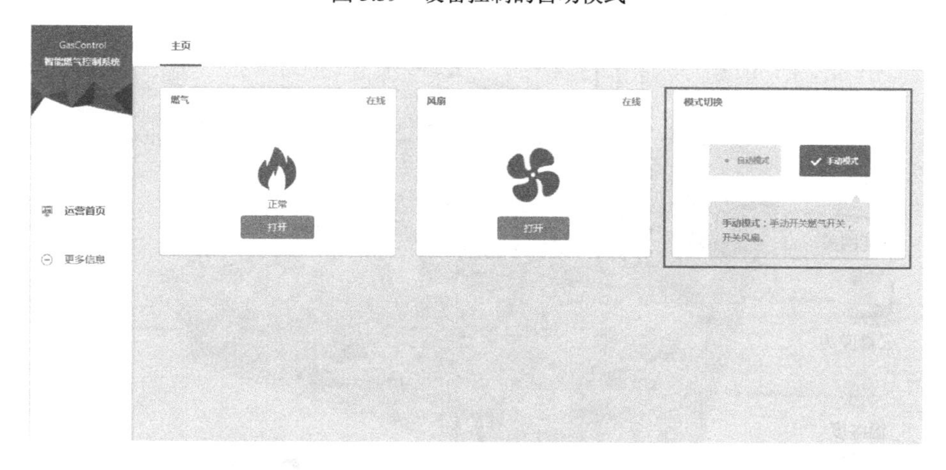

图 3.40　设备控制的手动模式

2．Android 端应用测试

Android 端应用测试同 Web 端应用测试基本一致，智能燃气控制系统的 Android 端运营首页如图 3.41 所示。

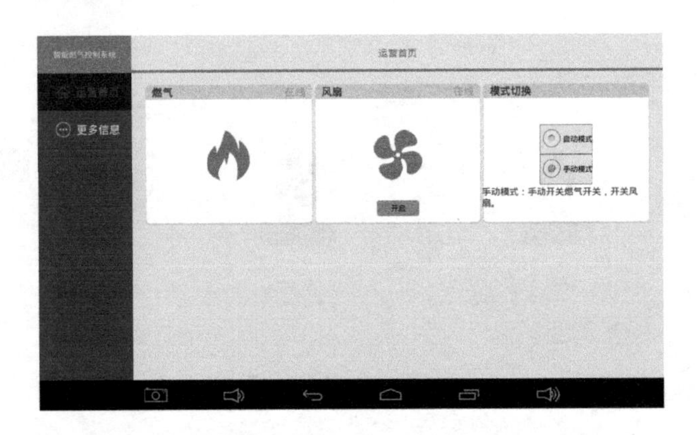

图 3.41　智能燃气控制系统的 Android 端运营首页

3.3.5　总结与拓展

本节基于 ZigBee 实现了燃气传感器的数据采集和风扇的控制，通过 Android 和 HTML5 技术实现了 Android 端和 Web 端的应用设计，可根据实时获取的燃气数据来控制风扇的开关，实现了基于 ZigBee 的智能燃气控制系统。

3.4　基于 ZigBee 的家庭安防监控系统

随着社会的发展，人们对于安全防范的需求却日益提高，家庭安防监控系统应运而生。家庭安防监控系统如图 3.42 所示，可以随时监控家庭的安全情况。

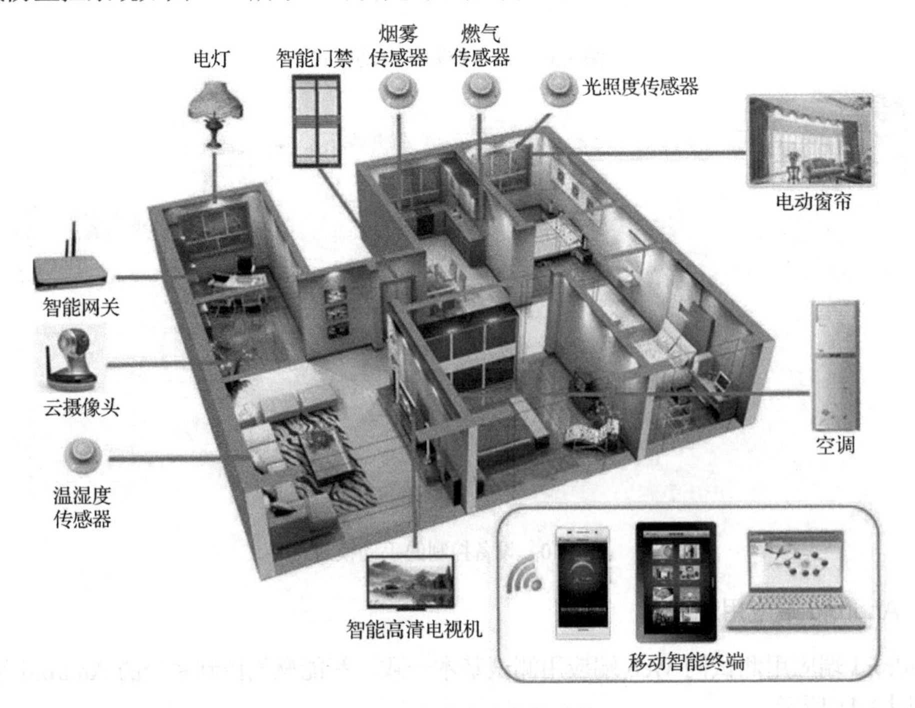

图 3.42　家庭安防监控系统

3.4.1 系统开发目标

（1）熟悉燃气传感器、火焰传感器、人体红外传感器、振动传感器和声光报警设备等的硬件原理和数据通信协议，基于 CC2530 微处理器和 ZigBee 实现传感器的驱动开发，通过对燃气、火焰器、人体红外、振动等状态的监控来控制声光报警设备的开关，从而实现家庭安防监控系统的设计。

（2）实现家庭安防监控系统的 Android 端应用开发和 Web 端应用开发。

3.4.2 系统设计分析

1. 系统的功能设计

家庭安防监控系统的主要功能是实时采集燃气、火焰、人体红外、振动等状态，并将采集到的数据主动推送到智云平台，如果监测到状态异常，则打开声光报警设备。从系统功能的角度出发，家庭安防监控系统可以分为两个模块：安防设备控制管理模块和系统设置模块，如图 3.43 所示。

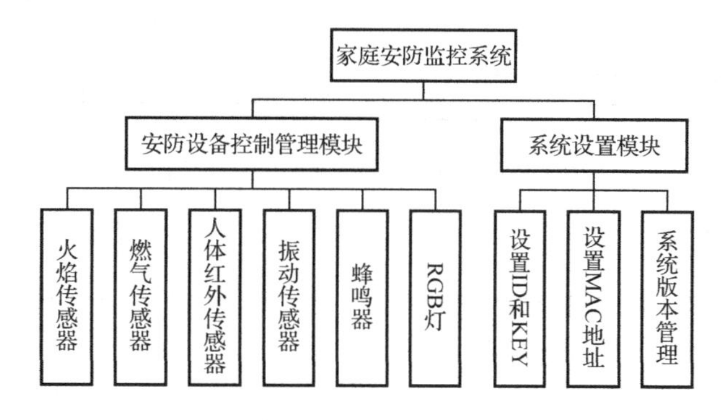

图 3.43 家庭安防监控系统的组成模块

安防设备控制管理模块的主要功能是将燃气传感器、火焰传感器、人体红外传感器、振动传感器等采集的数据主动推送到智云平台，如有异常，则通过声光报警设备（本系统使用 RGB 灯和蜂鸣器）进行报警。

系统设置模块的主要功能是设置智云服务器的 ID 和 KEY；设置 MAC 地址；查询与显示系统版本。

家庭安防监控系统的功能需求分析如表 3.11 所示。

表 3.11 家庭安防监控系统的功能需求分析

功 能	功 能 说 明
采集类传感器状态	在上层应用页面中实时更新并显示燃气、火焰、人体红外、振动等状态
模式设置	布防模式：开启全部传感器；撤防模式：关闭全部传感器
设备联动	当设备异常时引发报警
智云连接设置	设置智云服务器的参数和设备的 MAC 地址

2. 系统的总体架构设计

家庭安防监控系统基于物联网项目四层架构进行设计，其总体架构如图 3.44 所示。

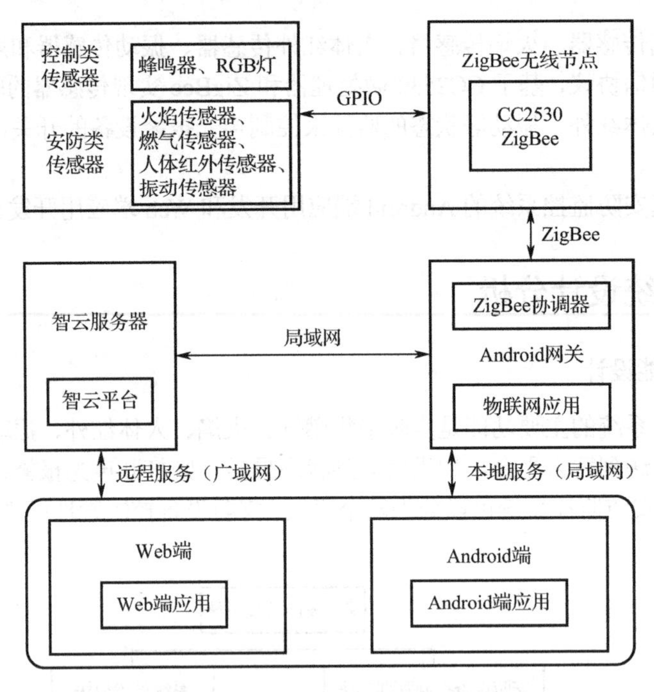

图 3.44　家庭安防监控系统的总体架构

3. 系统的数据传输

家庭安防监控系统的数据传输是在传感器节点、智能网关以及客户端（包括 Web 端和 Android 端）之间进行的，如图 3.45 所示。

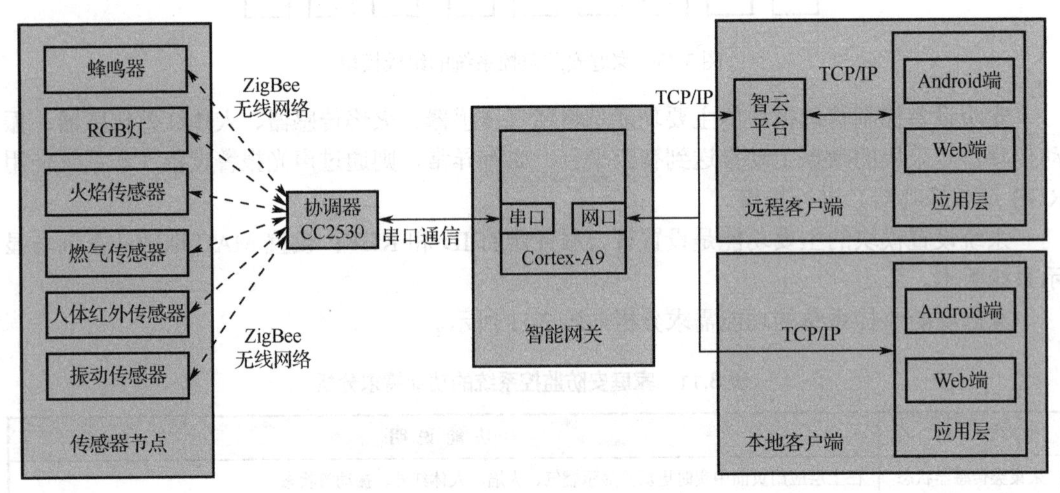

图 3.45　家庭安防监控系统的数据传输

3.4.3 系统的软硬件开发：家庭安防监控系统

1．系统底层软硬件设计

1）感知层硬件设计

家庭安防监控系统的感知层硬件主要包括 xLab 未来开发平台的智能网关、经典型无线节点 ZXBeeLiteB、控制类开发平台 Sensor-B 和安防类开发平台 Sensor-C。其中，智能网关负责汇集传感器采集的数据；ZigBee 无线节点（由经典型无线节点 ZXBeeLiteB 实现）通过无线通信的方式向智能网关发送传感器数据，接收智能网关发送的命令；控制类开发平台 Sensor-B 和安防类开发平台 Sensor-C 连接到 ZigBee 无线节点，由其中的 CC2530 微处理器对相关设备进行控制。

本系统使用的传感器主要有燃气传感器、火焰传感器、人体红外传感器、振动传感器、RGB 灯和蜂鸣器，燃气传感器的硬件接口电路如图 3.34 所示（见 3.3.3 节），火焰传感器的硬件接口电路如图 3.46 所示，人体红外传感器的硬件接口电路如图 3.47 所示，振动传感器的硬件接口电路如图 3.48 所示，RGB 灯的硬件接口电路如图 3.24 所示（见 3.2.3 节），蜂鸣器的硬件接口电路如图 3.49 所示。

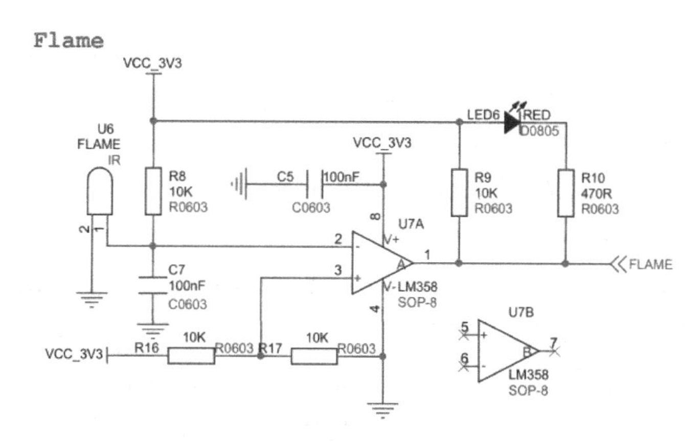

图 3.46　火焰传感器的硬件接口电路

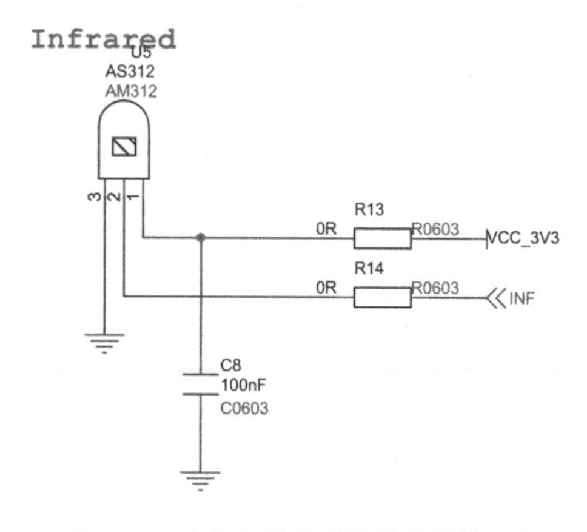

图 3.47　人体红外传感器的硬件接口电路

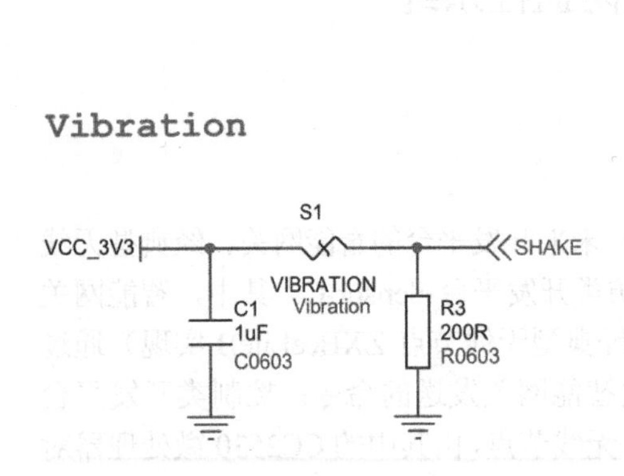

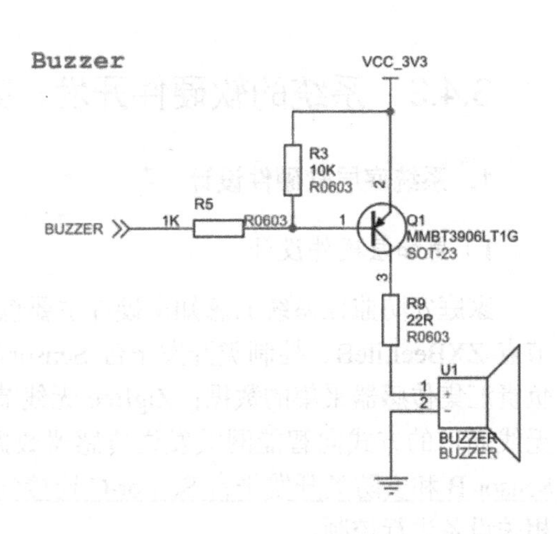

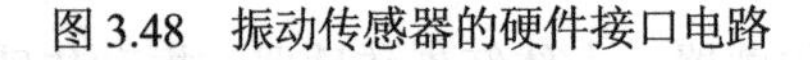
图 3.48　振动传感器的硬件接口电路　　　　图 3.49　蜂鸣器的硬件接口电路

2）系统底层开发

基于 ZigBee 的家庭安防监控系统的底层开发和基于 ZigBee 的城市环境信息采集系统的底层开发相同，详见 3.1.3 节。

3）传感器驱动设计

家庭安防监控系统的底层硬件主要包括控制类传感器和安防类传感器，传感器的驱动设计主要是针对这两类传感器进行的。控制类传感器的逻辑事件有 3 种，详见 3.2.3 节。安防类传感器的逻辑事件有 4 种，详见 3.3.3 节。

（1）数据通信协议的定义。本系统主要使用的是控制类开发平台 Sensor-B 和安防类开发平台 Sensor-C，其 ZXBee 协议定义如表 3.12 所示。

表 3.12　安防类开发平台和控制类开发平台 ZXBee 数据通信协议

开发平台	属　性	参　数	权限	说　明
Sensor-B（602）	RGB	D1(OD1/CD1)	R/W	D1 的 Bit0～Bit1 代表 RGB 灯的状态和三种颜色，00 表示关闭 RGB 灯，01 表示红色（R），10 表示绿色（G），11 表示蓝色（B）
	蜂鸣器	D1(OD1/CD1)	R/W	D1 的 Bit3 代表蜂鸣器的状态，0 表示关闭，1 表示打开
	数据上报时间间隔	V0	R/W	数据上报的时间间隔（循环上报）
Sensor-C（603）	火焰状态	A3	R	火焰状态值，0 表示未监测到火焰，1 表示监测到火焰
	人体红外状态	A0	R	人体红外状态值，0 表示未监测到人体，1 表示监测到人体
	振动状态	A1	R	振动状态值，0 表示未监测到振动，1 表示监测到振动
	燃气状态	A4	R	燃气泄漏状，0 表示未监测到燃气泄漏，1 表示监测到燃气泄漏
	数据上报时间间隔	V0	R/W	数据上报的时间间隔（循环上报）

（2）驱动程序的开发。本系统使用到控制类开发平台 Sensor-B 和安防类开发平台 Sensor-C，这两个开发平台是基于 CC2530 微处理器构建的。sensor.c 中 sensorInit()函数负责采集类传感器和控制类传感器的初始化，部分代码如下。

```
/*********************************************************************************
*名称：sensorInit()
*功能：传感器初始化
*********************************************************************************/
void sensorInit(void)
{
    //初始化传感器代码
    vibration_init();                //振动传感器初始化，根据该引脚的状态来判断传感器的跳线模式
    if(P0_1){                        //传感器跳线模式 1
        mode = 1;
        infrared_init();             //人体红外传感器初始化
        flame_init();                //火焰传感器初始化
        hall_init();                 //霍尔传感器初始化
    }
    else{                            //传感器跳线模式 2
        mode = 2;
        touch_init();                //触摸传感器初始化
        syn6288_init();              //语音合成传感器初始化
        syn6288_play("你好");
    }
    combustiblegas_init();           //燃气传感器初始化
    grating_init();                  //光栅传感器初始化
    relay_init();                    //继电器初始化

    //启动定时器，触发上报事件 MY_REPORT_EVT
    osal_start_timerEx(sapi_TaskID, MY_REPORT_EVT, (uint16)((osal_rand()%10) *1000));
    //启动定时器，触发报警事件 MY_CHECK_EVT
    osal_start_timerEx(sapi_TaskID, MY_CHECK_EVT, 100);
}
```

火焰传感器的初始化函数是 flame_init()，部分代码如下：

```
/*********************************************************************************
*名称：flame_init()
*功能：火焰传感器初始化
*********************************************************************************/
void flame_init(void)
{
    P0SEL &= ~0x08;                  //配置引脚为 GPIO 模式
    P0DIR &= ~0x08;                  //配置引脚为输入模式
}
/*********************************************************************************
*名称：unsigned char get_flame_status(void)
*功能：获取火焰传感器状态
*********************************************************************************/
unsigned char get_flame_status(void)
{
```

```
    if(P0_3)                        //监测 IO 口电平
        return 1;
    else
        return 0;
}
```

人体红外传感器的初始化函数是 infrared_init()，部分代码如下：

```
/*********************************************************************************************
*名称：infrared_init()
*功能：人体红外传感器初始化
*********************************************************************************************/
void infrared_init(void)
{
    P0SEL &= ~0x01;                 //配置引脚为 GPIO 模式
    P0DIR &= ~0x01;                 //配置引脚为输入模式
}
/*********************************************************************************************
*名称：unsigned char get_infrared_status(void)
*功能：获取人体红外传感器状态
*********************************************************************************************/
unsigned char get_infrared_status(void)
{
    if(P0_0)                        //人体红外传感器监测引脚
        return 1;                   //监测到信号返回 1
    else
        return 0;                   //没有监测到信号返回 0
}
```

振动传感器的初始化函数是 vibration_init()，部分代码如下：

```
/*********************************************************************************************
*名称：vibration_init()
*功能：振动传感器初始化
*********************************************************************************************/
void vibration_init(void)
{
    P0SEL &= ~0x02;                 //配置引脚为 GPIO 模式
    P0DIR &= ~0x02;                 //配置引脚为输入模式
}
/*********************************************************************************************
*名称：unsigned char get_vibration_status(void)
*功能：获取振动传感器状态
*********************************************************************************************/
unsigned char get_vibration_status(void)
{
    if(P0_1)                        //振动传感器监测引脚
        return 0;                   //没有监测到信号返回 0
```

```
    else
        return 1;                    //监测到信号返回 1
}
```

燃气传感器的初始化函数是 combustiblegas_init()，部分代码如下：

```
/*********************************************************************************************
*名称：combustibleGas_init()
*功能：燃气传感器初始化
*********************************************************************************************/
void combustiblegas_init(void)
{
    APCFG |= 0x20;                   //模拟 I/O 使能
    P0SEL |= 0x20;                   //端口 P0_5 功能选择外设功能
    P0DIR &= ~0x20;                  //设置输入模式
}
/*********************************************************************************************
*名称：unsigned int get_infrared_status(void)
*功能：获取燃气传感器状态
*********************************************************************************************/
unsigned int get_combustiblegas_data(void)
{
    unsigned int   value;
    value = HalAdcRead (HAL_ADC_CHANNEL_5,HAL_ADC_RESOLUTION_12);
    if(value <= 1500)
        return 0;
    else
        return 1;
}
```

蜂鸣器的初始化函数是 beep_init()，部分代码如下：

```
/*********************************************************************************************
*名称：beep_init()
*功能：蜂鸣器初始化
*********************************************************************************************/
void beep_init(void)
{
    P0SEL &= ~0x08;                  //配置引脚为 GPIO 模式
    P0DIR |= 0x08;                   //配置引脚为输出模式
}
```

RGB 灯的初始化函数是 rgb_init()，部分代码如下：

```
/*********************************************************************************************
*名称：rgb_init()
*功能：RGB 灯初始化
*********************************************************************************************/
void rgb_init(void)
```

```
{
    APCFG &= ~0x01;               //模拟 I/O 失能
    P0SEL &= ~0x07;               //配置引脚（P0_4 和 P0_5）为 GPIO 模式
    P0DIR |= 0x07;                //配置引脚（P0_4 和 P0_5）为输出模式
    RGB_R = OFF;                  //初始状态为关闭
    RGB_G = OFF;                  //初始状态为关闭
    RGB_B = OFF;                  //初始状态为关闭
}
/*********************************************************************************************
*名称：rgb_on()
*功能：打开 RGB 灯
*参数：rgb—RGB 灯号，在 rgb.h 中宏定义为 RGB_R、RGB_G、RGB_B
*返回：0 表示打开 RGB 灯，-1 表示参数错误
*********************************************************************************************/
signed char rgb_on(unsigned char rgb)
{
    if(rgb == RGB_R){             //打开 RGB_R
        RGB_R = ON;
        return 0;
    }
    if(rgb == RGB_G){             //打开 RGB_G
        RGB_G = ON;
        return 0;
    }
    if(rgb == RGB_B){             //打开 RGB_B
        RGB_B = ON;
        return 0;
    }
    return -1;                    //参数错误，返回-1
}
/*********************************************************************************************
*名称：rgb_off()
*功能：关闭 RGB 灯
*参数：rgb—RGB 灯号，在 rgb.h 中宏定义为 RGB_R、RGB_G、RGB_B
*返回：0 表示关闭 RGB 灯，-1 表示参数错误
*********************************************************************************************/
signed char rgb_off(unsigned char rgb)
{
    if(rgb == RGB_R){             //关闭 RGB_R
        RGB_R = OFF;
        return 0;
    }
    if(rgb == RGB_G){             //关闭 RGB_G
        RGB_G = OFF;
        return 0;
    }
    if(rgb == RGB_B){             //关闭 RGB_B
```

```
        RGB_B = OFF;
        return 0;
    }
    return -1;                          //参数错误，返回-1
}
```

安防类传感器的状态更新主要是通过 sensorCheck()函数实现的，部分代码如下：

```
/*********************************************************************************************
*名称：sensorCheck()
*功能：传感器监测
*********************************************************************************************/
void sensorCheck(void)
{
    static char lastA0 = 0,lastA1=0,lastA2=0,lastA3=0,lastA4=0,lastA5=0;
    static uint32 ct0=0, ct1=0, ct2=0, ct3=0, ct4=0, ct5=0;
    char pData[16];
    char *p = pData;
    updateA0();  updateA1();  updateA2();  updateA3();  updateA4();  updateA5();
    ZXBeeBegin();
    if (lastA0 != A0
      || ( //模式 1：人体红外传感器选通，每 3 s 上报一次数据
          (mode == 1)&&ct0 != 0 && osal_GetSystemClock() > (ct0+3000))          //每 3 s 上报一次
      ) {
        if (D0 & 0x01) {
            sprintf(p, "%u", A0);
            ZXBeeAdd("A0", p);
            ct0 = osal_GetSystemClock();
        }
        if (A0 == 0) {
            ct0 = 0;
        }
        lastA0 = A0;
    }
    if (mode == 1) {    //模式 1：振动传感器和火焰传感器选通
        if (lastA1 != A1 || (ct1 != 0 && osal_GetSystemClock() > (ct1+3000))) {        //振动传感器监测
            if (D0 & 0x02) {
                sprintf(p, "%u", A1);
                ZXBeeAdd("A1", p);
                ct1 = osal_GetSystemClock();
            }
            if (A1 == 0) {
                ct1 = 0;
            }
            lastA1 = A1;
        }
        if (lastA3 != A3 || (ct3 != 0 && osal_GetSystemClock() > (ct3+3000))) {        //火焰传感器监测
            if (D0 & 0x08) {
```

```
                    sprintf(p, "%u", A3);
                    ZXBeeAdd("A3", p);
                    ct3 = osal_GetSystemClock();
                }
                if (A3 == 0) {
                    ct3 = 0;
                }
                lastA3 = A3;
            }
        }
        if (lastA4 != A4 || (ct4 != 0 && osal_GetSystemClock() > (ct4+3000))) {          //燃气传感器监测
            if (D0 & 0x10) {
                sprintf(p, "%u", A4);
                ZXBeeAdd("A4", p);
                ct4 = osal_GetSystemClock();
            }
            if (A4 == 0) {
                ct4 = 0;
            }
            lastA4 = A4;
        }
        p = ZXBeeEnd();
        if (p != NULL) {
            int len = strlen(p);
            ZXBeeInfSend(p, len);
        }
    }
```

2. Android 端应用设计

1）Android 工程设计框架

打开 Android Studio 开发环境，可以看到家庭安防监控系统的工程目录，如图 3.50 所示，系统的工程框架如表 3.13 所示。

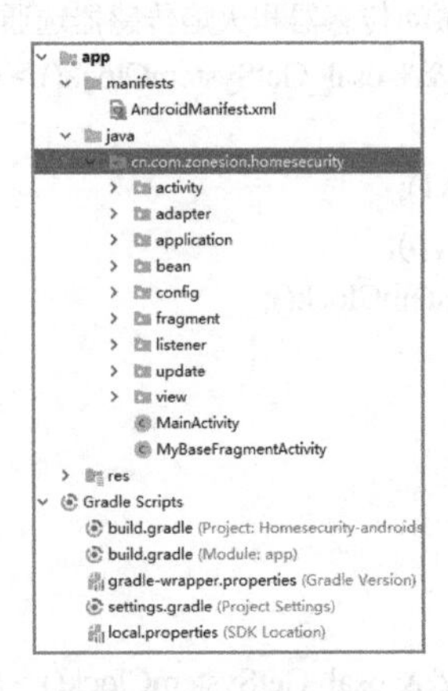

图 3.50　家庭安防监控系统的工程目录

表 3.13 家庭安防监控系统的工程框架

类 名	说 明
activity	
IdKeyShareActivity.java	在 IDKey 页面单击“分享”按钮时，可弹出 activity，用于分享二维码图片
adapter	
HdArrayAdapter.java	历史数据显示适配器
ViewPagerAdapter.java	对 ViewPager 进行适配，可以处理多个 Fragment 的横向滑动
application	
LCApplication.java	LCApplication 继承 application 类，使用单例模式（Singleton Pattern）创建 WSNRTConnect 对象
bean	
HistoricalData.java	历史数据的 bean 类，用于将从智云服务器获得的历史数据记录（JSON 形式）转换成该类对象
IdKeyBean.java	IdKeyBean 用来描述用户设备的 ID、KEY，以及智云服务器的地址 SERVER
config	
Config.java	config 用于修改用户的 ID、KEY，以及智云服务器的地址和 MAC 地址
fragment	
BaseFragment.java	页面基础 Fragment 定义类
BasicsFragment.java：下面一些 Fragment 的基类，定义共有的属性以及 getter 和 setter 方法	
HomepageFragment.java	展示首页页面的 Fragment
IDKeyFragment.java	IDKey 选项的页面
MacSettingFragment.java	当用户设置被监测项的 MAC 地址时显示的页面
MoreInformationFragment.java	更多信息显示页面
RunHomePageFragment.java	运营首页显示页面
VersionInformationFragment.java	显示版本等相关信息的页面
listener	
IOnWSNDataListener.java	传感器数据监听器接口
update	
UpdateService.java	应用下载服务类
view	
APKVersionCodeUtils.java	获取当前本地 apk 的版本
CustomRadioButton.java	自定义按钮类
PagerSlidingTabStrip.java	自定义滑动控件类
MainActivity.java：主页面类	
MyBaseFragmentActivity.java：系统 Fragment 通信类	

2）软件设计

根据智云 Android 端应用程序接口的定义，家庭安防监控系统的应用设计主要采用实时数据 API 接口（和城市环境信息采集系统相同），其流程见 3.1.3 节中的图 3.11。

（1）LCApplication.java 程序代码剖析。家庭安防监控系统中的 LCApplication.java 程序代码和城市环境信息采集系统的 LCApplication.java 程序代码相同，详见 3.1.3 节的相关内容。

（2）HomepageFragment.java 程序代码剖析。下面的代码通过(LCApplication) getActivity().getApplication()获取 LCApplication 类中的 WSNRTConnect 对象。

```
private void initInstance(){
    config = Config.getConfig();
    lcApplication = (LCApplication) getActivity().getApplication();
    lcApplication.registerOnWSNDataListener(this);
    wsnrtConnect = lcApplication.getWSNRConnect();
    preferences = getActivity().getSharedPreferences("user_info", Context.MODE_PRIVATE);
    editor = preferences.edit();
}
```

下面的代码通过复写 onMessageArrive 方法来处理节点接收到的无线数据包，实现了设备的 MAC 地址获取，并在当前的页面显示设备的状态。

```
@Override
    public void onMessageArrive(String mac, String tag, String val) {
        if (sensorBMac == null && sensorCMac == null) {
            wsnrtConnect.sendMessage(mac, "{TYPE=?}".getBytes());
        }
        if ("TYPE".equals(tag) && "602".equals(val.substring(2, val.length()))) {
            sensorBMac = mac;
        }
        if ("TYPE".equals(tag) && "603".equals(val.substring(2, val.length()))) {
            sensorCMac = mac;
        }
        if (tag.equals("D1") && mac.equals(sensorBMac)) {
            textLinkageState.setText("在线");
            textLinkageState.setTextColor(getResources().getColor(R.color.line_text_color));
            int numResult = Integer.parseInt(val);
            if (isSecurityMode == true) {
                if ((numResult & 8) == 8) {
                    imageLinkageState.setImageResource(R.drawable.alarm_on);
                 }else {
                     imageLinkageState.setImageResource(R.drawable.alarm);
                }
            }else {
                imageFlameState.setImageResource(R.drawable.alarm);
            }
        }
        if (tag.equals("A3") && mac.equals(sensorCMac)) {
            textFlameState.setText("在线");
            textFlameState.setTextColor(getResources().getColor(R.color.line_text_color));
            numResult1 = Integer.parseInt(val);
            if (isSecurityMode == true) {
```

```
            textFalmeDisabled.setText("已启用");
        }else {
            textFalmeDisabled.setText("已禁用");
        }
        if (isSecurityMode == true && Flame == true) {
            if ((numResult1 & 1) == 1) {
                imageFlameState.setImageResource(R.drawable.fire_on);
            }else {
                imageFlameState.setImageResource(R.drawable.fire);
            }
        }else {
            imageFlameState.setImageResource(R.drawable.fire);
        }
    }
    if (tag.equals("A4") && mac.equals(sensorCMac)) {
        textGasState.setText("在线");
        textGasState.setTextColor(getResources().getColor(R.color.line_text_color));
        numResult4 = Integer.parseInt(val);
        if (isSecurityMode == true) {
            textGasDisabled.setText("已启用");
        }else {
            textGasDisabled.setText("已禁用");
        }
        if (isSecurityMode == true && Gas == true) {
            if ((numResult4 & 1) == 1) {
                imageGasState.setImageResource(R.drawable.gas_on);
            }else {
                imageGasState.setImageResource(R.drawable.gas);
            }
        }else {
            imageGasState.setImageResource(R.drawable.gas);
        }
    }
    if (tag.equals("A0") && mac.equals(sensorCMac)) {
        textInfraredState.setText("在线");
        textInfraredState.setTextColor(getResources().getColor(R.color.line_text_color));
        numResult2 = Integer.parseInt(val);
        if (isSecurityMode == true) {
            textInfraredDisabled.setText("已启用");
        }else {
            textInfraredDisabled.setText("已禁用");
        }
        if (isSecurityMode == true && kHumanInfrared == true) {
            if ((numResult2 & 1) == 1) {
                imageInfraredState.setImageResource(R.drawable.body_on);
            }else {
                imageInfraredState.setImageResource(R.drawable.body);
```

```
                }
            }else {
                imageInfraredState.setImageResource(R.drawable.body);
            }
        }
        if (tag.equals("A1") && mac.equals(sensorCMac)) {
            textVibrationState.setText("在线");
            textVibrationState.setTextColor(getResources().getColor(R.color.line_text_color));
            numResult3 = Integer.parseInt(val);
            if (isSecurityMode == true) {
                textVibrationDisabled.setText("已启用");
            }else {
                textVibrationDisabled.setText("已禁用");
            }
            if (isSecurityMode == true && Vibration == true) {
                if ((numResult3 & 1) == 1) {
                    imageVibrationState.setImageResource(R.drawable.vibrate_on);
                }else {
                    imageVibrationState.setImageResource(R.drawable.vibrate);
                }
            }else {
                imageVibrationState.setImageResource(R.drawable.vibrate);
            }
        }
    }
```

实现设备联动功能的代码如下：

```
checkGas.setOnCheckedChangeListener(new OnCheckedChangeListener() {
    @Override
        public void onCheckedChanged(CompoundButton buttonView, boolean isChecked) {
            //TODO Auto-generated method stub
            if (isChecked) {
                Gas = true;
                if (sensorBMac != null) {
                    if (numResult4 == 1) {
                        new Thread(new Runnable() {
                        @Override
                            public void run() {
                                wsnrtConnect.sendMessage(sensorBMac, "{OD1=8,D1=?}".getBytes());
                            }
                        }).start();
                    }else if(numResult1 == 0 && numResult2 == 0 && numResult1 == 0 &&
                                                        numResult3 == 0 && numResult4 == 0){
                        new Thread(new Runnable() {
                            @Override
                                public void run() {
```

```
                                        wsnrtConnect.sendMessage(sensorBMac, "{CD1=8,D1=?}".
getBytes());
                                    }
                            }).start();
                        }
                    }else{
                        Toast.makeText(lcApplication, "请等待 MAC 地址上线", Toast.LENGTH_SHORT).
show();
                    }
                }else {
                    Gas = false;
                    if (Gas == false && Flame == false && kHumanInfrared == false && Vibration == false) {
                        new Thread(new Runnable() {
                            @Override
                                public void run() {
                                    wsnrtConnect.sendMessage(sensorBMac, "{CD1=8,D1=?}".getBytes());
                                }
                        }).start();
                    }
                }
            }
        });
```

3. Web 端应用设计

1）页面功能结构分析

家庭安防监控系统的 Web 端默认显示的是“运营首页”页面，在“运营首页”页面设计了 6 个模块，分别是设备联动设置模块、模式切换设置模块、火焰状态显示模块、燃气状态显示模块、人体红外状态显示模块、振动状态显示模块。家庭安防监控系统 Web 端的“运营首页”页面如图 3.51 所示。

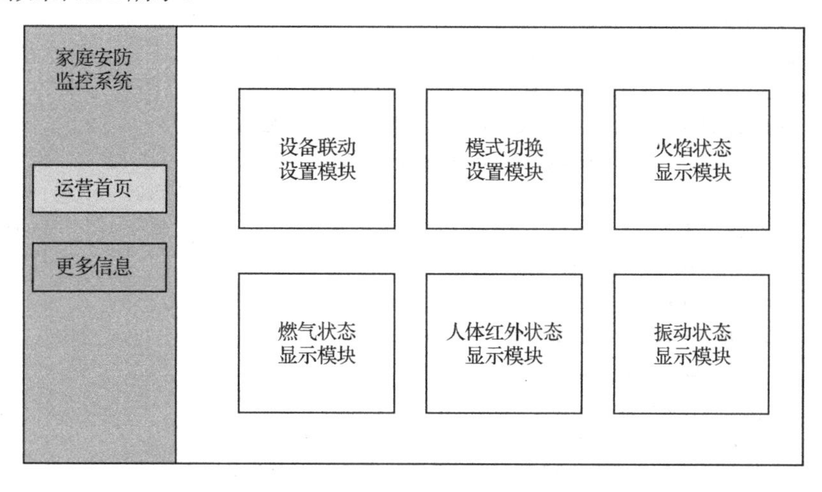

图 3.51 家庭安防监控系统 Web 端的“运营首页”页面

“更多信息”页面的主要功能是进行智云服务器的连接配置，和城市环境信息采集系统中的 Web 端“更多信息”页面类似，参见图 3.14。

2）软件设计

家庭安防监控系统 Web 端的 JS 开发逻辑与 Android 端的开发逻辑相似，首先通过配置 ID 和 KEY 与智云服务器进行连接，再通过实时监听数据的方法来获取相关传感器的数据并进行处理。JS 开发的部分代码如下。

在 getConnect()函数中定义了实时连接对象 rtc，连接成功回调函数是 rtc.onConnect，数据服务掉线回调函数是 rtc.onConnectLost，消息处理回调函数是 rtc.onmessageArrive。

```
function getConnect() {
    config.id = config.id ? config.id : $("#id").val();
    config.key = config.key ? config.key : $("#key").val();
    config.server = config.server ? config.server : $("#server").val();
    //创建数据连接服务对象
    rtc = new WSNRTConnect(config["id"], config["key"]);
    rtc.setServerAddr(config["server"] + ":28080");
    rtc.connect();
    rtc._connect = false;
    //连接成功回调函数
    rtc.onConnect = function() {
        rtc._connect = 1;
        console.log(rtc._connect);
        message_show("数据服务连接成功！");
        $("#idkeyInput").text("断开").addClass("btn-danger");
        $("#id,#key,#server").attr('disabled',true);
    };
    //数据服务掉线回调函数
    rtc.onConnectLost = function() {
        rtc._connect = 0;
        console.log(rtc._connect);
        $("#idkeyInput").text("连接").removeClass("btn-danger");
        $(".online_602").text("离线").css("color", "#e75d59");
        message_show("数据服务连接失败，检查网络或 ID、KEY");
        $("#id,#key,#server").removeAttr('disabled');
    };
    //消息处理回调函数
    rtc.onmessageArrive = function (mac, dat) {
        //console.log(mac+" >>> "+dat);
        if (dat[0]=='{' && dat[dat.length-1]=='}') {
            dat = dat.substr(1, dat.length-2);
            var its = dat.split(',');
            for (var i=0; i<its.length; i++) {
                var it = its[i].split('=');
                if (it.length == 2) {
                    var tag = it[0];
```

```
                var val = it[1];
                process_tag(mac, tag, val);
            }
        }
        if (!mac2type[mac]) { //如果没有获取到 TYPE 值，主动去查询
            rtc.sendMessage(mac, "{TYPE=?,A0=?,A1=?,A2=?,A3=?,A4=?,A5=?,A6=?,A7=?,D1=?}");
        }
    }
  }
}
```

下面的 JS 开发代码的功能是根据设备连接的情况，在页面中更新设备是否在线的状态，显示布防模式下安防类传感器的状态。

```
var wsn_config = {
    "602" : {
        "online" : function() {
            $(".online_602").text("在线").css("color", "#96ba5c");
        },
        "pro" : function (tag, val) {
            console.log(val);
            if(tag=="D1"){
                var flag = (val & 0x03 || val & 0x02 || val & 0x01) && val & 0x04;
                if(flag){
                    $("#alarmState").addClass("div-alarm-on Sensor-Active");
                }else{
                    $("#alarmState").removeClass("div-alarm-on Sensor-Active");
                }
                state.alarm = flag;
            }
        }
    },
    "603" : {
        "online" : function() {
            $(".online_603").text("在线").css("color", "#96ba5c");
        },
        "pro" : function (tag, val) {
            console.log(val);
            //仅当布防模式，才获取安防类传感器的状态
            if(config["mode"] == "secutiry-mode"){
                //A0=人体，A1=振动，A3=火焰，A4=燃气
                var sensor = channel2name[tag];
                if(val == 1){
                    $(".div-"+sensor).addClass(sensor+"-on");
                } else{
                    $(".div-"+sensor).removeClass(sensor+"-on");
                }
```

```
                    checkAlarm(sensor, val);
                } else {
                    rtc.sendMessage(config.mac_602, '{CD1=8,D1=?}');
                }
            }
        }
};
```

布防模式和撤防模式切换的代码如下：

```
//模式切换
$("#mode-switch input").on("click", function() {
    config["mode"] = $(this).val();
    console.log("切换到："+config["mode"]);
    var isManualMode = config["mode"]=="homeout-mode";
    console.log(isManualMode);
    if(isManualMode){
        $("#mode-txt-2").removeClass("hidden").siblings("span").addClass("hidden");
        $("#mode-text").addClass("mode-right");
        //关闭所有传感器
        $(".sensor-status").addClass("sensor-disable");
        console.log("禁用所有传感器");
    }
    //撤防模式下禁用开关按钮
    else{
        $(".sensor-status").removeClass("sensor-disable");
        $("#mode-txt-1").removeClass("hidden").siblings("span").addClass("hidden");
        $("#mode-text").removeClass("mode-right");
    }
    storeStorage();
})
```

3.4.4 开发验证

1. Web 端应用测试

在 Web 端打开家庭安防监控系统，连接服务器成功后切换到系统的主页，可看到设备状态更新为“在线”。家庭安防监控系统的 Web 端主页如图 3.52 所示。

在智能燃气控制系统的 Web 端主页中，选中“布防模式”后，系统会开启传感器，这些传感器的状态会变为“在线”，在“设备联动”勾选在线的设备，如果该设备出现异常，设备图标的颜色就会变成红色，并且会引发报警。联动报警测试如图 3.53 所示。

在智能燃气控制系统的 Web 端主页中，还可以进行模式的切换，在布防模式下系统会开启全部传感器，在撤防模式下系统会禁用全部传感器。模式切换测试如图 3.54 所示。

图 3.52　家庭安防监控系统的 Web 端主页

图 3.53　联动报警测试

图 3.54　模式切换测试

2. Android 端应用测试

Android 端应用测试同 Web 端应用测试的流程基本一致，可参考本系统的 Web 端应用测试流程进行操作。家庭安防监控系统的 Android 端“运营首页”页面如图 3.55 所示。

图 3.55　家庭安防监控系统的 Android 端“运营首页”页面

3.4.5　总结与拓展

本节基于 ZigBee 实现了燃气传感器、火焰传感器、人体红外传感器、振动传感器等的数据采集以及声光报警设备的控制，通过 Android 和 HTML5 技术实现了 Android 端和 Web 端的应用设计，能根据传感器实时采集的数据来控制声光报警设备的开关，实现了基于 ZigBee 的家庭安防监控系统。

第4章 BLE高级应用开发

蓝牙技术是一种支持设备之间进行短距离无线通信的技术，工作在全球通用的 2.4 GHz 的 ISM（工业、科学、医学）频段，采用快跳频和短数据包技术，支持点对点及点对多点的通信方式，其数据传输速率为 1 Mbps，采用时分双工传输方案实现了全双工传输。低功耗蓝牙（BLE）技术具有低成本、跨厂商互操作性、3 ms 低时延、100 m 以上超长传输距离、采用 AES-128 加密等诸多特色，使无线连接具有超低的功耗和较高的稳定性。有关更详尽的 BLE 内容请参考《物联网短距离无线通信技术应用与开发》。

本章通过基于 BLE 的家庭灯光控制系统和智能门禁管理系统这两个贴近生活的开发案例，详细地介绍了 BLE 物联网系统的架构和软硬件开发，实现了控制类节点和识别类节点的驱动程序，进行了 Android 端和 Web 端的应用开发。

4.1 基于 BLE 的家庭灯光控制系统

家庭灯光控制系统可以感知光线的强弱，并根据光线的强弱来对灯光进行合理的控制，其控制方式更加符合人的行为模式。家庭灯光控制系统如图 4.1 所示。

图 4.1 家庭灯光控制系统

4.1.1 系统开发目标

（1）熟悉 RGB 灯和 LED 等硬件原理和数据通信协议，实现基于 CC2540 和 BLE 的 RGB 灯和 LED 驱动开发，通过控制 RGB 灯和 LED 来实现家庭灯光控制系统的设计。

（2）实现家庭灯光控制系统的 Android 端应用开发和 Web 端应用开发。

4.1.2 系统设计分析

1．系统的功能设计

从系统功能的角度出发，家庭灯光控制系统可以分为设备控制和系统设置两个模块，如图 4.2 所示。

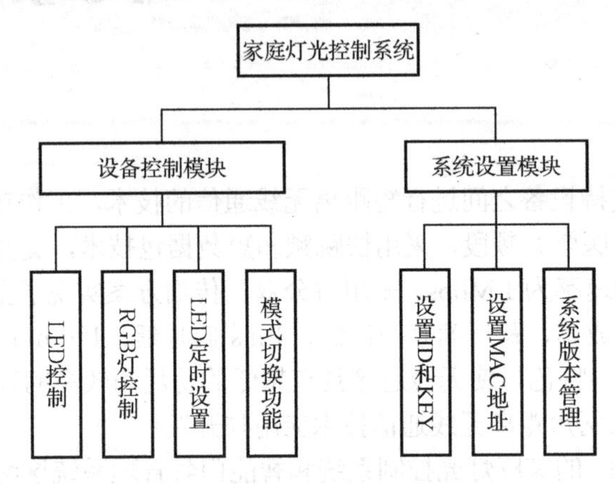

图 4.2 家庭灯光控制系统的组成模块

设备控制模块：分为 LED 控制、RGB 灯控制、LED 定时设置和模式切换功能，在安防模式下可以定时开关指定的灯具，模拟有人在家效果，在离家模式下可以一键关闭所有的灯具。

系统设置模块：设置智云服务器的 ID 和 KEY；设置 MAC 地址；系统软件版本查询与显示。

家庭灯光控制系统的功能需求分析如表 4.1 所示。

表 4.1 家庭灯光控制系统的功能需求分析

功　能	功 能 说 明
灯光状态显示	在上层应用页面中实时更新并显示 LED 和 RGB 灯的状态，如设备是否在线
灯光实时控制	通过应用程序，实时控制 LED 和 RGB 灯的开关
定时开发功能	设置 LED 和 RGB 灯的开启及关闭时间
模式设置	安防模式：定时开启指定的灯具，模拟家中有人的效果。离家模式：关闭所有的灯具
智云连接设置	设置智云服务器的参数和设备的 MAC 地址

2．系统的总体架构设计

家庭灯光控制系统是基于物联网四层架构模型来设计的，其总体架构如图 4.3 所示。

感知层：由控制类传感器构成，本系统使用 LED 和 RGB 灯。

网络层：感知层的采集类传感器和智能网关（Android 网关）之间是通过 BLE 连接的，智能网关和智云服务器、上层应用设备之间是通过局域网（互联网）来传输数据的。

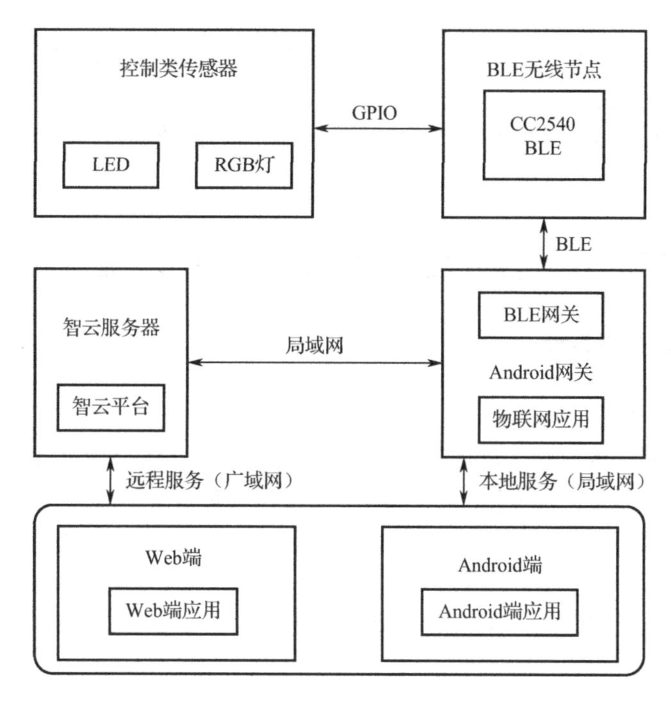

图 4.3 家庭灯光控制系统的总体架构

平台层：平台层提供物联网设备之间的基于互联网的存储、访问、控制。

应用层：提供物联网系统的人机交互接口，通过 Web 端、Android 端来提供页面友好、操作交互性强的应用。

3. 系统的数据传输

家庭灯光控制系统的数据传输是在传感器节点、智能网关以及客户端（包括 Web 端和 Android 端）之间进行的，如图 4.4 所示，具体通信流程如下所述。

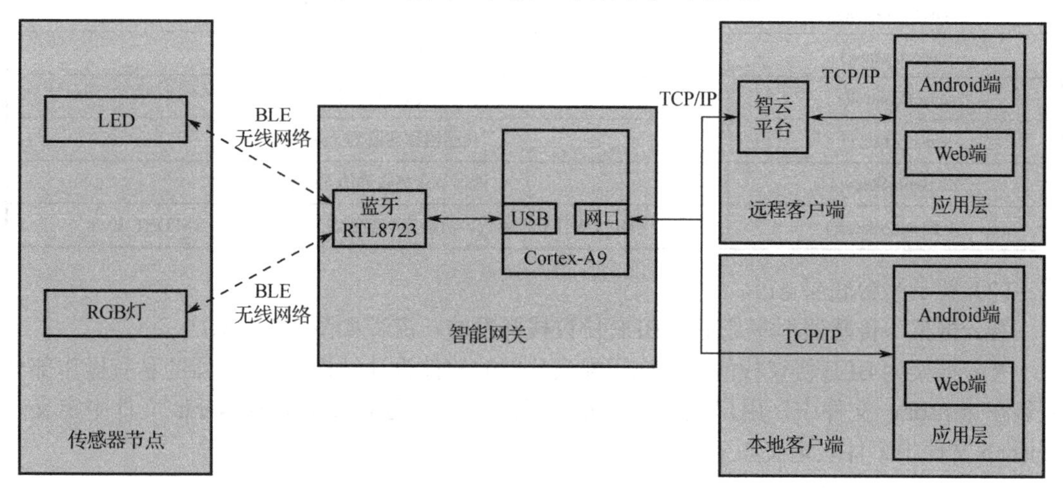

图 4.4 家庭灯光控制系统的数据传输

（1）传感器通过 BLE 无线网络与智能网关进行组网。

（2）传感器采集的数据通过 BLE 无线网络发送到智能网关，智能网关将接收到的数据推送给所有连接智能网关的客户端。

（3）客户端（Android 端和 Web 端）应用通过调用智云数据接口，实现数据实时采集的功能。

4.1.3 系统的软硬件开发：家庭灯光控制系统

1. 系统底层软硬件设计

1）感知层硬件设计

家庭灯光控制系统的感知层硬件包括 xLab 未来开发平台的智能网关、经典型无线节点 ZXBeeLiteB、控制类开发平台 Sensor-B。其中，智能网关负责汇集传感器采集的数据；BLE 无线节点（由经典型无线节点 ZXBeeLiteB 实现）通过无线通信的方式向智能网关发送传感器数据，接收智能网关发送的命令；控制类开发平台 Sensor-B 连接到 BLE 无线节点，由其中的 CC2540 微处理器对相关设备进行控制。LED 和 RGB 灯的硬件接口电路如图 3.24 所示（见 3.2.3 节）。

2）系统底层开发

本系统使用 BLE 无线网络进行开发。

（1）BLE 开发框架。开发框架是在 BLE 协议栈接口基础上搭建起来的，通过合理调用这些接口，使 BLE 的开发形成了一套系统的开发逻辑。传感器应用程序接口函数是在 sensor.c 文件中实现的，如表 4.2 所示。

表 4.2 传感器应用程序接口函数

函数名称	函数说明
sensorInit()	传感器初始化
sensorLinkOn()	传感器节点入网成功调用的函数
sensorUpdate()	传感器数据定时上报
sensorControl()	传感器控制函数
sensorCheck()	传感器预警监测及处理函数
ZXBeeInfRecv()	处理节点接收到的无线数据包
MyEventProcess()	自定义事件处理函数，启动定时器触发上报事件 MY_REPORT_EVT

（2）智云平台底层 API。

智云框架下传感器程序是基于 BLE 协议栈开发的，流程如图 4.5 所示。

智云框架为 BLE 协议栈的上层应用提供分层的软件设计结构，将传感器的私有操作部分封装在 sensor.c 文件中，用户任务中的处理事件和节点类型选择在 sensor.h 文件中定义。sensor.h 文件中事件宏定义如下：

```
#define MY_REPORT_EVT    0x0010
#define MY_CHECK_EVT     0x0020
#define NODE_NAME        "601"
extern uint8 simpleBLEPeripheral_TaskID;
```

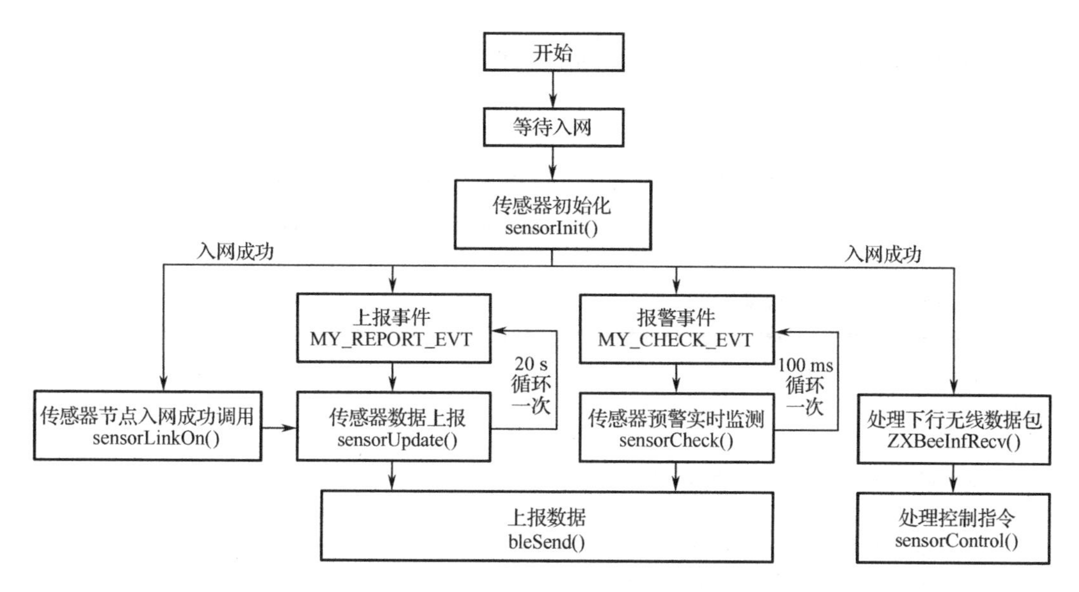

图4.5 智云框架下传感器程序流程

用户事件中定义的内容分别是上报事件（MY_REPORT_EVT）和报警事件（MY_CHECK_EVT），上报事件用于上报传感器采集到的数据，报警事件用于对传感器监测到的危险信息进行响应。在sensor.h文件中，通过节点类型的宏定义可以将将节点设置为路由节点（NODE_ROUTER）或终端节点（NODE_ENDDEVICE），同时还声明了智云框架下的传感器应用文件sensor.c中的函数。

sensorInit()函数用于进行传感器的初始化，以及触发上报事件MY_REPORT_EVT和报警事件MY_CHECK_EVT的定义，相关代码如下：

```
/*********************************************************************************
*名称：sensorInit()
*功能：传感器初始化
*********************************************************************************/
void sensorInit(void)
{
    printf("sensor->sensorInit(): Sensor init!\r\n");
    //传感器初始化
    //启动定时器，触发上报事件 MY_REPORT_EVT 和报警事件 MY_CHECK_EVT
    osal_start_timerEx(simpleBLEPeripheral_TaskID, MY_REPORT_EVT, (uint16)((osal_rand()%10) *1000));
    osal_start_timerEx(simpleBLEPeripheral_TaskID, MY_CHECK_EVT, 100));
}
```

传感器节点入网成功后会调用sensorLinkOn()函数来执行相关的操作，相关代码如下：

```
/*********************************************************************************
*名称：sensorLinkOn()
*功能：传感器节点入网成功后调用函数
*********************************************************************************/
void sensorLinkOn(void)
{
```

```
    printf("sensor->sensorLinkOn(): Sensor Link on!\r\n");
    sensorUpdate();                                    //入网成功后上报一次传感器数据
}
```

sensorUpdate()函数用于更新传感器的数据，并将更新后的数据打包上报，相关代码如下：

```
/*********************************************************************************************
*名称：sensorUpdate()
*功能：处理主动上报的数据
*********************************************************************************************/
void sensorUpdate(void)
{
    ......
    sprintf(p, "humidity=%.1f", humidity);
    bleSend(p, strlen(p));
    HalLedSet( HAL_LED_1, HAL_LED_MODE_OFF );
    HalLedSet( HAL_LED_1, HAL_LED_MODE_BLINK );
    printf("sensor->sensorUpdate(): humidity=%.1f\r\n", humidity);
}
```

MyEventProcess()函数用于启动和处理用户定义事件，相关代码如下：

```
/*********************************************************************************************
*名称：MyEventProcess()
*功能：自定义事件处理
*参数：event—事件编号
*********************************************************************************************/
void MyEventProcess( uint16 event )
{
    if (event & MY_REPORT_EVT) {
        sensorUpdate();                                //传感器数据定时上报
        //启动定时器，触发事件 MY_REPORT_EVT
        osal_start_timerEx(simpleBLEPeripheral_TaskID, MY_REPORT_EVT, 20*1000);
    }
    if (event & MY_CHECK_EVT) {
        sensorCheck();                                 //传感器状态实时监测
        //启动定时器，触发事件 MY_CHECK_EVT
        osal_start_timerEx(simpleBLEPeripheral_TaskID, MY_CHECK_EVT, 100);
    }
}
```

ZXBeeInfRecv()函数用于处理节点接收到的无线数据包，相关代码如下：

```
/*********************************************************************************************
*名称：ZXBeeInfRecv()
*功能：处理节点接收到的无线数据包
*参数：*pkg—收到的无线数据包；len—无线数据包的长度
*********************************************************************************************/
void ZXBeeInfRecv(char *pkg, int len)
```

```
{
    uint8 val;
    char pData[16];
    char *p = pData;
    char *ptag = NULL;
    char *pval = NULL;
    printf("sensor->ZXBeeInfRecv(): Receive ZigBee Data!\r\n");
    HalLedSet(HAL_LED_1, HAL_LED_MODE_BLINK);
    ptag = pkg;
    p = strchr(pkg, '=');
    if (p != NULL) {
        *p++ = 0;
        pval = p;
    }
    val = atoi(pval);
    //控制命令解析
    if (0 == strcmp("cmd", ptag)){                    //对D0的位进行操作，CD0表示位清零操作
        sensorControl(val);
    }
}
```

sensorControl()函数用于控制传感器，相关代码如下：

```
/*********************************************************************************************
*名称：sensorControl()
*功能：控制传感器
*参数：cmd—控制命令
*********************************************************************************************/
void sensorControl(uint8 cmd)
void sensorControl(uint8 cmd)
{
    //根据cmd参数执行对应的控制程序
}
```

3）传感器驱动设计

家庭灯光控制系统的底层硬件主要是控制类传感器，传感器驱动设计主要是针对控制类传感器进行的。对于控制类节点，主要关注的是它对远程设备的控制是否有效，以及控制的结果。控制类传感器的逻辑事件可分为3种，详见3.2.3节。

（1）数据通信协议的定义。本系统主要使用的是控制类开发平台Sensor-B，其ZXBee数据通信协议如表4.3所示。

表4.3 控制类开发平台ZXBee数据通信协议

开发平台	属　性	参　数	权限	说　明
Sensor-B（602）	RGB灯	D1(OD1/CD1)	R/W	D1的Bit0和Bit1代表RGB灯的状态和颜色，00表示关闭RGB灯，01表示红色，10表示绿色，11表示蓝色

续表

开发平台	属　性	参　数	权限	说　　明
Sensor-B（602）	LED	D1(OD1/CD1)	R/W	D1 的 Bit4 和 Bit5 分别代表 LED1 和 LED2 状态，0 表示关闭，1 表示打开
	上报使能	D0(OD0/CD0)	R/W	上报状态，Bit0～Bit3 分别对应 A0～A1；0 表示不允许上报，1 表示允许上报，通过 OD0/CD0 进行状态控制
	数据上报时间间隔	V0	R/W	定时上报数据的时间间隔，单位为 s
	设备类型	TYPE	R	12602

（2）驱动程序的开发。在智云框架下不仅可以很容易地实现传感器驱动程序的开发，还可以省略无线传感器节点的组网和用户任务的创建等烦琐过程。例如，调用 sensorInit()函数可以实现传感器的初始化；调用 ZXBeeInfRecv()函数可以处理节点接收到的无线数据包；设备状态的定时上报使用 MyEventProcess()作为 sensorUpdate()函数的定时进入接口来反馈设备状态信息。

在 sensor.c 中，需要在 sensorInit()函数中添加传感器初始化的内容，并定义上报事件和报警事件来实现设备工作状态的定时反馈。部分代码如下：

```
/*********************************************************************************
*名称：sensorInit()
*功能：传感器初始化
*********************************************************************************/
void sensorInit(void)
{
    //初始化传感器代码
    //通过检测 P0_3 引脚来判断传感器跳线模式，并初始化对应的传感器

    P0SEL &= ~0x08;                                    //配置引脚为 GPIO 模式
    P0DIR &= !0x08;

    if(P0_3 == 0){
        mode = 1;
        stepmotor_init();                              //步进电机初始化
        fan_init();                                    //风扇初始化
    }else{
        mode = 2;
        rgb_init();                                    //RGB 灯初始化
        beep_init();                                   //蜂鸣器初始化
    }

    led_init();                                        //LED 初始化
    relay_init();                                      //继电器初始化

    //启动定时器，触发上报事件 MY_REPORT_EVT
    osal_start_timerEx(simpleBLEPeripheral_TaskID, MY_REPORT_EVT, (uint16)((osal_rand()%10) *1000));
}
```

RGB 灯的初始化函数是 rgb_init()，代码如下。

```
/*********************************************************************************
*名称：rgb_init()
*功能：RGB 灯初始化
*********************************************************************************/
void rgb_init(void)
{
    APCFG &= ~0x01;                              //模拟 I/O 使能
    P0SEL &= ~0x07;                              //配置控制引脚（P0_4 和 P0_5）为 GPIO 模式
    P0DIR |= 0x07;                               //配置控制引脚（P0_4 和 P0_5）为输出模式
    RGB_R = OFF;                                 //初始状态为关闭
    RGB_G = OFF;                                 //初始状态为关闭
    RGB_B = OFF;                                 //初始状态为关闭
}
/*********************************************************************************
*名称：rgb_on()
*功能：打开 RGB 灯
*参数：rgb—RGB 灯号，在 rgb.h 中宏定义为 RGB_R、RGB_G、RGB_B
*返回：0 表示打开 RGB 灯，-1 表示参数错误
*注释：参数只能填入 RGB_R、RGB_G、RGB_B，否则会返回-1
*********************************************************************************/
signed char rgb_on(unsigned char rgb)
{
    if(rgb == RGB_R){                            //打开 RGB_R
        RGB_R = ON;
        return 0;
    }
    if(rgb == RGB_G){                            //打开 RGB_G
        RGB_G = ON;
        return 0;
    }
    if(rgb == RGB_B){                            //打开 RGB_B
        RGB_B = ON;
        return 0;
    }
    return -1;                                   //参数错误，返回-1
}
/*********************************************************************************
*名称：rgb_off()
*功能：关闭 RGB 灯
*参数：rgb—RGB 灯号，在 rgb.h 中宏定义为 RGB_R、RGB_G、RGB_B
*返回：0 表示关闭 RGB 灯，-1 表示参数错误
*********************************************************************************/
signed char rgb_off(unsigned char rgb)
{
    if(rgb == RGB_R){                            //关闭 RGB_R
```

```
        RGB_R = OFF;
        return 0;
    }
    if(rgb == RGB_G){                                   //关闭 RGB_G
        RGB_G = OFF;
        return 0;
    }
    if(rgb == RGB_B){                                   //关闭 RGB_B
        RGB_B = OFF;
        return 0;
    }
    return -1;                                          //参数错误，返回-1
}
```

LED 的初始化函数是 led_init()，代码如下：

```
/*********************************************************************************************
*名称：led_init()
*功能：LED 控制引脚初始化
*********************************************************************************************/
void led_init(void)
{
    P0SEL &= ~0x30;                                     //配置控制引脚（P0_4 和 P0_5）为 GPIO 模式
    P0DIR |= 0x30;                                      //配置控制引脚（P0_4 和 P0_5）为输出模式
    LED1 = OFF;                                         //初始状态为关闭
    LED2 = OFF;                                         //初始状态为关闭
}
/*********************************************************************************************
*名称：led_on()
*功能：打开 LED
*参数：led—LED 号，在 led.h 中宏定义为 LED1、LED2
*********************************************************************************************/
signed char led_on(unsigned char led)
{
    if(led == LED1){                                    //开启 LED1
        LED1 = ON;
        return 0;
    }
    if(led == LED2){                                    //开启 LED2
        LED2 = ON;
        return 0;
    }
    return -1;                                          //参数错误，返回-1
}
/*********************************************************************************************
*名称：led_off()
*功能：关闭 LED
*参数：led—LED 号，在 led.h 中宏定义为 LED1、LED2
```

```
*****************************************************************************************/
signed char led_off(unsigned char led)
{
    if(led == LED1){                                        //关闭 LED1
        LED1 = OFF;
        return 0;
    }
    if(led == LED2){                                        //关闭 LED2
        LED2 = OFF;
        return 0;
    }
    return -1;                                              //参数错误，返回-1
}
```

家庭灯光控制系统是通过参数 cmd 来控制 RGB 灯和 LED 的，部分代码如下。

```
/*****************************************************************************************
*名称：sensorControl()
*功能：传感器控制
*参数：cmd—控制命令
*****************************************************************************************/
void sensorControl(uint8 cmd)
{
    static uint8 stepmotor_flag = 0;
    //根据参数 cmd 执行对应的控制程序
    ……
    if(mode == 2){                                          //传感器模式二跳线
        if ((cmd & 0x01) == 0x01){                          //RGB 灯控制位：Bit0~Bit1
            if ((cmd & 0x02) == 0x02){                      //cmd 为 3，打开蓝灯
                RGB_R = OFF;    RGB_G = OFF;    RGB_B = ON;
            }else{                                          //cmd 为 1，打开红灯
                RGB_R = ON;    RGB_G = OFF;     RGB_B = OFF;
            }
        }else if ((cmd & 0x02) == 0x02){                    //cmd 为 2，打开绿灯
            RGB_R = OFF;    RGB_G = ON; RGB_B = OFF;
        }else{                                              //cmd 为 1，关闭所有的 RGB 灯
            RGB_R = OFF; RGB_G = OFF; RGB_B = OFF;
        }
        if(cmd & 0x08) {                                    //蜂鸣器控制位：Bit3
            BEEPIO = ON;                                    //开启蜂鸣器
        } else {
            BEEPIO = OFF;                                   //关闭蜂鸣器
        }
    }
    if(cmd & 0x20){                                         //LED2 灯控制位：Bit5
        LED2 = ON;    //开启 LED2
    }else{
        LED2 = OFF;   //关闭 LED2
```

```
    }
    if(cmd & 0x10){                                    //LED1 灯控制位：Bit4
        LED1 = ON;                                     //开启 LED1
    }else{
        LED1 = OFF;                                    //关闭 LED1
    }

    if(cmd & 0x80){                                    //继电器 2 控制位：Bit7
        RELAY2 = ON;                                   //开启继电器 2
    }else{
        RELAY2 = OFF;                                  //关闭继电器 2
    }
    if(cmd & 0x40){                                    //继电器 1 控制位：Bit6
        RELAY1 = ON;                                   //开启继电器 1
    }else{
        RELAY1 = OFF;                                  //关闭继电器 1
    }
}
```

2. Android 端应用设计

1）Android 工程设计框架

打开 Android Studio 开发环境，可以看到家庭灯光控制系统的工程目录，如图 4.6 所示，系统的工程框架如表 4.4 所示。

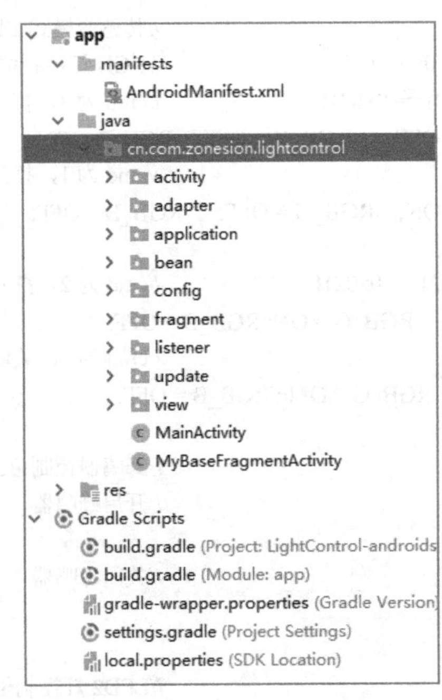

图 4.6　家庭灯光控制系统的工程目录

表 4.4　家庭灯光控制系统的工程框架

类　名	说　明
activity	
IdKeyShareActivity.java	在 IDKey 页面单击“分享”按钮时，可弹出 activity，用于分享二维码图片
TimePickerActivity.java	自动控制中的时间选择器
adapter	
HdArrayAdapter.java	历史数据显示适配器
ViewPagerAdapter.java	对 ViewPager 进行适配，可以处理多个 Fragment 的横向滑动
application	
LCApplication.java	LCApplication 继承 application 类，使用单例模式（Singleton Pattern）创建 WSNRTConnect 对象
bean	
HistoricalData.java	历史数据的 bean 类，用于将从智云服务器获得的历史数据记录（JSON 形式）转换成该类对象
IdKeyBean.java	IdKeyBean 用来描述用户设备的 ID、KEY，以及智云服务器的地址 SERVER
config	
Config.java	config 用于修改用户的 ID、KEY，以及智云服务器的地址和 MAC 地址
fragment	
BaseFragment.java	页面基础 Fragment 定义类
BasicsFragment.java：下面一些 Fragment 的基类，用于定义共有的属性以及 getter 和 setter 方法	
HomepageFragment.java	展示首页页面的 Fragment
IDKeyFragment.java	IDKey 选项的页面
MacSettingFragment.java	当用户设置被监测项的 MAC 地址时显示的页面
MoreInformationFragment.java	更多信息显示页面
RunHomePageFragment.java	运营首页显示页面
VersionInformationFragment.java	显示版本等相关信息的页面
listener	
IOnWSNDataListener.java	传感器数据监听器接口
update	
UpdateService.java	应用下载服务类
view	
APKVersionCodeUtils.java	APKVersionCodeUtils.java
CustomRadioButton.java	CustomRadioButton.java
PagerSlidingTabStrip.java	PagerSlidingTabStrip.java
MainActivity.java：主页面类	
MyBaseFragmentActivity.java：系统 Fragment 通信类	

2）软件设计

根据智云 Android 端应用程序接口的定义，家庭灯光控制系统的 Android 端应用设计主要采用实时数据 API 接口，其流程如图 4.7 所示。

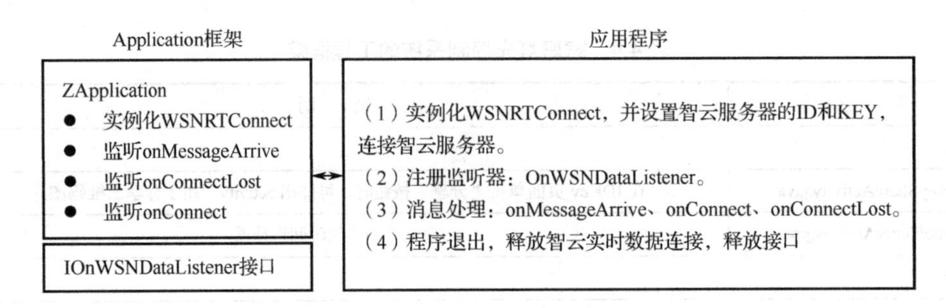

图 4.7　实时数据 API 接口的流程

（1）LCApplication.java 程序代码剖析。家庭灯光控制系统中的 LCApplication.java 程序代码和城市环境信息采集系统的 LCApplication.java 程序代码相同，详见 3.1.3 节的相关内容。

（2）HomepageFragment.java 程序代码剖析。下面的代码通过（LCApplication）getActivity().getApplication()获取 LCApplication 类中的 WSNRTConnect 对象。

```
private void initInstance(){
    config = Config.getConfig();
    lcApplication = (LCApplication) getActivity().getApplication();
    lcApplication.registerOnWSNDataListener(this);
    wsnrtConnect = lcApplication.getWSNRConnect();
    preferences = getActivity().getSharedPreferences("user_info", Context.MODE_PRIVATE);
    editor = preferences.edit();
}
```

下面的代码通过 RGB 灯和 LED 的开关按钮的 setOnClickListener 监听器调用 wsnrtConnect.sendMessage 接口，实现了对节点设备的控制。

```
open1OrCloseLamp.setOnClickListener(new OnClickListener() {
    @Override
        public void onClick(View v) {
            //TODO Auto-generated method stub
            if (totalMac != null) {
                if (open1OrCloseLamp.getText().equals("开启")) {
                    new Thread(new Runnable() {
                        @Override
                            public void run() {
                                wsnrtConnect.sendMessage(totalMac, "{OD1=16,D1=?}".getBytes());
                            }
                    }).start();
                }
                if (open1OrCloseLamp.getText().equals("关闭")) {
                    new Thread(new Runnable() {
                        @Override
                            public void run() {
                                wsnrtConnect.sendMessage(totalMac, "{CD1=16,D1=?}".getBytes());
                            }
                    }).start();
```

```
                }
            }else {
                Toast.makeText(lcApplication, "请等待 MAC 地址上线", Toast.LENGTH_SHORT).show();
            }
        }
});
```

下面的代码通过 onCheckedChanged 方法实现了模式切换的功能。

```
@Override
    public void onCheckedChanged(CompoundButton compoundButton, boolean isChecked) {
        switch (compoundButton.getId()) {
            case R.id.security_module:
                if (isChecked) {
                    isSecurityMode = true;
                    securityModuleTip.setVisibility(View.VISIBLE);
                    leaveHomeModuleTip.setVisibility(View.GONE);
                    wsnrtConnect.sendMessage(totalMac, "{OD1=16，D1=?}".getBytes());
                    wsnrtConnect.sendMessage(totalMac, "{OD1=32，D1=?}".getBytes());
                }
            break;
            case R.id.leave_home_module:
                if (isChecked) {
                    isSecurityMode = false;
                    securityModuleTip.setVisibility(View.GONE);
                    leaveHomeModuleTip.setVisibility(View.VISIBLE);
                    wsnrtConnect.sendMessage(totalMac, "{CD1=16，D1=?}".getBytes());
                    wsnrtConnect.sendMessage(totalMac, "{CD1=32，D1=?}".getBytes());
                    wsnrtConnect.sendMessage(totalMac, "{CD1=3,D1=?}".getBytes());
                }
            break;
        }
}
```

3. Web 端应用设计

1）页面功能结构分析

家庭灯光控制系统的 Web 端默认显示的是“运营首页”页面，在“运营首页”页面上设计了 6 个模块，分别是 LED1 显示模块、LED2 显示模块、模式切换设置模块、LED1 自动控制设置模块、LED2 自动控制设置模块、RGB 灯显示模块。家庭灯光控制系统 Web 端的“运营首页”页面如图 4.8 所示。

“更多信息”页面的主要功能是进行智云服务器的连接配置，和城市环境信息采集系统 Web 端的“更多信息”页面类似，参见图 3.14。

2）软件设计

家庭灯光控制系统 Web 端的 JS 开发逻辑与 Android 端的开发逻辑相似，首先通过配置

ID 和 KEY 与智云服务器进行连接，再通过实时监听数据的方法来获取相关传感器的数据并进行处理。JS 开发的部分代码如下。

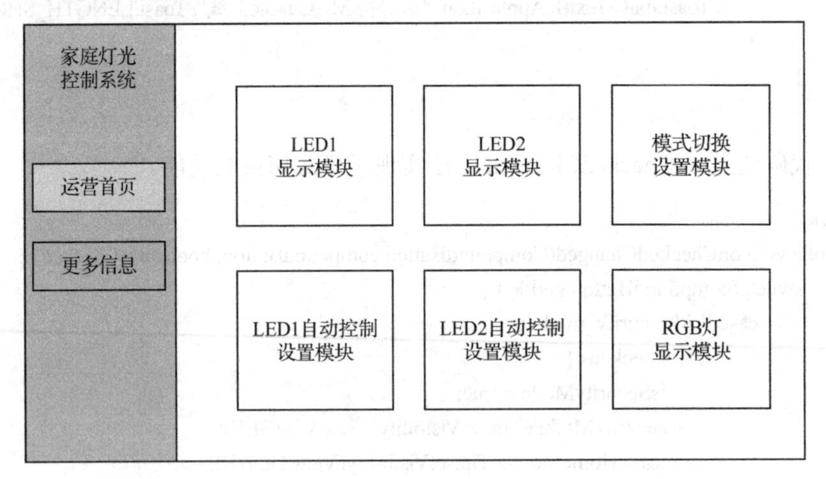

图 4.8　家庭灯光控制系统 Web 端的“运营首页”页面

在 getConnect()函数中定义了实时连接对象 rtc，连接成功回调函数是 rtc.onConnect，数据服务掉线回调函数是 rtc.onConnectLost，消息处理回调函数是 rtc.onmessageArrive。

```
//连接服务器
function getConnect() {
    config["id"] = config["id"] ? config["id"] : $("#id").val();
    config["key"] = config["key"] ? config["key"] : $("#key").val();
    config["server"] = config["server"] ? config["server"] : $("#server").val();
    //创建数据连接服务对象
    rtc = new WSNRTConnect(config["id"], config["key"]);
    rtc.setServerAddr(config["server"] + ":28080");
    rtc.connect();
    rtc._connect = false;
    //连接成功回调函数
    rtc.onConnect = function() {
        rtc._connect = 1;
        console.log(rtc._connect);
        message_show("数据服务连接成功！");
        $("#idkeyInput").text("断开").addClass("btn-danger");
        $("#id,#key,#server").attr('disabled',true);
    };
    //数据服务掉线回调函数
    rtc.onConnectLost = function() {
        rtc._connect = 0;
        console.log(rtc._connect);
        $("#idkeyInput").text("连接").removeClass("btn-danger");
        $(".online_602").text("离线").css("color", "#e75d59");
        message_show("数据服务连接失败，检查网络或 ID、KEY");
        $("#id,#key,#server").removeAttr('disabled');
```

```
        };
        //消息处理回调函数
        rtc.onmessageArrive = function (mac, dat) {
            //console.log(mac+" >>> "+dat);
            if (dat[0]=='{' && dat[dat.length-1]=='}') {
                dat = dat.substr(1, dat.length-2);
                var its = dat.split(',');
                for (var i=0; i<its.length; i++) {
                    var it = its[i].split('=');
                    if (it.length == 2) {
                        var tag = it[0];
                        var val = it[1];
                        process_tag(mac, tag, val);
                    }
                }
                if (!mac2type[mac]) { //如果没有获取到 TYPE 值，则主动去查询
                    rtc.sendMessage(mac, "{TYPE=?,A0=?,A1=?,A2=?,A3=?,A4=?,A5=?,A6=?,A7=?,D1=?}");
                }
            }
        }
    }
```

下面 JS 开发代码的功能是当设备连接到智云服务器后，在“运营首页”页面中更新设备的状态为“在线”，RGB 的状态有红色（R）、绿色（G）、蓝色（B）和关闭，LED 的状态有打开和关闭。

```
    var wsn_config = {
        "602" : {
            "online" : function() {
                $(".online_602").text("在线").css("color", "#96ba5c");
            },
            "pro" : function (tag, val) {
                if(tag=="D1"){
                    //Bit 0～1 代表 RGB 灯的状态及颜色
                    var colorname;
                    if(val & 0x03){
                        colorname = "blue";
                    }else if(val & 0x02){
                        colorname = "green";
                    }else if(val & 0x01){
                        colorname = "red";
                    }else{
                        colorname = "off";
                    }
                    activeColor(colorname);
                    //Bit 4 代表 LED1
                    var led1_state = val & 16;
```

```
                    if(led1_state){
                        $("#led1_light").addClass("bulb-on");
                    }else{
                        $("#led1_light").removeClass("bulb-on");
                    }
                    $("#ledSwitcher1").attr("checked", led1_state);
                    //Bit 5 代表 LED2
                    var led2_state = val & 32;
                    if(led2_state){
                        $("#led2_light").addClass("bulb-on");
                    }else{
                        $("#led2_light").removeClass("bulb-on");
                    }
                    $("#ledSwitcher2").attr("checked", led2_state);
                }
            }
        }
    }
```

对 LED 和 RGB 灯的控制是通过 rtc.sendMessage 实现的，代码如下：

```
//开关 LED
$(".led-switcher").on("click", function(){
    if(page.checkOnline() && page.checkMac("mac_602")){
        var cmd;
        var index = $(this).data("index");
        var index2cmd = {
            1 : 16,
            2 : 32
        }
        if(this.checked == true){
            cmd = "{OD1="+index2cmd[index]+",D1=?}";
        }else{
            cmd = "{CD1="+index2cmd[index]+",D1=?}";
        }
        rtc.sendMessage(config["mac_602"], cmd);
    }
});
//切换 RGB 灯的颜色
var color2cmd = {
    "red" : "CD1=3,OD1=1",
    "green" : "CD1=1,OD1=2",
    "blue" : "OD1=3",
    "off": "CD1=3"
};
$(".color-btn").on("click", function() {
    var color = $(this).data("title");
    var cmd = "{"+color2cmd[color]+"}";
```

```
        if(page.checkOnline() && page.checkMac("mac_602")){
            rtc.sendMessage(config["mac_602"], cmd);
            activeColor(color);
        }
});
```

离家模式和安防模式对灯具（LED 和 RGB 灯）的控制方式是不同的，具体实现代码如下：

```
//模式切换
$("#mode-switch input").on("click", function() {
    config["mode"] = $(this).val();
    var isHomeoutMode = config["mode"]=="homeout-mode";
    //在离家模式下关闭所有灯具
    if(isHomeoutMode){
        $("#mode-txt-2").removeClass("hidden").siblings("span").addClass("hidden");
        $("#mode-text").addClass("mode-right");
        console.log("发送关闭灯具指令");
        if(page.checkOnline() && page.checkMac("mac_602")){
            rtc.sendMessage(config["mac_602"], "{CD1=16,CD1=32,CD1=3,D1=?}");
            $(".switch-box .switch-box-input").prop('checked',false);
        }
    }else{
        //在安防模式下可自动开关指定的灯具
        $("#mode-txt-1").removeClass("hidden").siblings("span").addClass("hidden");
        $("#mode-text").removeClass("mode-right");
    }
    //storeStorage();
});
```

家庭灯光控制系统是通过 setTimeout 方法实现定时功能的，具体代码如下：

```
function getTime(){
    var nowdate = new Date();
    //获取年、月、日、时、分、秒
    var hours = nowdate.getHours(), minutes = nowdate.getMinutes();
    var hour_info = hours >=10 ? hours : "0"+hours;
    var minute_info = minutes >=10 ? minutes : "0"+minutes;
    var cur_time = hour_info + ":" + minute_info;
    //设置一个全局变量来缓存上一次的时间字符串，如果最新的时间字符串与保存的时间字符串不
同，则更新为最新的时间字符串，并执行相应的动作
    if(last_time != cur_time){
        last_time = cur_time;
        console.log("每分钟更新：当前时间："+cur_time);
        if(config["mode"]=="secutiry-mode" && $.inArray(cur_time, config["setTime"])>-1){
            if(page.checkOnline() && page.checkMac("mac_602")){
                if(cur_time==config["setTime"][0]){
                    message_show("定时器执行 LED1 开启！");
```

```
                $("#led1_switch").prop('checked',true);
                rtc.sendMessage(config["mac_602"], "{OD1=16,D1=?}");
            }
            else if(cur_time == config["setTime"][1]){
                message_show("定时器执行 LED1 关闭！");
                $("#led1_switch").prop('checked',false);
                rtc.sendMessage(config["mac_602"], "{CD1=16,D1=?}");
            }
        } else{
            setTimeout(function() {
                message_show("定时器控制失败！");
            },4000)
        }
    }
    if(config["mode"] == "secutiry-mode" && $.inArray(cur_time, config["setTime2"])>-1){
        if(page.checkOnline() && page.checkMac("mac_602")){
            if(cur_time == config["setTime2"][0]){
                message_show("定时器执行 LED2 开启！");
                $("#led2_switch").prop('checked',true);
                rtc.sendMessage(config["mac_602"], "{OD1=32,D1=?}");
            } else if(cur_time == config["setTime2"][1]) {
                message_show("定时器执行 LED2 关闭！");
                $("#led2_switch").prop('checked',false);
                rtc.sendMessage(config["mac_602"], "{CD1=32,D1=?}");
            }
        }
    }
  }
  setTimeout(getTime, 1000);
}
```

家庭灯光控制系统 Web 端其他部分的代码请查看本书配套资料中的项目源文件。

4.1.4 开发验证

1. Web 端应用测试

打开家庭灯光控制系统的 Web 端后，在“运营首页”下可以看到 Web 端的主页，如图 4.9 所示。

当设备在线后，在安防模式下，既可以通过 LED 和 RGB 灯的按钮来实时控制灯具，也可以通过定时控制功能来控制指定的灯具。家庭灯光控制系统的定时控制功能如图 4.10 所示。

在模式切换功能中选中“离家模式”后，可以关闭所有的灯具，如图 4.11 所示。

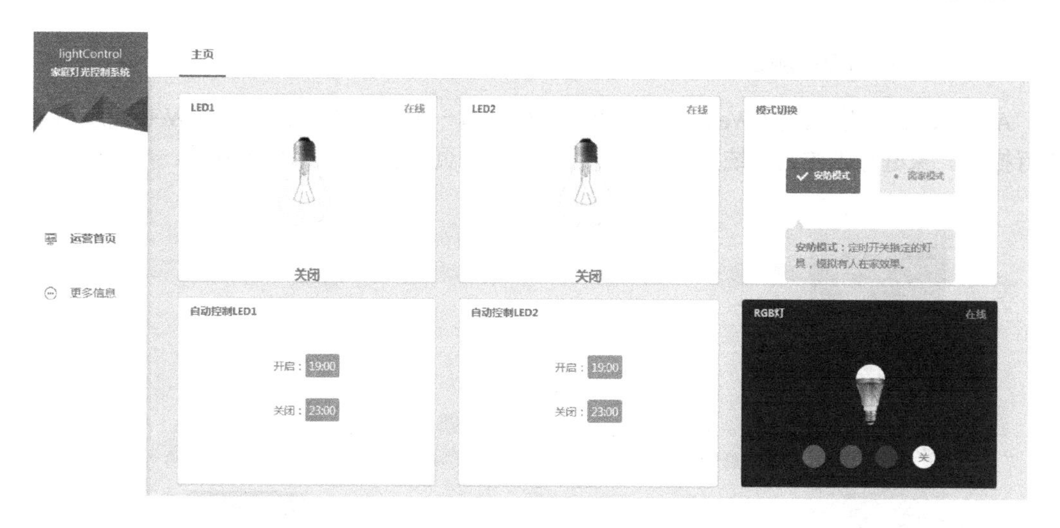

图 4.9　家庭灯光控制系统 Web 端的主页

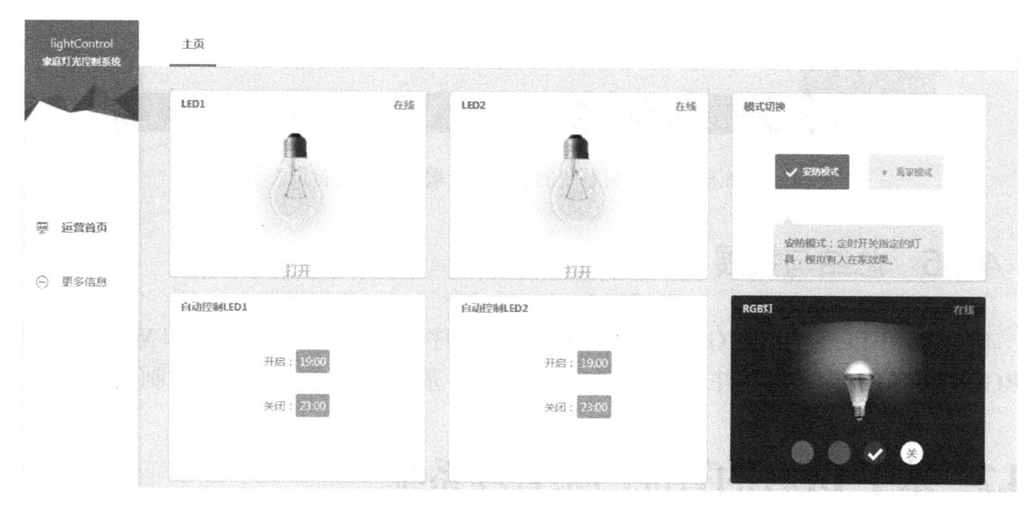

图 4.10　家庭灯光控制系统的定时控制功能

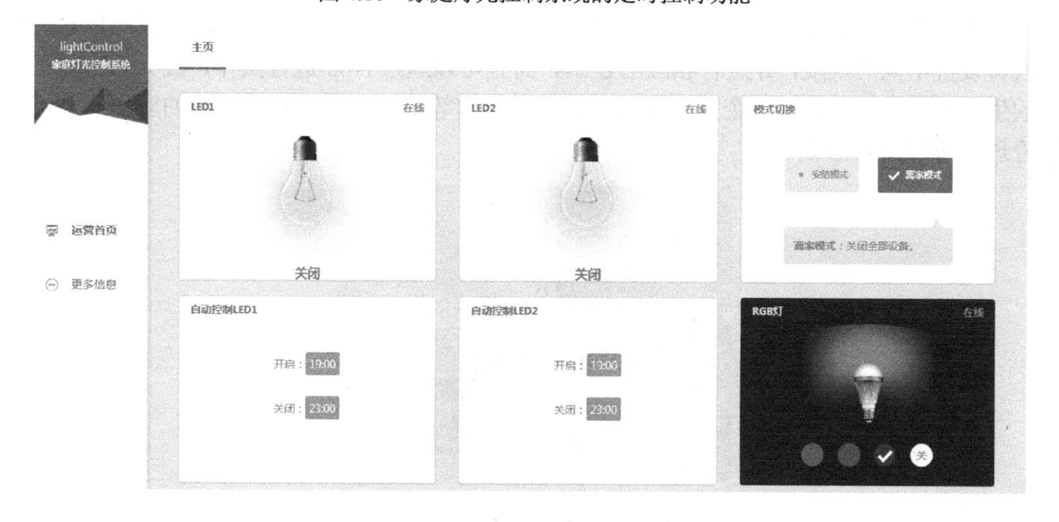

图 4.11　家庭灯光控制系统的离家模式

2. Android 端应用测试

Android 端应用测试同 Web 端应用测试流程基本一致，可参考本系统的 Web 端应用测试进行操作。家庭灯光控制系统 Android 端的“运营首页”页面如图 4.12 所示。

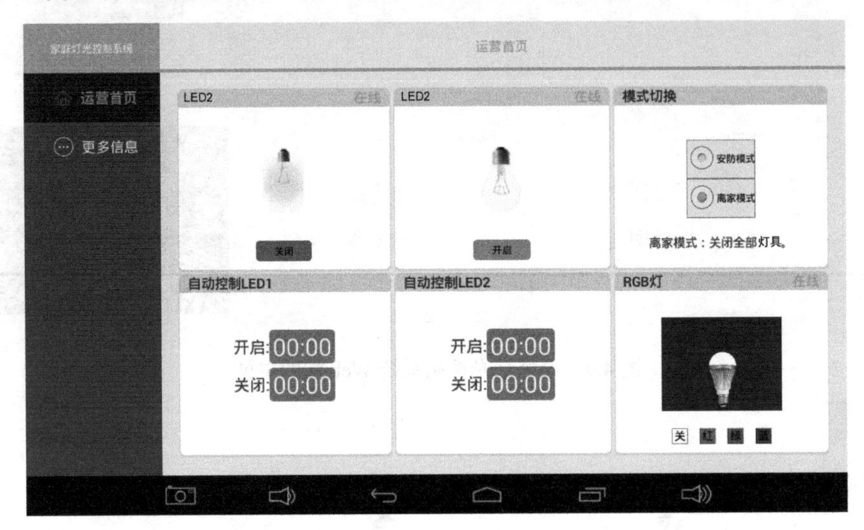

图 4.12　家庭灯光控制系统的 Android 端的“运营首页”页面

4.1.5　总结与拓展

本节基于 BLE 设计了 RGB 灯和 LED 的驱动程序，可以在 Android 端和 Web 端实时获取 RGB 灯和 LED 的状态，并对其进行控制，实现了基于 BLE 的家庭灯光控制系统。

4.2　基于 BLE 的智能门禁管理系统

智能门禁管理系统将信息技术、电子技术和机械锁有机地结合在一起，不仅可以对进出的人员设置不同的权限，并进行相应的控制，而且可以通过 App 进行操作，极大地方便了人们的生活和工作，在住宅、企业、园区、酒店、公寓等多个领域得到了广泛的应用。智能门禁管理系统如图 4.13 所示。

图 4.13　智能门禁管理系统

4.2.1 系统开发目标

(1) 熟悉射频传感器和继电器等硬件原理和数据通信协议，并基于 CC2540 和 BLE 进行射频传感器和继电器的驱动开发，识别合法的门禁卡后可以打开电磁锁，实现智能门禁管理系统的设计。

(2) 实现智能门禁管理系统的 Android 端应用开发和 Web 端应用开发。

4.2.2 系统设计分析

1. 系统的功能设计

从系统功能的角度出发，智能门禁管理系统可以分为两个模块：设备识别和控制模块，以及系统设置模块，其组成模块如图 4.14 所示。

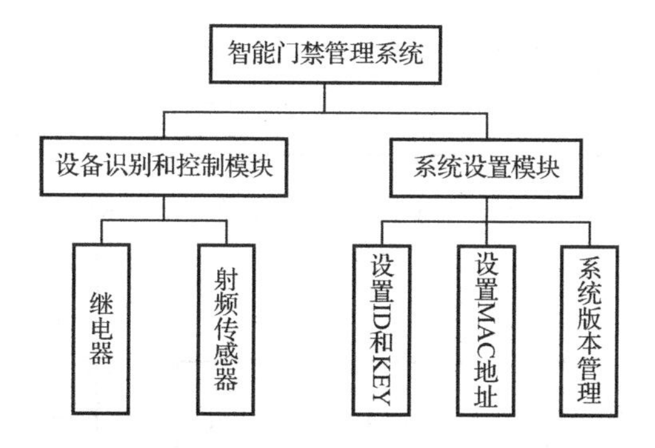

图 4.14 智能门禁管理系统的组成模块

智能门禁管理系统的功能需求如表 4.5 所示。

表 4.5 智能门禁管理系统的功能需求

功 能	功 能 说 明
门禁卡数据显示	在上层应用页面中实时更新并显示门禁卡信息
继电器实时控制	通过上层应用程序控制继电器
智云连接设置	设置智云服务器的参数和设备的 MAC 地址

2. 系统的总体架构设计

智能门禁管理系统是基于物联网四层架构模型来设计的，其总体架构如图 4.15 所示。

3. 系统的数据传输

智能门禁管理系统的数据传输是在传感器节点、智能网关以及客户端（包括 Web 端和 Android 端）之间进行的，如图 4.16 所示。

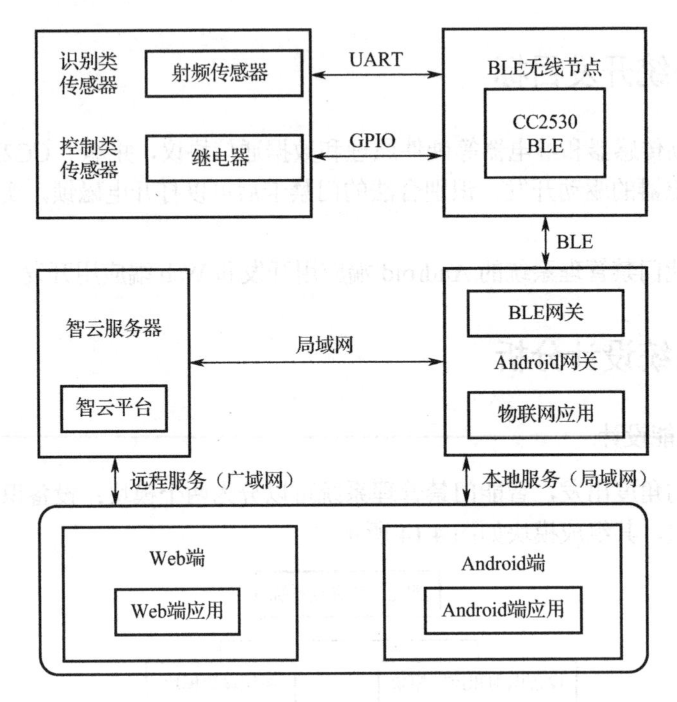

图 4.15　智能门禁管理系统的总体架构

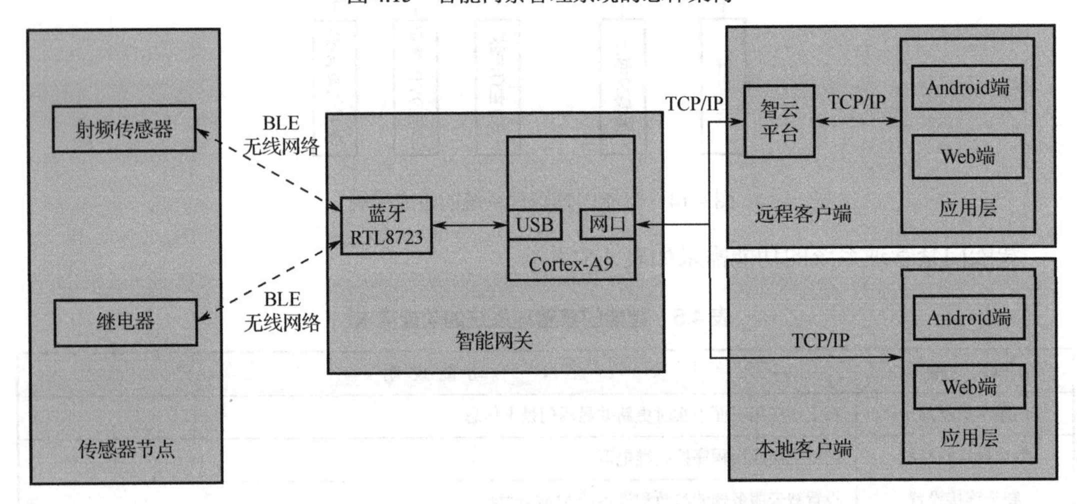

图 4.16　智能门禁管理系统的数据传输

4.2.3　系统的软硬件开发：智能门禁管理系统

1．系统底层软硬件设计

1）感知层硬件设计

智能门禁管理系统的底层硬件包括 xLab 未来开发平台的智能网关、经典型无线节点 ZXBeeLiteB、控制类开发平台 Sensor-B、识别类传感器 Sensor-EL。其中，智能网关负责汇集传感器采集的数据；BLE 无线节点（由经典型无线节点 ZXBeeLiteB 实现）通过无线通信的方式向智能网关发送传感器采集的数据，接收智能网关发送的命令；控制类开发平台 Sensor-B 和识别类传感器 Sensor-EL 连接到 BLE 无线节点，由其中的 CC2540 对相关设备进行控制。智能门禁管理系统使用的传感器包括 125 kHz 和 13.56 MHz 的射频传感器以及继电器，射频传感器的硬件接口电路如图 4.17 所示，继电器的硬件接口电路如图 4.18 所示。

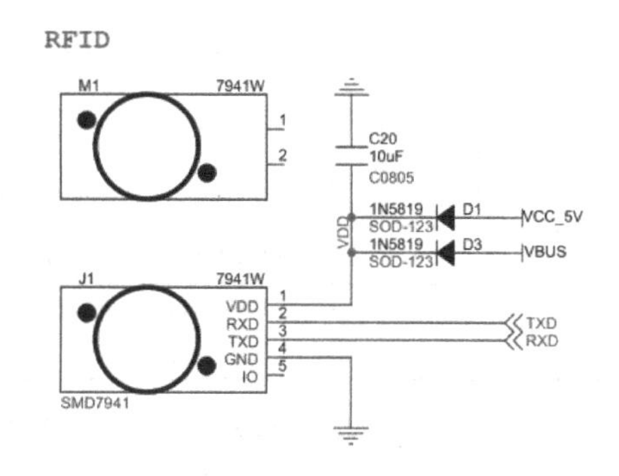

图 4.17 射频传感器的硬件接口电路

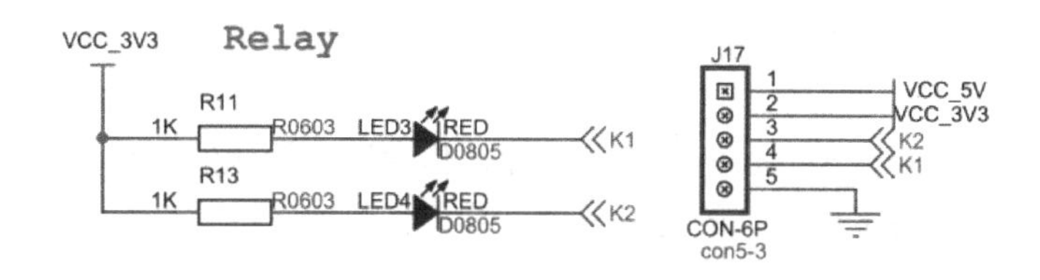

图 4.18 继电器的硬件接口电路

2）系统底层开发

智能门禁管理系统的底层开发与 4.1 节介绍的家庭灯光控制系统的底层开发类似，详见 4.1.3 节。

3）传感器驱动设计

智能门禁管理系统的底层硬件主要是控制类传感器与识别类传感器，主要关注控制类传感器对设备的控制是否有效，以及控制结果。控制类传感器的逻辑事件有 3 种，详见 3.2.3 节。

（1）数据通信协议的定义。智能门禁管理系统使用的是控制类开发平台 Sensor-B 和识别类开发平台 Sensor-EL，其 ZXBee 数据通信协议如表 4.6 所示。

表 4.6 控制类开发平台和识别类开发平台的 ZXBee 数据通信协议

开发平台	属性	参数	权限	说明
Sensor-EL（605）	卡号	A0	—	字符串（主动上报，不可查询）
	卡类型	A1	R	整型数据，0 表示 125K 型(频率为 125 kHz)，1 表示 13.56M 型（频率为 13.56 MHz）
	卡余额	A2	R	整型数据，范围 0～800000，手动查询
	设备余额	A3	R	浮点型数据，设备余额

续表

开发平台	属性	参数	权限	说明
Sensor-EL（605）	设备单次消费金额	A4	R	浮点型数据，设备本次消费扣款金额
	设备累计消费	A5	R	浮点型数据，设备累计扣款金额
	门锁和设备状态	D1(OD1/CD1)	R/W	D1 的 Bit0～Bit1 表示门锁和设备的状态，0 表示关闭，1 表示打开
	充值金额	V1	R/W	返回充值状态，0 表示操作未成功，1 表示操作成功
	扣款金额	V2	R/W	返回扣款状态，0 表示操作未成功，1 表示操作成功
	充值金额（设备）	V3	R/W	返回充值状态，0 表示操作未成功，1 表示操作成功
	扣款金额（设备）	V4	R/W	返回扣款状态，0 表示操作未成功，1 表示操作成功
Sensor-B（602）	继电器	D1(OD1/CD1)	R/W	D1 的 Bit6、Bit7 分别代表继电器 K1、K2 的状态，0 表示断开，1 表示吸合

（2）驱动程序的开发。在智云框架下不仅可以很容易地实现传感器驱动程序的开发，还可以省略无线传感器节点的组网和用户任务的创建等烦琐过程。例如，调用 sensorInit()函数可以实现传感器的初始化；调用 ZXBeeInfRecv()函数可以节点接收到的无线数据包；设备状态的定时上报使用 MyEventProcess()作为 sensorUpdate()函数的定时进入接口来反馈设备状态信息。

在 sensor.c 中，需要在 sensorInit()函数中添加传感器初始化的内容，并通过定义上报事件和报警事件来实现设备工作状态的定时反馈。部分代码如下：

```
void sensorInit(void)
{
    //初始化传感器代码
    RFID7941Init();                                         //射频传感器初始化
    OLED_ShowString(21+13*3,2,(unsigned char *)buf,12);
    //启动定时器，触发上报事件 MY_REPORT_EVT
    osal_start_timerEx(simpleBLEPeripheral_TaskID, MY_REPORT_EVT, (uint16)((osal_rand()%10) *1000));
    //启动定时器，触发报警事件 MY_CHECK_EVT
    osal_start_timerEx(simpleBLEPeripheral_TaskID, MY_CHECK_EVT, (uint16)((osal_rand()%10) *300));
}
```

射频传感器的初始化函数是 RFID7941Init()，代码如下：

```
void RFID7941Init(void)
{
    uart_init();
    uart_set_input_call(uartRecvCallBack);
}
```

继电器的初始化函数是 relay_init()，代码如下：

```
void relay_init(void)
{
    P0SEL &= ~0xC0;                                  //配置引脚为 GPIO 模式
    P0DIR |= 0xC0;                                   //配置控制引脚为输入模式
}
```

sensorControl()函数是根据其参数 cmd 来执行对应的继电器控制程序的，代码如下：

```
void sensorControl(uint8 cmd)
{
    //根据 cmd 参数执行对应的控制程序
    if(cmd & 0x80){                                   //继电器 2 控制位：Bit7
        RELAY2 = ON;                                  //开启继电器 2
    } else{
        RELAY2 = OFF;                                 //关闭继电器 2
    }
    if(cmd & 0x40){                                   //继电器 1 控制位：Bit6
        RELAY1 = ON;                                  //开启继电器 1
    }else{
        RELAY1 = OFF;                                 //关闭继电器 1
    }
}
```

SensorUpdate()函数用于主动上报数据，代码如下：

```
/*********************************************************************************************
*名称：sensorUpdate()
*功能：主动上报的数据
*********************************************************************************************/
void sensorUpdate(void)
{
    char pData[16];
    char *p = pData;

    if (A3 > 0) {
        A4 = 1 + rand()%500;                          //设备本次消费扣款金额，通过随机数产生
        if (A4 > A3) A4 = A3;

        A3 = (A3 - A4);                               //设备余额
        A5 = (A5 + A4);                               //设备累计消费金额

        ZXBeeBegin();
        sprintf(p, "%ld.%ld", A3/100, A3%100);        //余额
        ZXBeeAdd("A3", p);
        sprintf(p, "%ld.%ld", A4/100, A4%100);        //本次消费扣款金额
        ZXBeeAdd("A4", p);
        sprintf(p, "%ld.%ld", A5/100, A5%100);        //累积消费金额
        ZXBeeAdd("A5", p);
        p = ZXBeeEnd();                               //智云数据帧格式包尾
        if (p != NULL) {
            ZXBeeInfSend(p, strlen(p));      //将需要上传的数据打包后通过 zb_SendDataRequest()发送
到协调器
        }
        sprintf(p, "%ld.%ld    ", A3/100, A3%100);
```

```
        OLED_ShowString(21+13*3,2,(unsigned char *)p,12);                    //通过 OLED 显示余额
        if(A3 == 0) {      //余额不足扣费，继电器断开
            D1 &= ~2;
            RELAY2 = OFF;
        }
    }
}
```

2. Android 端应用设计

1）Android 工程设计框架

打开 Android Studio 开发环境，可以看到智能门禁管理系统的工程目录，如图 4.19 所示，智能门禁管理系统的工程框架如表 4.7 所示。

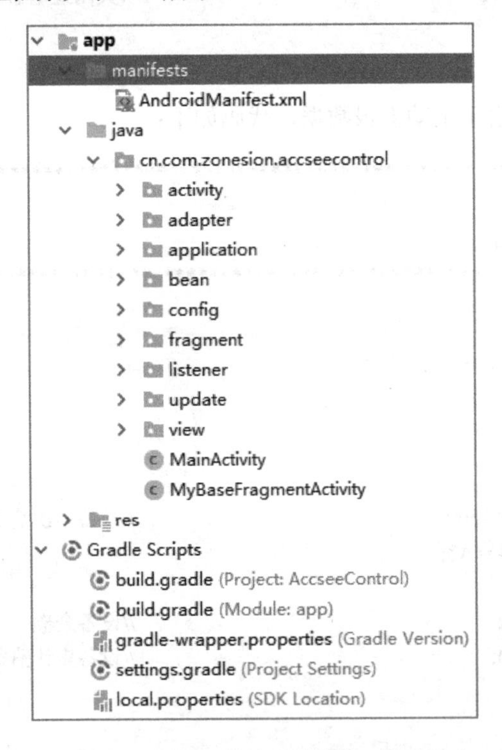

图 4.19 智能门禁管理系统的工程目录

表 4.7 智能门禁管理系统的工程框架

类 名	说 明
activity	
IdKeyShareActivity.java	在 IDKey 页面单击“分享”按钮时，可弹出 activity，用于分享二维码图片
adapter	
HdArrayAdapter.java	历史数据显示适配器
ViewPagerAdapter.java	对 ViewPager 进行适配，可以处理多个 Fragment 的横向滑动

续表

类　名	说　明
application	
LCApplication.java	LCApplication 继承 application 类，使用单例模式（Singleton Pattern）创建 WSNRTConnect 对象
bean	
HistoricalData.java	历史数据的 bean 类，用于将从智云服务器获得的历史数据记录（JSON 形式）转换成该类对象
IdKeyBean.java	IdKeyBean 用来描述用户设备的 ID、KEY，以及智云服务器的地址 SERVER
config	
Config.java	该类用于对用户的 ID、KEY 和智云服务器地址以及 MAC 地址进行修改
fragment	
BaseFragment.java	页面基础 Fragment 定义类
BasicsFragment.java：下面一些 Fragment 的基类，定义共有的属性以及 getter 和 setter 方法	
HomepageFragment.java	展示首页页面的 Fragment
IDKeyFragment.java	IDKey 选项的页面
MacSettingFragment.java	当用户设置被监测项的 MAC 地址时显示的页面
MoreInformationFragment.java	更多信息显示页面
RunHomePageFragment.java	运营首页显示页面
VersionInformationFragment.java	显示版本等相关信息的页面
listener	
IOnWSNDataListener.java	传感器数据监听器接口
update	
UpdateService.java	应用下载服务类
view	
APKVersionCodeUtils.java	获取当前本地 apk 的版本
CustomRadioButton.java	自定义按钮类
PagerSlidingTabStrip.java	自定义滑动控件类
MainActivity.java：主页面类	
MyBaseFragmentActivity.java：系统 Fragment 通信类	

2）软件设计

根据智云 Android 端应用程序接口的定义，智能门禁管理系统的 Android 端应用设计主要采用实时数据 API 接口，其流程和家庭灯光控制系统相同（见图 4.9）。

（1）LCApplication.java 程序代码剖析。家庭灯光控制系统中的 LCApplication.java 程序代码和城市环境信息采集系统的 LCApplication.java 程序代码相同，详见 3.1.3 节的相关内容。

（2）HomepageFragment.java 程序代码剖析。下面的代码通过(LCApplication) getActivity().getApplication()获取 LCApplication 类中的 WSNRTConnect 对象。

```
private void initInstance(){
    config = Config.getConfig();
```

```
        lcApplication = (LCApplication) getActivity().getApplication();
        lcApplication.registerOnWSNDataListener(this);
        wsnrtConnect = lcApplication.getWSNRConnect();
        preferences = getActivity().getSharedPreferences("user_info", Context.MODE_PRIVATE);
        editor = preferences.edit();
}
```

下面的代码通过 onMessageArrive 的方法实现了对节点设备状态的更新显示。

```
@Override
    public void onMessageArrive(String mac, String tag, String val) {
        if (sensorELMac == null) {
            wsnrtConnect.sendMessage(mac, "{TYPE=?}".getBytes());
        }
        if (tag.equals("TYPE")) {
            if (val.substring(2, val.length()).equals("605")) {
                sensorELMac = mac;
            }
        }
        if (mac.equals(sensorELMac) && "A0".equals(tag)) {
            entranceGuardState.setText("在线");
            entranceGuardState.setTextColor(getResources().getColor(R.color.line_text_color));
        }
        if (mac.equals(sensorELMac) && "D1".equals(tag)) {
            doorState.setText("在线");
            doorState.setTextColor(getResources().getColor(R.color.line_text_color));
        }

    }
```

3. Web 端应用设计

1）页面功能结构分析

智能门禁管理系统的 Web 端默认显示的是“运营首页”页面，在“运营首页”页面上设计了 4 个模块，分别是注册用户列表显示模块、刷卡记录显示模块、门禁 ID 显示模块、门锁控制模块。智能门禁管理系统 Web 端的“运营首页”面面如图 4.20 所示。

“更多信息”页面的主要功能是进行智云服务器的连接配置，和城市环境信息采集系统中的 Web 端“更多信息”页面类似，参见图 3.14。

2）软件设计

智能门禁管理系统 Web 端的 JS 开发逻辑与 Android 端的开发逻辑相似，首先通过配置 ID 和 KEY 与智云服务器进行连接，再通过实时监听数据的方法来获取相关传感器的数据并进行处理。JS 开发的部分代码如下。

在 getConnect()函数中定义了实时连接对象 rtc，连接成功回调函数是 rtc.onConnect，数据服务掉线回调函数是 rtc.onConnectLost，消息处理回调函数是 rtc.onmessageArrive。

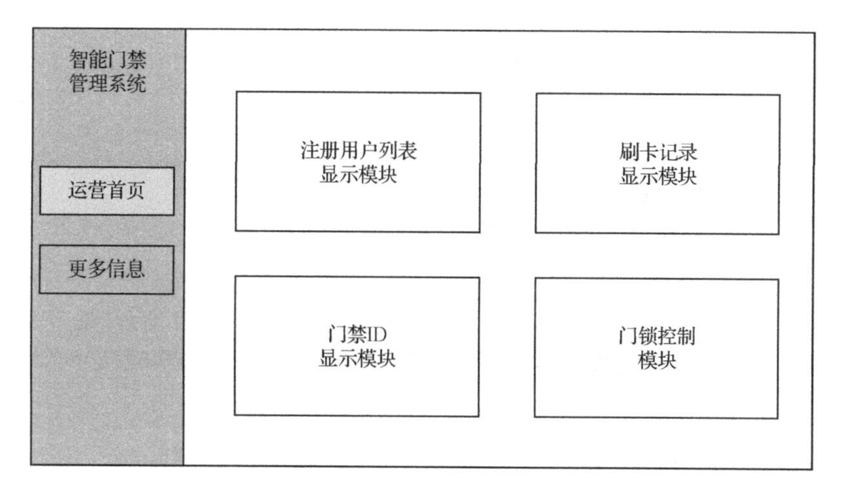

图 4.20　智能门禁管理系统的 Web 端“运营首页”页面

```
function getConnect() {
    config["id"] = config["id"] ? config["id"] : $("#id").val();
    config["key"] = config["key"] ? config["key"] : $("#key").val();
    config["server"] = config["server"] ? config["server"] : $("#server").val();
    //创建数据连接服务对象
    rtc = new WSNRTConnect(config["id"], config["key"]);
    rtc.setServerAddr(config["server"] + ":28080");
    rtc.connect();
    rtc._connect = false;
    //连接成功回调函数
    rtc.onConnect = function() {
        $("#ConnectState").text("数据服务连接成功！");
        rtc._connect = 1;
        message_show("数据服务连接成功！");
        $("#idkeyInput").text("断开").addClass("btn-danger");
        $("#id,#key,#server").attr('disabled',true);
    };
    //数据服务掉线回调函数
    rtc.onConnectLost = function() {
        rtc._connect = 0;
        $("#ConnectState").text("数据服务连接掉线！");
        $("#idkeyInput").text("连接").removeClass("btn-danger");
        message_show("数据服务连接失败，检查网络或 ID、KEY");
        $("#RFIDLink").text("离线").css("color", "#e75d59");
        $("#doorLink").text("离线").css("color", "#e75d59");
        $("#id,#key,#server").removeAttr('disabled',true);
    };
    //消息处理回调函数
    rtc.onmessageArrive = function (mac, dat) {
        //console.log(mac+" >>> "+dat);
        if (dat[0]=='{' && dat[dat.length-1]=='}') {
```

```
                dat = dat.substr(1, dat.length-2);
                var its = dat.split(',');
                for (var i=0; i<its.length; i++) {
                    var it = its[i].split('=');
                    if (it.length == 2) {
                        process_tag(mac, it[0], it[1]);
                    }
                }
                if (!mac2type[mac]) {          //如果没有获取到 TYPE 值，则主动去查询
                    rtc.sendMessage(mac, "{TYPE=?,A0=?,A1=?,A2=?,A3=?,A4=?,A5=?,A6=?,A7=?,D1=?}");
                }
            }
        }
    }
```

下面 JS 开发代码的功能是当设备连接到智云服务器后，在“运营首页”页面中将设备的状态更新为“在线”。

```
var wsn_config = {
    "602" : {
        "online" : function() {
            $(".online_602").text("在线").css("color", "#96ba5c");
        },
        "pro" : function (tag, val) {
            if(tag=="D1"){
                if(val & 0x20){
                    $("#doorStatus").text("关闭");
                    $("#doorImg").attr("src", "img/door-on.png");
                }else{
                    $("#doorStatus").text("打开");
                    $("#doorImg").attr("src", "img/door-off.png");
                }
            }
        }
    },
    "605" : {
        "online" : function() {
            $(".online_605").text("在线").css("color", "#96ba5c");
        },
        "pro" : function (tag, val) {
            if(tag=="A0"){
                if(val !=0){
                    testRfid(val);
                }
            }
        }
    }
}
```

下面的代码通过 rtc.sendMessage(config["mac_602"], cmd)对门锁进行控制。

```
//门锁控制-打开、关闭
$("#doorStatus").click(function(){
    if (page.checkOnline() && page.checkMac("mac_602")){
        var curState = $(this).text(), cmd;
        console.log(curState);
        if(curState=="打开"){
            cmd = "{OD1=1,D1=?}";
        }else{
            cmd = "{CD1=1,D1=?}";
        }
        console.log("cmd="+cmd);
        rtc.sendMessage(config["mac_602"], cmd);
    }
});
```

注册用户列表的增加、删除、修改功能通过下面的代码实现。

```
//弹出模态框，添加新的 ID 并确认
$("#newConfirm").on("click",function(){
    var data = {};
    var new_rfid = $("#newRfidId").val();
    var new_name = $("#newRfidName").val();
    data[new_rfid] = {
        "name" : new_name
    };
    console.log("add : " + data+"----"+JSON.stringify(data));
    if(checkRFID(data)){
        console.log("add rfid data="+JSON.stringify(data));
        var tr = '<tr><td class="name">'+data[new_rfid]['name']+'</td><td class="rfid">'+new_rfid+
                '</td>><td class="actionTd"><a class="edit-id" href="#"data-toggle="modal" data-target=
                "#editModal"  data-type="rfid">编辑</a>    <a class=
                "del-id"  href="#"data-toggle="modal"  data-target="#delModal"  data-type="rfid">删除</a>
</td></tr>';
        $("#rfidTable").append(tr);
        var len = $("#rfidTable").find("tr").length;
        console.log("len="+len);
        config['rfid'][new_rfid] = data[new_rfid];
        page.storeStorage();
        //console.log(" 读 取  localStorage :  localStorage.FwsSmartHome="+JSON.stringify(localStorage.
FwsSmartHome));
        $("#newModal").modal("hide");
        console.log("添加后的 length="+$("#rfidTable").find("tr").length);
    }
})
$("#newModal").on("hide.bs.modal", function() {
    $("#newModal").find("input").val("");
```

```
})
//验证新增的 ID 是否已存在
function isExist(n){
    if(config.idList && $.inArray(n, config.idList)>-1){
        $("#id_check").text("ID 已存在！")
        console.log("ID 已存在");
        return false;
    }else{
        return true;
    }
}
function storeIdTable(){
    var tr = $("#historyIdTable tbody").find("tr");
    var configTableTxt ="";
    for(var i=0;i<tr.length;i++){
        var td =$("#historyIdTable tbody tr:eq("+i+")").find("td");
        for(var j=0;j<td.length;j++){
            var text = $("#historyIdTable tbody tr:eq("+i+")").find("td:eq("+j+")").text();
            if(text.indexOf("单击")>-1){
                text = $("#historyIdTable tbody tr:eq("+i+")").find("td:eq("+j+")").find("a").data("href");
            }
            configTableTxt += text+",";
        }
    }
    //console.log("configTableTxt="+configTableTxt);
    localStorage["accessControlTableId"] = configTableTxt;
}
//弹出编辑模态框
$("#editModal").on("show.bs.modal", function (e) {
    var $dom = $(e.relatedTarget);
    modal.curIndex = $dom.parents("tr").index();
    //获取编辑按钮对应行的数据并填入模态框
    var obj = {};
    $dom.parents("tr").find("td").each(function (index,element) {
        obj[index] = $(this).text();
    })
    console.log("obj="+JSON.stringify(obj));
    $("#editRfidName").val(obj[0]);
    $("#editRfidId").val(obj[1]);
});
//在模态框内编辑确认
$("#editConfirm").on("click",function(){
    var data = {};
    var cur_rfid = $("#editRfidId").val();
    var cur_name = $("#editRfidName").val();
    data[cur_rfid] = {
        "name" : cur_name
```

```
        };
        if(checkRFID(data)){
            //编辑 ID：data={"0D69CFF0":{"name":"王小虎 2","sex":"男"}}
            console.log("编辑 ID：data="+JSON.stringify(data)+"----index="+modal.curIndex);
            $("#rfidTable tr:eq("+ (modal.curIndex+1 )+")").find(".name").text(cur_name);
            for(var k in config['rfid']){
                if(k == cur_rfid){
                    config['rfid'][k] = data[k];
                }
            }
            page.storeStorage();
            console.log(" 读 取  localStorage ： localStorage.FwsSmartHome="+localStorage.getItem("course_
accessControl"));
            $("#editModal").modal("hide");
        }
    });
    function checkRFID() {
        return true;
    }
    //删除模态框
    $("#delModal").on("show.bs.modal", function (e) {
        console.log("del ready");
        var $del_tr =  $(e.relatedTarget).parents("tr")
        modal.del_index = $del_tr.index();
        modal.del_rfid = $del_tr.find(".rfid").text();
        $("#delInfo").text(modal.del_rfid);
    })
    $("#delConfirm").on("click", function() {
        console.log("del rfid = "+modal.del_rfid);
        $("#rfidTable").find("tr:eq("+(modal.del_index+1)+")").remove();
        delete  config['rfid'][(modal.del_rfid)];
        console.log(config['rfid'][(modal.del_rfid)]);
        page.storeStorage();
        $('#delModal').modal('hide');
    })
```

智能门禁管理系统 Web 端其他部分的代码请查看本书配套资料中的项目源文件。

4.2.4 开发验证

1. Web 端应用测试

在 Web 端打开智能门禁管理系统后，在“运营首页”下可以看到 Web 端的主页，如图 4.21 所示。

当设备在线后，可以手动地控制门锁，如图 4.22 所示。

在“运营首页”页面的主页上还可以显示门禁 ID 和刷卡记录，如图 4.23 所示。

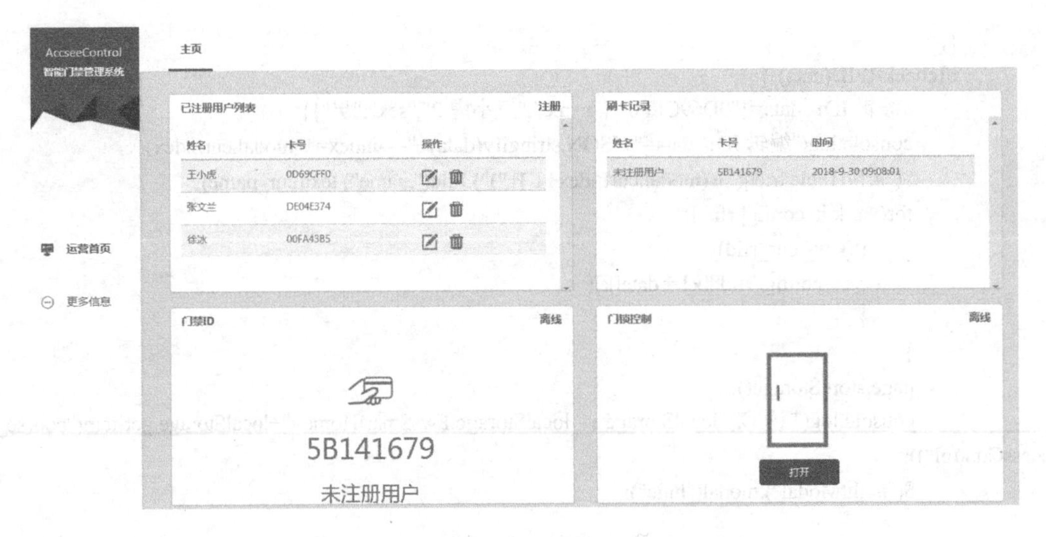

图 4.21　智能门禁管理系统的 Web 端主页面

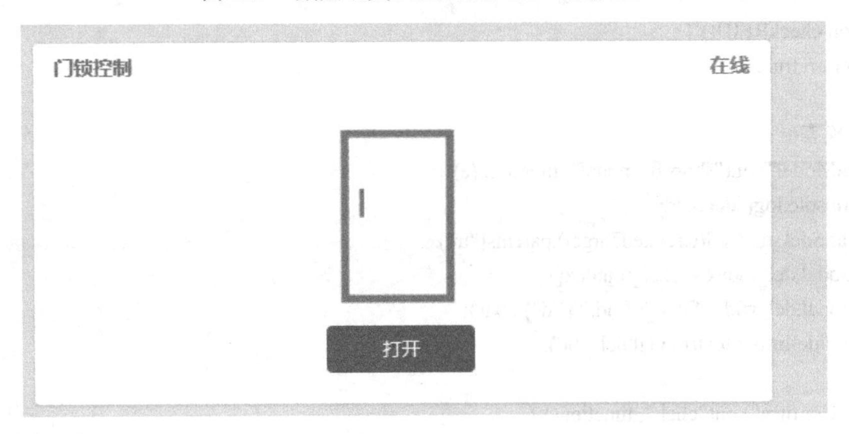

图 4.22　手动控制门锁

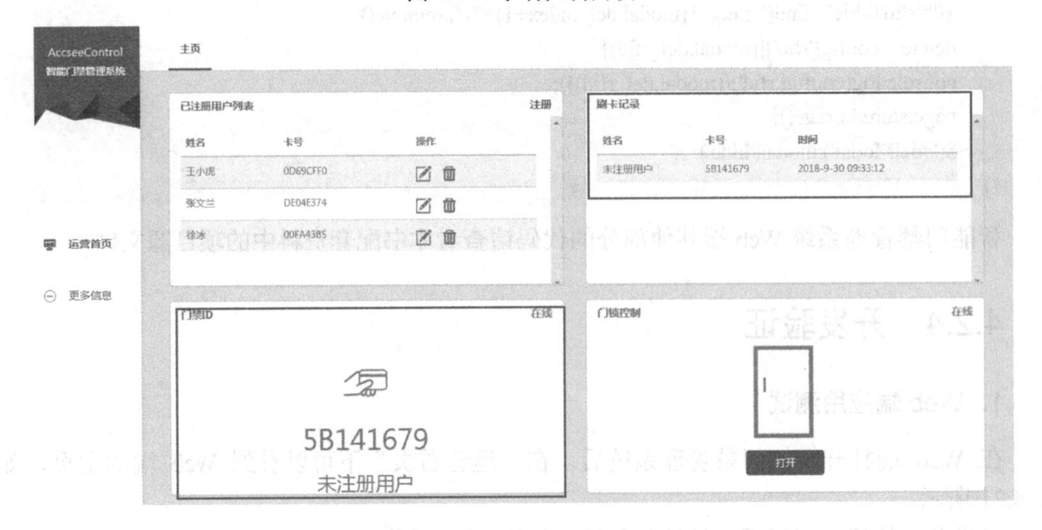

图 4.23　门禁 ID 和刷卡记录

2. Android 端应用测试

Android 端应用测试同 Web 端应用测试流程基本一致，可参考本系统的 Web 端应用测试进行操作。智能门禁管理系统 Android 端的“运营首页”页面如图 4.24 所示。

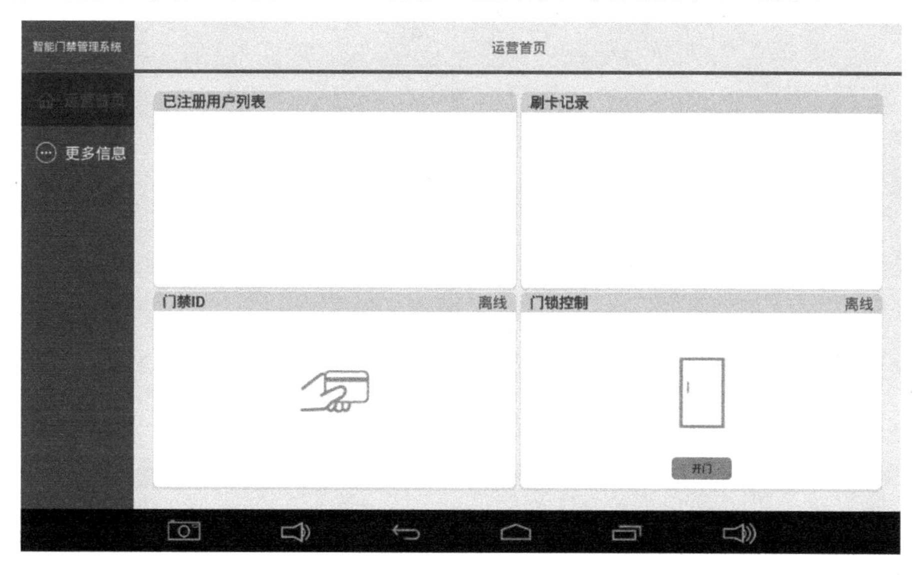

图 4.24 智能门禁管理系统 Android 端的“运营首页”页面

4.2.5 总结与拓展

本节基于 BLE 实现了射频传感器的识别和继电器的控制，通过 Android 和 HTML5 技术实现了 Android 端和 Web 端的应用设计，能够根据射频传感器和继电器的状态来实时控制继电器的开关，实现了基于 BLE 的智能门禁管理系统。

第5章 Wi-Fi高级应用开发

Wi-Fi（Wireless Fidelity，无线保真技术）符合IEEE 802.11b标准，可以将个人电脑、手持设备(如PDA、手机)等终端以无线方式互相连接起来。Wi-Fi是为了改善基于IEEE 802.11b标准的无线网络产品之间的互通性而提出的，其主要特点为数据传输速率快、可靠性高，方便与现有的网络整合，组网成本低。有关更详尽的Wi-Fi内容请参考《物联网短距离无线通信技术应用与开发》。

本章通过基于Wi-Fi的楼宇消防控制系统和楼宇通风控制系统这两个贴近生活的开发案例，详细地介绍了Wi-Fi物联网系统的架构和软硬件开发，实现了采集类传感器、控制类传感器和安防类传感器的驱动程序，进行了Android端和Web端的应用开发。

5.1 基于Wi-Fi的楼宇消防控制系统

楼宇消防控制系统（见图5.1）是智能消防系统中一个重要的组成部分，是现代建筑与信息技术相结合的产物，在智能消防系统中起到了重要的作用。楼宇消防控制系统通过温湿度传感器、空气质量传感器和火焰传感器采集环境数据，并与设置的阈值进行比较，当数据高于阈值时启动火灾报警器（由RGB灯与蜂鸣器实现）和喷淋系统。

5.1.1 系统开发目标

（1）熟悉温湿度传感器、空气质量传感器、火焰传感器、蜂鸣器、RGB灯、继电器等硬件原理和数据通信协议，基于CC3200和Wi-Fi实现蜂鸣器、RGB灯、继电器的驱动程序的开发，通过比较温湿度传感器、空气质量传感器和火焰传感器采集的数据与设定的阈值来控制火灾报警器和喷淋系统，实现楼宇消防控制系统的设计。

（2）实现楼宇消防控制系统的Android端应用开发和Web端应用开发。

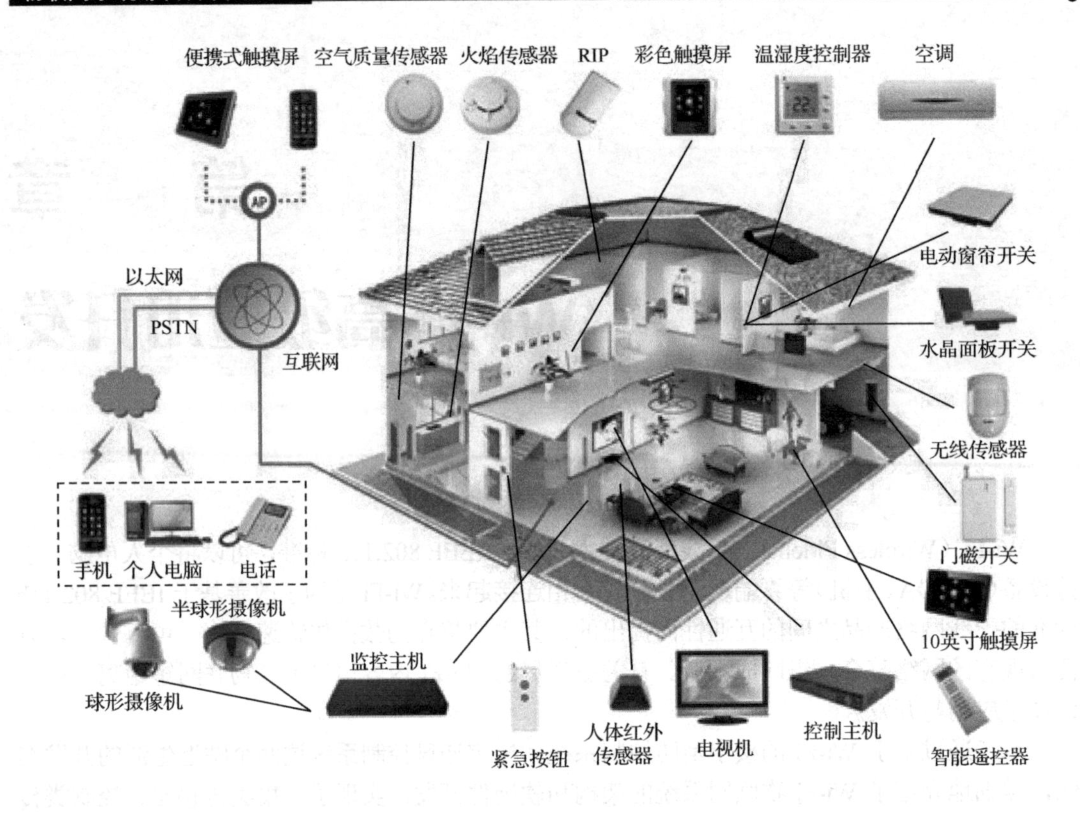

图 5.1　楼宇消防控制系统

5.1.2　系统设计分析

1. 系统的功能设计

从系统功能的角度出发，楼宇消防控制系统可以分为两个模块：设备采集和控制模块以及系统设置模块，如图 5.2 所示。

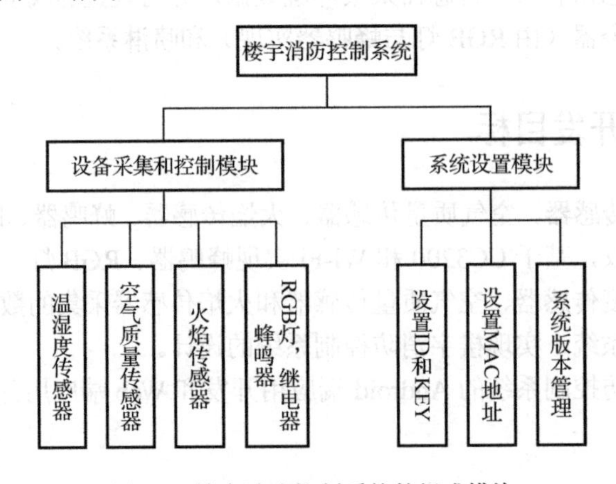

图 5.2　楼宇消防控制系统的组成模块

设备采集和控制模块的主要功能是监测温湿度传感器、空气质量传感器数据和火焰传感器的状态，控制继电器、RGB 灯和蜂鸣器。

系统设置模块的主要功能是设置智云服务器的 ID 和 KEY；设置 MAC 地址；系统软件版本查询与显示。

楼宇消防控制系统的功能需求如表 5.1 所示。

表 5.1　楼宇消防控制系统的功能需求

功　能	功 能 说 明
采集数据显示	在上层应用页面中实时更新显示温湿度传感器、空气质量传感器、火焰传感器采集的数据
喷淋系统实时控制	通过上层应用程序，对喷淋系统进行控制
声光报警控制	通过比较传感器采集的数据与设定的阈值，控制 RGB 灯与蜂鸣器
模式设置	自动模式：当传感器采集的数据超出设定的阈值时将自动报警。手动模式：通过页面控制
智云连接设置	设置智云服务器的参数和设备的 MAC 地址

2. 系统的总体架构设计

楼宇消防控制系统是基于物联网四层架构模型来设计的，其总体架构如图 5.3 所示。

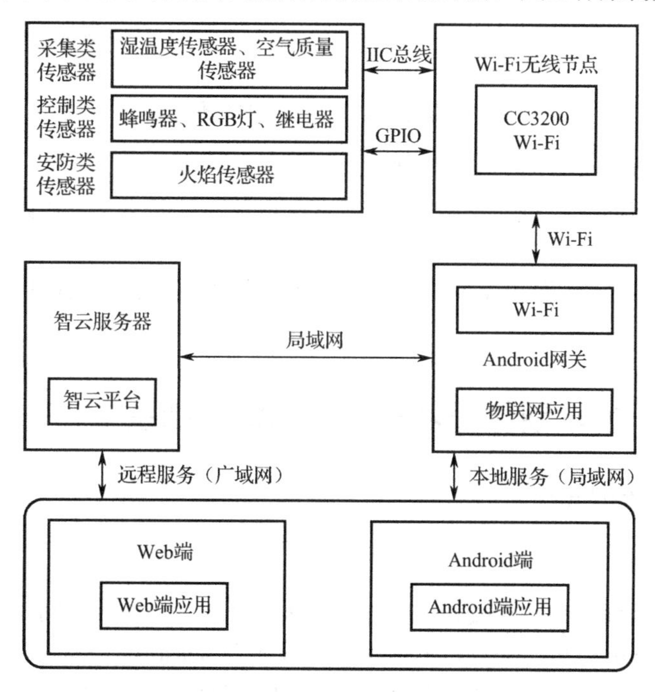

图 5.3　楼宇消防控制系统的总体架构

感知层：通过采集类和安防类传感器对温度、空气质量和火焰等进行监测，通过 CC3200 控制安防类传感器（如蜂鸣器、RGB 灯、继电器）的操作。

网络层：感知层节点和智能网关之间的通信是通过 Wi-Fi 来实现的，智能网关和智云服务器、上层应用设备间是通过局域网（互联网）进行数据传输的。

平台层：平台层提供物联网设备之间的基于互联网的存储、访问、控制。

应用层：应用层主要是物联网系统的人机交互接口，通过 Web 端、Android 端提供页面友好、操作交互性强的应用。

3. 系统的数据传输

楼宇消防控制系统的数据传输是在传感器节点、智能网关以及客户端（包括 Web 端和 Android 端）之间进行的，如图 5.4 所示。

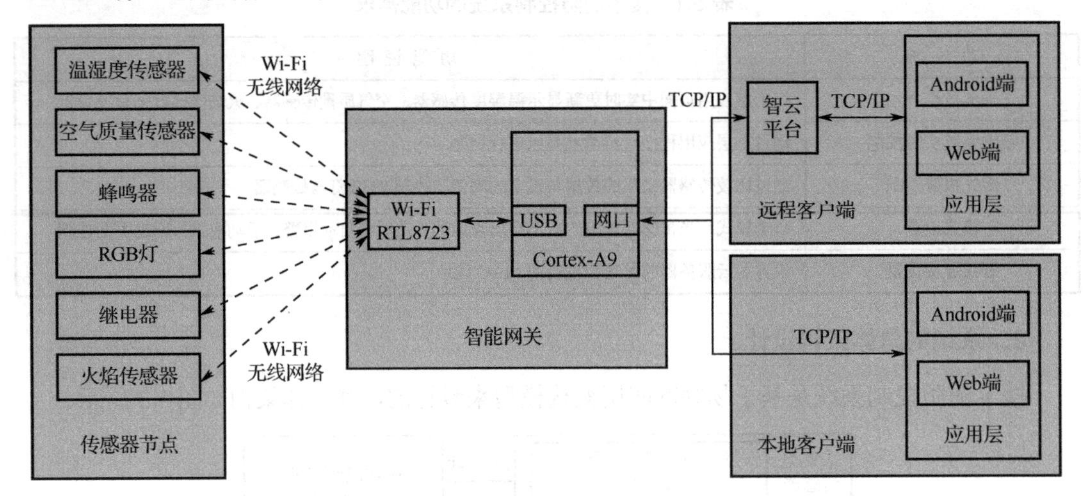

图 5.4　楼宇消防控制系统的数据传输

（1）传感器节点通过 Wi-Fi 与智能网关进行组网。

（2）传感器节点的数据通过 Wi-Fi 发送给智能网关，智能网关将数据推送给所有连接智能网关的客户端。

（3）客户端（Android 端和 Web 端）的应用是通过调用智云数据接口来实现实时数据采集等功能的。

5.1.3　系统的软硬件开发：楼宇消防控制系统

1. 系统底层软硬件设计

1）感知层硬件设计

楼宇消防控制系统的感知层硬件包括 xLab 未来开发平台的智能网关、经典型无线节点 ZXBeeLiteB、采集类开发平台 Sensor-A、控制类开发平台 Sensor-B、安防类开发平台 Sensor-C。其中智能网关负责汇集传感器采集的数据；Wi-Fi 无线节点（由经典型无线节点 ZXBeeLiteB 实现）通过无线通信的方式向智能网关发送传感器数据，接收智能网关发送的命令；控制类传感器连接到 Wi-Fi 无线节点，由 CC3200 对相关设备进行控制。本系统中的传感器包括温湿度传感器、空气质量传感器、火焰传感器、继电器、蜂鸣器和 RGB 灯，温湿度传感器的硬件接口电路如图 3.5 所示，空气质量传感器的硬件接口电路如图 3.7 所示，火焰传感器的硬件接口电路如图 3.46 所示，继电器的硬件接口电路如图 4.18 所示，蜂鸣器的硬件接口电路如图 3.49 所示，RGB 灯的硬件接口电路如图 3.24 所示。

2）系统底层开发

本系统使用 Wi-Fi 无线网络进行开发。

（1）Wi-Fi 智云开发框架。智云框架是在传感器应用程序接口和 SAPI 框架的基础上搭建起来的，通过合理调用这些接口，可以使 Wi-Fi 的开发形成一套系统的开发逻辑。传感器应用程序接口函数是在 sensor.c 文件中实现的，具体如表 5.2 所示。

表 5.2 传感器应用程序接口函数

函数名称	函数说明
sensorInit()	传感器初始化
sensorLinkOn()	传感器节点入网成功调用的函数
sensorUpdate()	传感器数据定时上报
sensorControl()	传感器控制函数
sensorCheck()	传感器预警监测及处理函数
ZXBeeInfRecv()	处理节点接收到的无线数据包
sensorLoop()	启动定时器触发事件

（2）智云平台底层 API。智云框架下传感器程序执行流程如图 5.5 所示。

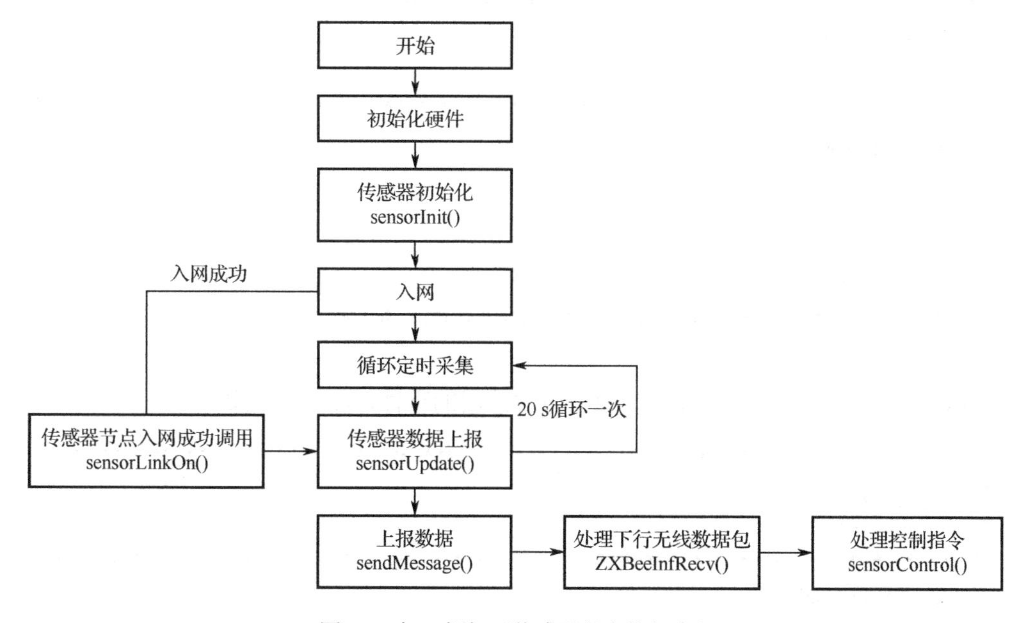

图 5.5 智云框架下传感器程序执行流程

智云框架为 Wi-Fi 协议栈的上层应用提供分层的软件设计结构，将传感器的私有操作部分封装到 sensor.c 文件中，用户任务中的处理事件和节点类型选择在 sensor.h 文件中定义。

sensorInit()函数用于对传感器进行初始化，相关代码如下：

```
/*********************************************************************************************
*名称：sensorInit()
*功能：传感器初始化
```

```
*****************************************************************************/
void sensorInit(void)
{
    DebugMsg("sensor->sensorInit(): Sensor init!\r\n");
    //温湿度传感器初始化
    relay_init();  //继电器初始化
}
```

在传感器节点入网成功后调用 sensorLinkOn()函数进行相关操作，该函数的代码如下：

```
/*****************************************************************************
*名称：sensorLinkOn()
*功能：传感器节点入网成功调用函数
*****************************************************************************/
void sensorLinkOn(void)
{
    printf("sensor->sensorLinkOn(): Sensor Link on!\r\n");
    sensorUpdate();       //入网成功后上报一次传感器数据
}
```

sensorUpdate()函数用于对传感器采集的数据进行更新和打包上报，相关代码如下：

```
/*****************************************************************************
*名称：sensorUpdate()
*功能：处理主动上报的数据
*****************************************************************************/
void sensorUpdate(void)
{
    ……
    sprintf(p, "temperature=%.1f", temperature);
    sendMessage(p, strlen(p));
    DebugMsg("sensor->sensorUpdate(): temperature=%.1f\r\n", temperature);
}
```

sensorLoop()函数用于进行循环定时触发传感器的事件，相关代码如下：

```
/*****************************************************************************
*名称：sensorLoop()
*功能：定时触发功能
*****************************************************************************/
void sensorLoop(void)
{
    static unsigned long ct_update = 0;
    if (t4exp(ct_update)) {
        sensorUpdate();
    ct_update = t4ms()+20*1000;
    }
}
```

ZXBeeInfRecv()用于对节点接收到的无线数据包进行处理，相关代码如下：

```
/*********************************************************************************************
*名称：ZXBeeInfRecv()
*功能：处理节点接收到的无线数据包
*参数：*pkg—收到的无线数据包；len—无线数据包的长度
*********************************************************************************************/
void ZXBeeInfRecv(char *pkg, int len)
{
    ……
    DebugMsg("sensor->ZXBeeInfRecv(): Receive Wi-Fi Data!\r\n");
    ptag = pkg;
    p = strchr(pkg, '=');
    if (p != NULL) {
        *p++ = 0;
        pval = p;
    }
    val = atoi(pval);
    //控制命令解析
    if (0 == strcmp("cmd", ptag)){                    //对 D0 的位进行操作，CD0 表示位清零操作
        sensorControl(val);
    }
}
```

sensorControl()函数用于控制传感器，相关代码如下：

```
/*********************************************************************************************
*名称：sensorControl()
*功能：传感器控制
*参数：cmd—控制命令
*********************************************************************************************/
void sensorControl(uint8 cmd)
{
    //根据 cmd 参数执行对应的控制程序
}
```

通过在 sensor.c 文件中实现具体函数即可快速地完成 Wi-Fi 项目开发。

3）传感器驱动设计

楼宇消防控制系统中底层的传感器主要是安防类传感器和控制类传感器，主要关注两个方面，一是安防类传感器的报警信息要及时可靠地上报，二是要了解控制类传感器对设备的控制是否有效以及控制结果。安防类传感器的逻辑事件可以分为 4 种，详见 3.3.3 节；控制类传感器的逻辑事件可以分为 3 种，详见 3.2.3 节。

（1）数据通信协议的定义。本系统主要使用的是采集类开发平台 Sensor-A、控制类开发平台 Sensor-B 和安防类开发平台 Sensor-C，三种开发平台的 ZXBee 数据通信协议如表 5.3 所示。

表 5.3　三种开发平台的 ZXBee 数据通信协议

开发平台	属　性	参　数	权限	说　明
Sensor-A（601）	温度	A0	R	温度值，浮点型数据，精度为 0.1，范围为-40.0～105.0
	空气质量	A3	R	空气质量值
	上报状态	D0（OD0/CD0）	R/W	D0 的 Bit0～Bit7 分别代表 A0～A7 的上报
	数据上报时间间隔	V0	R/W	A0～A7 传感器值的循环上报时间间隔
Sensor-B（602）	RGB 灯的状态和颜色	D1（OD1/CD1）	R/W	D1 的 Bit0～Bit1 代表 RGB 灯的状态和颜色，00 表示关闭 RGB 灯，01 表示红色（R），10 表示绿色（G），11 表示蓝色（B）
	蜂鸣器状态	D1（OD1/CD1）	R/W	D1 的 Bit3 代表蜂鸣器的开关状态，0 表示关闭，1 表示打开
	继电器状态	D1（OD1/CD1）	R/W	D1 的 Bit6、Bit7 分别代表继电器 K1、K2 的状态，0 表示断开，1 表示吸合
	数据上报时间间隔	V0	R/W	定时上报数据的时间间隔，单位为 s
Sensor-C（603）	火焰状态	A3	R	火焰状态值，0 或 1 变化；0 表示未监测到火焰，1 表示监测到火焰

（2）驱动程序的开发。在智云框架下不仅可以很容易地实现传感器驱动程序的开发，还可以省略无线传感器节点的组网和用户任务的创建等烦琐过程。例如，调用 sensorInit()函数可以实现传感器的初始化；调用 ZXBeeInfRecv()函数可以处理节点接收到的无线数据包；设备状态的定时上报使用 MyEventProcess()作为 sensorUpdate()函数的定时进入接口来反馈设备状态信息。

在 sensor.c 中，需要在 sensorInit()函数中添加传感器初始化的内容，并通过定义上报事件和报警事件来实现设备工作状态的定时反馈。部分代码如下：

```
/*********************************************************************************************
*名称：sensorInit()
*功能：传感器初始化
*********************************************************************************************/
void sensorInit(void)
{
    //初始化传感器代码
    htu21d_init();
    airgas_init();
    ……
}
```

温湿度传感器的初始化函数是 htu21d_init()，该函数是通过 IIC 总线写寄存器地址来初始化温湿度传感器的。htu21d_read_reg()函数用于实现读寄存器的操作，htu21d_get_data()函数用于实现测量温湿度，代码详见 3.1.3 节中 htu21d.c 文件的代码。

空气质量传感器的初始化函数是 airgas_init()，代码如下：

```
/*********************************************************************************************
*名称：airgas_init()
*功能：空气质量传感器初始化
*********************************************************************************************/
```

```
void airgas_init(void)
{
    PinTypeADC(PIN_60,PIN_MODE_255);
    ADCTimerConfig(ADC_BASE, 2^17);                    //配置 ADC 内部定时器
    ADCTimerEnable(ADC_BASE);                          //使能内部定时器
    ADCChannelEnable(ADC_BASE,ADC_CH_3);
    ADCEnable(ADC_BASE);
}
/*********************************************************************************
*名称：unsigned int get_airgas_data(void)
*功能：获取空气质量传感器状态
*********************************************************************************/
unsigned int get_airgas_data(void)
{
    unsigned int    value;
    while(!ADCFIFOLvlGet(ADC_BASE, ADC_CH_3));
      value = ((ADCFIFORead(ADC_BASE,ADC_CH_3)>>2)&0xfff)>>4;
    return value;                                      //返回有效值
}
```

蜂鸣器的初始化函数是 Beep _init()，该函数是通过将对应引脚配置为输出模式来实现蜂鸣器的初始化的，在初始化后关闭蜂鸣器，代码如下：

```
/*********************************************************************************
*名称：Beep_init()
*功能：蜂鸣器初始化
*********************************************************************************/
void Beep_init(void)
{
    PRCMPeripheralClkEnable(PRCM_GPIOA1, PRCM_RUN_MODE_CLK);        //使能时钟
    PinTypeGPIO(PIN_01,PIN_MODE_0,0);                              //选择引脚为 GPIO 模式
    GPIODirModeSet(GPIOA1_BASE, G10_UCPINS, GPIO_DIR_MODE_OUT);     //设置 GPIO14 为输
出模式
    Beep_off(0x01);
}
/*********************************************************************************
*名称：Beep_on()
*功能：打开蜂鸣器
*返回：0 表示打开成功，-1 表示参数错误
*********************************************************************************/
signed char Beep_off(unsigned char beep)
{
    if(beep & 0x01){                                               //打开蜂鸣器
        GPIOPinWrite(GPIOA1_BASE, G10_UCPINS, 0xff);
        return 0;
    }
    return -1;                                         //参数错误，返回-1
```

```
}
/**************************************************************************************
*名称：Beep_off()
*功能：关闭蜂鸣器
*返回：0 表示关闭成功，-1 表示参数错误
**************************************************************************************/
signed char Beep_on(unsigned char beep)
{
    if(beep &0x01){                                             //关闭蜂鸣器
        GPIOPinWrite(GPIOA1_BASE, G10_UCPINS, 0x00);
        return 0;
    }
    return -1;                                                  //参数错误，返回-1
}
```

RGB 灯的初始化函数是 rgb_init()，该函数是通过将对应引脚配置为输出模式来实现 RGB 灯的初始化的，在初始化后关闭 RGB 灯，代码如下：

```
void rgb_init(void)
{
    GPIO_InitTypeDef   GPIO_InitStructure;
    RCC_APB2PeriphClockCmd(RCC_APB2Periph_GPIOA, ENABLE);          //使能 PA 端口时钟
    RCC_APB2PeriphClockCmd(RCC_APB2Periph_GPIOB, ENABLE);          //使能 PB 端口时钟
    GPIO_InitStructure.GPIO_Pin = GPIO_Pin_1 | GPIO_Pin_0;
    GPIO_InitStructure.GPIO_Speed = GPIO_Speed_2MHz;
    GPIO_InitStructure.GPIO_Mode = GPIO_Mode_Out_PP;
    GPIO_Init(GPIOB, &GPIO_InitStructure);
    GPIO_InitStructure.GPIO_Pin = GPIO_Pin_3;
    GPIO_Init(GPIOA, &GPIO_InitStructure);
    rgb_off(0x01);                                                  //初始状态为关闭
    rgb_off(0x02);
    rgb_off(0x04);
}
```

继电器的初始化函数是 relay_init()，该函数也是通过将对应的引脚配置为输出模式来初始化继电器的，代码如下：

```
/**************************************************************************************
*名称：relay_init()
*功能：继电器初始化
**************************************************************************************/
void relay_init(void)
{
    PRCMPeripheralClkEnable(PRCM_GPIOA0, PRCM_RUN_MODE_CLK);          //使能时钟
    PinTypeGPIO(PIN_08,PIN_MODE_0,0);                                  //配置引脚为 GPIO 模式
    GPIODirModeSet(GPIOA2_BASE, G17_UCPINS, GPIO_DIR_MODE_OUT);      //配置 GPIO17 为输
出模式
    PRCMPeripheralClkEnable(PRCM_GPIOA3, PRCM_RUN_MODE_CLK);          //使能时钟
```

```
        PinTypeGPIO(PIN_18,PIN_MODE_0,0);                              //配置引脚为 GPIO 模式
        GPIODirModeSet(GPIOA3_BASE, G28_UCPINS, GPIO_DIR_MODE_OUT);    //配置 GPIO28 为输
出模式
        GPIOPinWrite(GPIOA2_BASE, G17_UCPINS, 0xFF);
        GPIOPinWrite(GPIOA3_BASE, G28_UCPINS, 0xFF);
    }
    signed int relay_on(char cmd)
    {
        if(cmd & 0x01){
            GPIOPinWrite(GPIOA2_BASE, G17_UCPINS, 0x00);
        }
        if(cmd & 0x02){
            GPIOPinWrite(GPIOA3_BASE, G28_UCPINS, 0x00);
        }
        return 0;
    }
    signed int relay_off(char cmd)
    {
        if(cmd & 0x01){
            GPIOPinWrite(GPIOA2_BASE, G17_UCPINS, 0xff);
        }
        if(cmd & 0x02){
            GPIOPinWrite(GPIOA3_BASE, G28_UCPINS, 0xff);
        }
        return 0;
    }
    void relay_control(char cmd)
    {
        relay_on(cmd);
        relay_off((~cmd)&0x03);
    }
```

火焰传感器的初始化函数是 flame_init()，该函数是通过将对应的引脚配置为输入模式来初始化火焰传感器的，程序是通过监测引脚的电平状态来判断火焰传感器的状态的。

```
/*********************************************************************************************
*名称：flame_init()
*功能：火焰传感器初始化
*********************************************************************************************/
void flame_init(void)
{
    PRCMPeripheralClkEnable(PRCM_GPIOA1, PRCM_RUN_MODE_CLK);  //使能时钟
    PinTypeGPIO(PIN_01,PIN_MODE_0,false);                     //配置引脚为 GPIO 模式
    GPIODirModeSet(GPIOA1_BASE, G10_UCPINS, GPIO_DIR_MODE_IN); //配置引脚为输入模式
    PinConfigSet(PIN_01,PIN_TYPE_STD_PD,PIN_MODE_0);          //下拉
}
/*********************************************************************************************
```

```
*名称：unsigned char get_flame_status(void)
*功能：获取火焰传感器状态
*********************************************************************************/
unsigned char get_flame_status(void)
{
    if((unsigned char)GPIOPinRead(GPIOA1_BASE,G10_UCPINS) > 0)          //监测引脚的电平状态
        return 1;
    else
        return 0;
}
```

2. Android 端应用设计

1）Android 工程设计框架

打开 Android Studio 开发环境，可以看到楼宇消防控制系统的工程目录，如图 5.6 所示，系统的工程框架如表 5.4 所示。

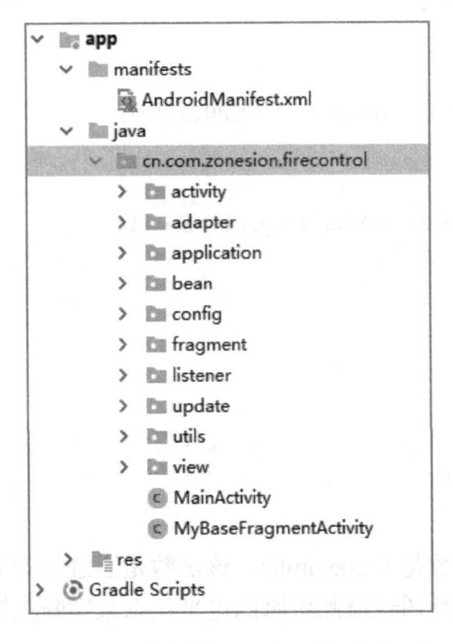

图 5.6　楼宇消防控制系统的工程目录

表 5.4　楼宇消防控制系统的工程框架

类　名	说　明
activity	
IdKeyShareActivity.java	在 IDKey 页面单击“分享”按钮时，可弹出 activity，用于分享二维码图片
adapter	
HdArrayAdapter.java	历史数据显示适配器
Application	
LCApplication.java	LCApplication 继承 application 类，使用单例模式（Singleton Pattern）创建 WSNRTConnect 对象

续表

类　名	说　明
bean	
HistoricalData.java	历史数据的 bean 类，用于将从智云服务器获得的历史数据记录（JSON 形式）转换成该类对象
IdKeyBean.java	IdKeyBean 用来描述用户设备的 ID、KEY，以及智云服务器的地址 SERVER
Config	
Config.java	config 用于修改用户的 ID、KEY，以及智云服务器的地址和 MAC 地址
fragment	
BaseFragment.java	页面基础 Fragment 定义类
HDFragment.java	历史数据页面
HistoricalDataFragment.java	历史数据显示页面
HomepageFragment.java	展示首页的 Fragment
IDKeyFragment.java	IDKey 选项的页面
MacSettingFragment.java	当用户设置被监测项的 MAC 地址时显示的页面
MoreInformationFragment.java	更多信息显示页面
RunHomePageFragment.java	运营首页显示页面
VersionInformationFragment.java	显示版本等相关信息的页面
listener	
IOnWSNDataListener.java	传感器数据监听器接口
update	
UpdateService.java	应用下载服务类
view	
APKVersionCodeUtils.java	获取当前本地 apk 的版本
CustomRadioButton.java	自定义按钮类
PagerSlidingTabStrip.java	自定义滑动控件类
MainActivity.java：主页面类	
MyBaseFragmentActivity.java：系统 Fragment 通信类	

2）软件设计

根据智云 Android 端应用程序接口的定义，楼宇消防控制系统的应用设计主要采用实时数据 API 接口，实时数据 API 接口的流程见 3.1.3 节的图 3.11。

（1）LCApplication.java 程序代码剖析。楼宇消防控制系统中的 LCApplication.java 程序代码和城市环境信息采集系统的 LCApplication.java 程序代码相同，详见 3.1.3 节的相关内容。

（2）HomepageFragment.java 程序代码剖析。下面的代码通过(LCApplication) getActivity().getApplication()获取 LCApplication 类中的 WSNRTConnect 对象。

```
private void initViewAndBindEvent() {
    preferences = getActivity().getSharedPreferences("user_info", Context.MODE_PRIVATE);
    lcApplication = (LCApplication) getActivity().getApplication();
```

```
        wsnrtConnect = lcApplication.getWSNRConnect();
        lcApplication.registerOnWSNDataListener(this);
        editor = preferences.edit();
    }
```

下面的代码通过复写 onMessageArrive 方法来处理节点接收到的无线数据包，实现了设备的 MAC 地址获取，并在当前的页面显示设备的状态。

```
    @Override
        public void onMessageArrive(String mac, String tag, String val) {
            if (sensorAMAC == null && sensorBMAC    == null && sensorCMAC == null) {
                wsnrtConnect.sendMessage(mac, "{TYPE=?}".getBytes());
            }
            if ("TYPE".equals(tag) && "601".equals(val.substring(2, val.length()))) {
                sensorAMAC = mac;
            }
            if ("TYPE".equals(tag) && "602".equals(val.substring(2, val.length()))) {
                sensorBMAC = mac;
            }
            if ("TYPE".equals(tag) && "603".equals(val.substring(2, val.length()))) {
                sensorCMAC = mac;
            }
            if (mac.equals(sensorAMAC) && "A0".equals(tag)) {
                textTemperaturestate.setText("在线");
                textTemperaturestate.setTextColor(getResources().getColor(R.color.line_text_color));
                temperatureText.setText(val+"°C");
                lightIntensity = Float.valueOf(val);
                if(seekbarThreshold.getProgress() != 0) {
                    limitofilluminationTooHigh();
                }
            }
            if (mac.equals(sensorAMAC) && "A3".equals(tag)) {
                textAirState.setText("在线");
                textAirState.setTextColor(getResources().getColor(R.color.line_text_color));
                airText.setText(val+"μg/m3");
            }
            if (mac.equals(sensorCMAC) && "A3".equals(tag)) {
                textFlameState.setText("在线");
                textFlameState.setTextColor(getResources().getColor(R.color.line_text_color));
                int numResult = Integer.parseInt(val);
                if (Spray==true) {
                    if ((numResult & 1) == 1) {
                        imageFlameState.setImageResource(R.drawable.fire_on);
                        FlameText.setText("监测到火焰");
                        wsnrtConnect.sendMessage(sensorBMAC, "{OD1=64,D1=?}".getBytes());
                    }else {
                        imageFlameState.setImageResource(R.drawable.fire);
```

```
                        wsnrtConnect.sendMessage(sensorBMAC, "{CD1=64,D1=?}".getBytes());
                        FlameText.setText("正常");
                    }
                }else {
                    if ((numResult & 1) == 1) {
                        imageFlameState.setImageResource(R.drawable.fire_on);
                        FlameText.setText("监测到火焰");
                    }else {
                        imageFlameState.setImageResource(R.drawable.fire);
                        FlameText.setText("正常");
                    }
                }
            }
            if (mac.equals(sensorBMAC) && "D1".equals(tag)) {
                textSprayState.setText("在线");
                textSprayState.setTextColor(getResources().getColor(R.color.line_text_color));
                textTipsState.setText("在线");
                textTipsState.setTextColor(getResources().getColor(R.color.line_text_color));
                int numResult = Integer.parseInt(val);
                if ((numResult & 0X40) == 0x40) {
              imageSprayState.setImageDrawable(getResources().getDrawable(R.drawable.spray_on));
                    openOrCloseSpray.setText("关闭");
                openOrCloseSpray.setBackground(getResources().getDrawable(R.drawable.close));
                }else {
              imageSprayState.setImageDrawable(getResources().getDrawable(R.drawable.spray));
                    openOrCloseSpray.setText("开启");
                openOrCloseSpray.setBackground(getResources().getDrawable(R.drawable.open));
                }
                if ((numResult & 0X8) == 0x8) {
            imageTipsState.setImageDrawable(getResources().getDrawable(R.drawable.alarm_on));
                    openOrCloseTips.setText("关闭");
                openOrCloseTips.setBackground(getResources().getDrawable(R.drawable.close));
                }else {
              imageTipsState.setImageDrawable(getResources().getDrawable(R.drawable.alarm));
                    openOrCloseTips.setText("开启");
              openOrCloseTips.setBackground(getResources().getDrawable(R.drawable.open));
                }
            }
        }
```

下面的代码实现了对喷淋系统的控制。

```
openOrCloseSpray.setOnClickListener(new OnClickListener() {
    @Override
    public void onClick(View v) {
        if (sensorBMAC != null) {
            if (openOrCloseSpray.getText().equals("开启")) {
                new Thread(new Runnable() {
```

```
                        @Override
                            public void run() {
                                wsnrtConnect.sendMessage(sensorBMAC,
"{OD1=64,D1=?}".getBytes());
                            }
                    }).start();
                }
                if (openOrCloseSpray.getText().equals("关闭")) {
                    new Thread(new Runnable() {
                        @Override
                            public void run() {
                                wsnrtConnect.sendMessage(sensorBMAC,
"{CD1=64,D1=?}".getBytes());
                            }
                    }).start();
                }
            }else {
                Toast.makeText(lcApplication, "请等待 MAC 地址上线", Toast.LENGTH_SHORT).show();
            }
        }
    });
```

3. Web 端应用设计

1）页面功能结构分析

楼宇消防控制系统的 Web 端默认显示的是“运营首页”页面，在“运营首页”页面上设计了 6 个模块，分别是温度数据显示模块、PM2.5 数据显示模块、火焰状态显示模块、喷淋设备控制模块、报警提示控制模块、阈值设置模块。楼宇消防控制系统 Web 端的“运营首页”页面如图 5.7 所示。

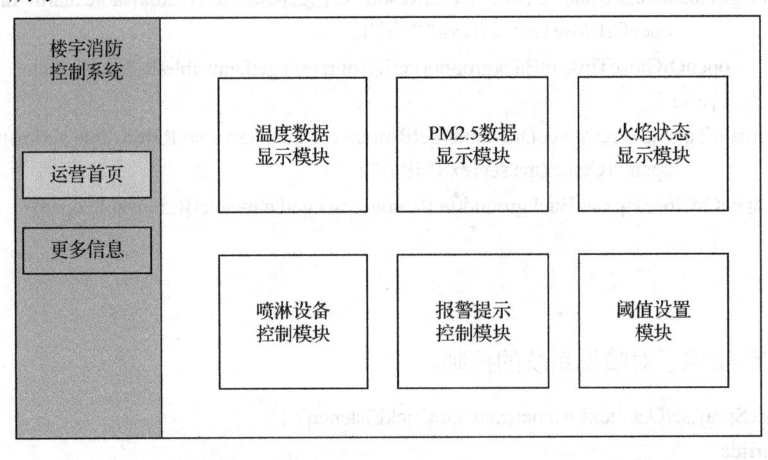

图 5.7　楼宇消防控制系统 Web 端的 “运营首页”页面

“更多信息”页面的主要功能是进行智云服务器的连接配置，和城市环境信息采集系统中的 Web 端“更多信息”页面类似，参见图 3.14。

2）软件设计

楼宇消防控制系统 Web 端的 JS 开发逻辑与 Android 端的开发逻辑相似，首先通过配置 ID 和 KEY 与智云服务器进行连接，再通过实时监听数据的方法来获取相关传感器的数据并进行处理。JS 开发的部分代码如下。

在 getConnect()函数中定义了实时连接对象 rtc，连接成功回调函数是 rtc.onConnect，数据服务掉线回调函数是 rtc.onConnectLost，消息处理回调函数是 rtc.onmessageArrive。

```
function getConnect() {
    config["id"] = config["id"] ? config["id"] : $("#id").val();
    config["key"] = config["key"] ? config["key"] : $("#key").val();
    config["server"] = config["server"] ? config["server"] : $("#server").val();
    //创建数据连接服务对象
    rtc = new WSNRTConnect(config["id"], config["key"]);
    rtc.setServerAddr(config["server"] + ":28080");
    rtc.connect();
    rtc._connect = false;
    //连接成功回调函数
    rtc.onConnect = function() {
        $("#ConnectState").text("数据服务连接成功！");
        rtc._connect = 1;
        message_show("数据服务连接成功！");
        $("#idkeyInput").text("断开").addClass("btn-danger");
        $("#id,#key,#server").attr('disabled',true);
    };
    //数据服务掉线回调函数
    rtc.onConnectLost = function() {
        rtc._connect = 0;
        $("#ConnectState").text("数据服务连接掉线！");
        $("#idkeyInput").text("连接").removeClass("btn-danger");
        message_show("数据服务连接失败，检查网络或 ID、KEY");
        $("#RFIDLink").text("离线").css("color", "#e75d59");
        $("#doorLink").text("离线").css("color", "#e75d59");
        $("#id,#key,#server").removeAttr('disabled',true);
    };
    //消息处理回调函数
    rtc.onmessageArrive = function (mac, dat) {
        //console.log(mac+" >>> "+dat);
        if (dat[0]=='{' && dat[dat.length-1]=='}') {
            dat = dat.substr(1, dat.length-2);
            var its = dat.split(',');
            for (var i=0; i<its.length; i++) {
                var it = its[i].split('=');
                if (it.length == 2) {
                    process_tag(mac, it[0], it[1]);
                }
            }
```

```
                    if (!mac2type[mac]) { //如果没有获取到 TYPE 值，主动去查询
                        rtc.sendMessage(mac,
"{TYPE=?,A0=?,A1=?,A2=?,A3=?,A4=?,A5=?,A6=?,A7=?,D1=?}");
                    }
                }
            }
        }
```

下述 JS 开发代码的功能是根据设备连接情况，在页面显示设备是否在线，如果设备在线，则更新设备状态和报警状态。

```
    var wsn_config = {
        "601" : {
            "online" : function() {
                $(".online_601").text("在线").css("color", "#96ba5c");
            },
            "pro" : function (tag, val) {
                if(tag=="A1"){
                    thermometer("temp","℃","100%","100%","#e47e77", -20, 100, val);
                    if(config["mac_602"]!="" && config.mode == "secutiry-mode"){
                        //温度超过阈值的上限，开启警报
                        if(val>config["threshold2"]){
                            //引发报警
                            checkOnAlarm();
                            message_show("温度阈值超限，警报开启！");
                        }
                    }
                }
                else if(tag=="A2"){
                    thermometer("pm","μg/m3","100%","100%","#229", 0, 800, val);
                    if(config["mac_602"]!="" && config.mode == "secutiry-mode"){
                        //PM2.5 超过阈值的上限，开启警报
                        if(val>config["threshold"]){
                            //引发报警
                            checkOnAlarm();
                            message_show("PM2.5 阈值超限，警报开启！");
                        }
                    }
                }
            }
        },
        "602" : {
            "online" : function() {
                $(".online_602").text("在线").css("color", "#96ba5c");
            },
            "pro" : function (tag, val) {
                if(tag=="D1" && config["alarm"]){
```

```
                }
            }
        },
        "603" : {
            "online" : function() {
                $(".online_603").text("在线").css("color", "#96ba5c");
            },
            "pro" : function (tag, val) {
                if(tag=="A3" && config["mac_603"]!="" && config.mode == "secutiry-mode"){
                    if(val == 1){
                        $("#fireStatus").addClass("fire-on");
                        $("#fire-text").text("监测到火焰");
                        //引发报警
                        checkOnAlarm();
                    }else{
                        $("#fireStatus").removeClass("fire-on");
                        $("#fire-text").text("正常");
                        //引发报警
                        checkCloseAlarm();
                    }
                    state.fire = val;
                }
            }
        }
    };
```

空气质量阈值（PM2.5 阈值）与温度阈值的设置及更新代码如下：

```
$('#nstSliderS').nstSlider({
    "left_grip_selector": "#leftGripS",
    "value_bar_selector": "#barS",
    "value_changed_callback": function(cause, leftValue, rightValue) {
        var $container = $(this).parent(),
        g = 255 - 127 + leftValue,
        r = 255 - g,
        b = 0;
        $container.find('#leftLabelS').text(rightValue);
        $container.find('#rightLabelS').text(leftValue);
        $(this).find('#barS').css('background', 'rgb(' + [r, g, b].join(',') + ')');
        console.log("PM2.5 阈值更新："+leftValue);
        config["threshold"] = leftValue;
        storeStorage();
    }
});
//温度阈值
$('#nstSliderS2').nstSlider({
    "left_grip_selector": "#leftGripS2",
    "value_bar_selector": "#barS2",
```

```
        "value_changed_callback": function(cause, leftValue, rightValue) {
            var $container = $(this).parent(),
            g = 255 - 0 + leftValue,
            r = 255 - g,
            b = 0;
            $container.find('#leftLabelS2').text(rightValue);
            $container.find('#rightLabelS2').text(leftValue);
            $(this).find('#barS2').css('background', 'rgb(' + [r, g, b].join(',') + ')');
            console.log("温度阈值更新："+leftValue);
            config["threshold2"] = leftValue;
            storeStorage();
        }
    });
```

在 Web 端“运营首页”中，灾情提示与喷淋系统中的按钮状态更新以及命令发送代码如下。

```
    //灾情提示中的按钮
    $("#alarmStatus").on("click", function() {
        if(page.checkOnline() && page.checkMac("mac_602")) {
            var state = $(this).text() == "已开", cmd;
            if (state) {
                $(this).text("已关");
                $("#alarmImg").attr('src', 'img/alarm.png');
                config["alarm"] = false;
                cmd = "{CD1=1,CD1=8,D1=?}"
            } else {
                $(this).text("已开");
                $("#alarmImg").attr('src', 'img/alarm-on.png');
                config["alarm"] = true;
                cmd = "{OD1=1,OD1=8,D1=?}"
            }
            console.log(cmd);
            rtc.sendMessage(config["mac_602"], cmd);
        }
    });
    //喷淋系统中的按钮
    $("#sprayStatus").on("click", function() {
        if(page.checkOnline() && page.checkMac("mac_602")){
            var state = $(this).text()=="已开", cmd;
            if(state){
                cmd = "{CD1=64,D1=?}";
                $("#sprayImg").attr('src','img/spray.png');
                $(this).text("已关");
            }else{
                cmd = "{OD1=64,D1=?}";
                $("#sprayImg").attr('src','img/spray-on.png');
```

```
                $(this).text("已开");
            }
            console.log(cmd);
            rtc.sendMessage(config["mac_602"], cmd);
        }
    });
```

楼宇消防控制系统 Web 端其他部分的代码请查看本书配套资料中的项目源文件。

5.1.4　开发验证

1. Web 端应用测试

在 Web 端打开楼宇消防控制系统后，在“运营首页”下可以看到 Web 端的主页，如图 5.8 所示。

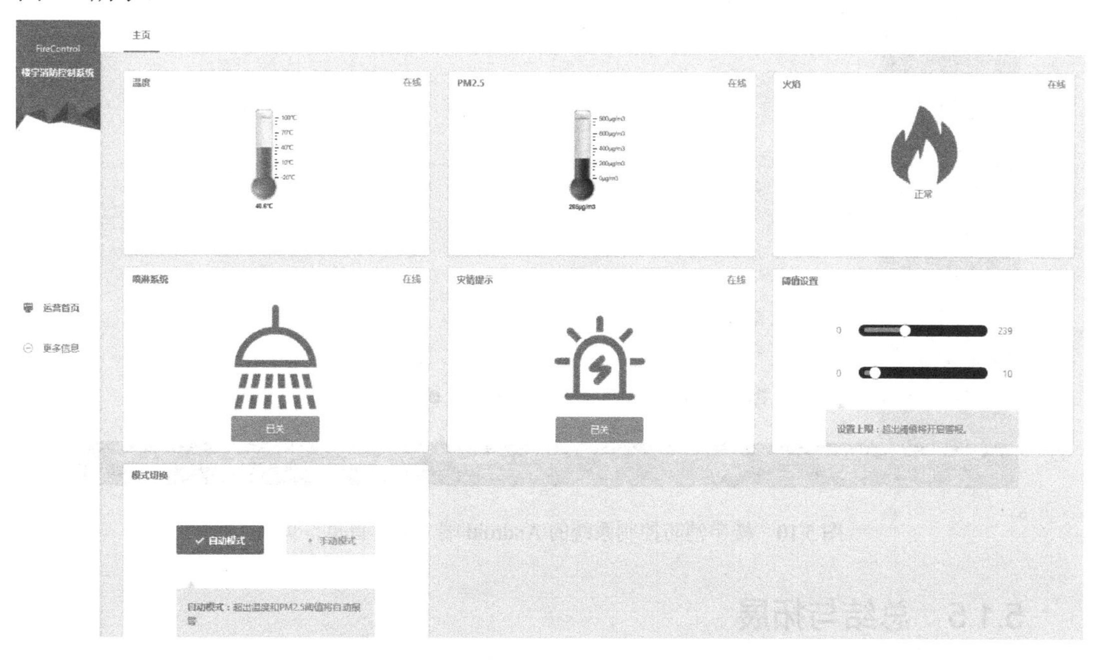

图 5.8　楼宇消防控制系统的 Web 端主页

当设备在线后，可以在楼宇消防控制系统的 Web 端手动控制喷淋系统，如图 5.9 所示。

2. Android 端应用测试

Android 端应用测试同 Web 端应用测试流程基本一致，可参考本系统的 Web 端应用测试进行操作。楼宇消防控制系统的 Android 端“运营首页”页面如图 5.10 所示。

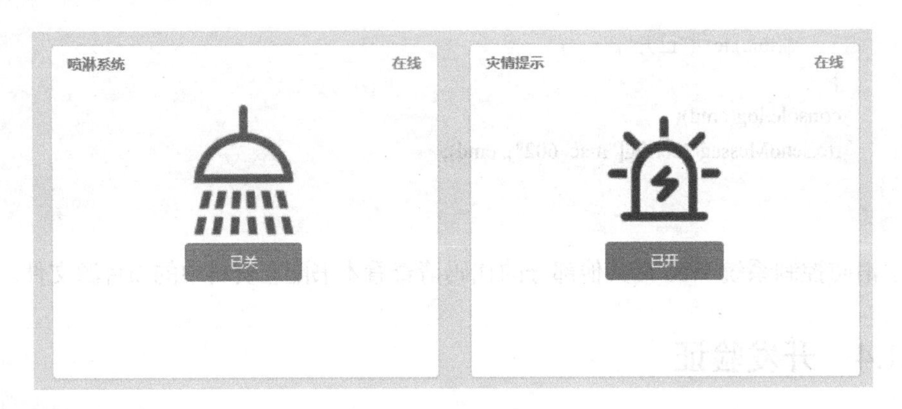

图 5.9　手动控制喷淋

图 5.10　楼宇消防控制系统的 Android 端“运营首页”页面

5.1.5　总结与拓展

本节基于 CC3200 和 Wi-Fi 实现了温湿度传感器、空气质量传感器、火焰传感器的数据采集以及蜂鸣器、RGB 灯、继电器的控制，通过 Android 和 HTML5 技术实现了 Android 端和 Web 端的应用设计，根据传感器实时获取的数据来控制蜂鸣器、RGB 灯和继电器，实现了基于 Wi-Fi 的楼宇消防控制系统。

5.2　基于 Wi-Fi 的楼宇通风控制系统

楼宇通风控制系统是智能楼宇系统中一个重要的组成部分，可以通过比较实时采集的空气质量数据和设定的阈值来控制风扇。

5.2.1 系统开发目标

（1）熟悉空气质量传感器和风扇等硬件原理和数据通信协议，基于 Wi-Fi 实现空气质量传感器和风扇的驱动程序开发，通过比较空气质量传感器采集的数据和设定的阈值来控制风扇，实现楼宇通风控制系统的设计。

（2）实现楼宇通风控制系统的 Android 端应用开发和 Web 端应用开发。

5.2.2 系统设计分析

1. 系统的功能设计

从系统功能的角度出发，楼宇通风控制系统可以分为两个模块：设备采集和控制模块以及系统设置模块，如图 5.11 所示。

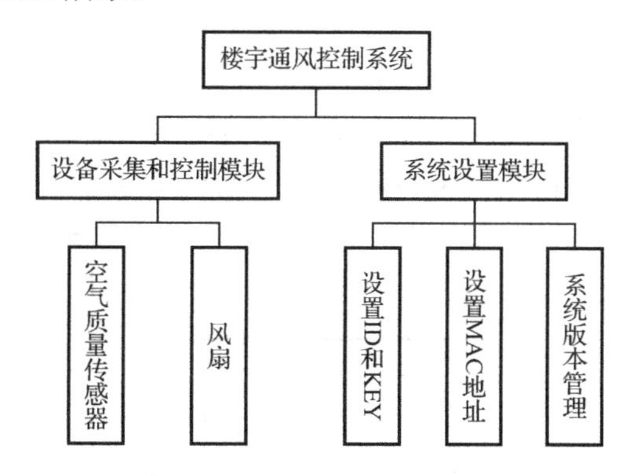

图 5.11 楼宇通风控制系统的组成模块

楼宇通风控制系统的功能需求如表 5.5 所示。

表 5.5 楼宇通风控制系统的功能需求

功 能	功能说明
采集数据显示	在上层应用页面中实时更新显示空气质量传感器采集的数据
风扇实时控制	通过上层应用程序，对风扇进行控制
模式设置	自动模式：定时或者通过比较空气质量传感器采集的数据和设定的阈值来对风扇进行控制。手动模式：通过页面控制风扇
智云连接设置	设置智云服务器的参数和设备的 MAC 地址

2. 系统的总体架构设计

楼宇通风控制系统是基于物联网四层架构模型来设计的，其总体架构如图 5.12 所示。

3. 系统的数据传输

楼宇通风控制系统的数据传输是在传感器节点、智能网关以及客户端（包括 Web 端和 Android 端）之间进行的，如图 5.13 所示。

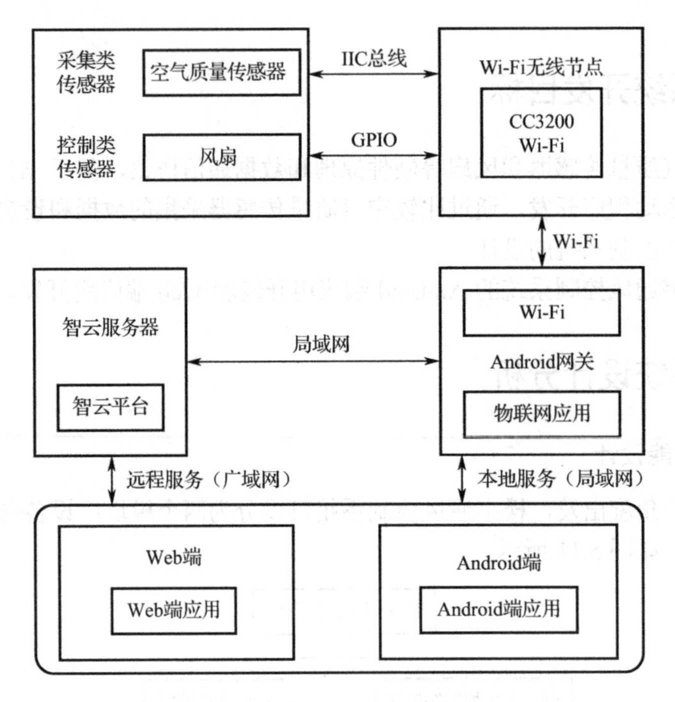

图 5.12　楼宇通风控制系统的总体架构

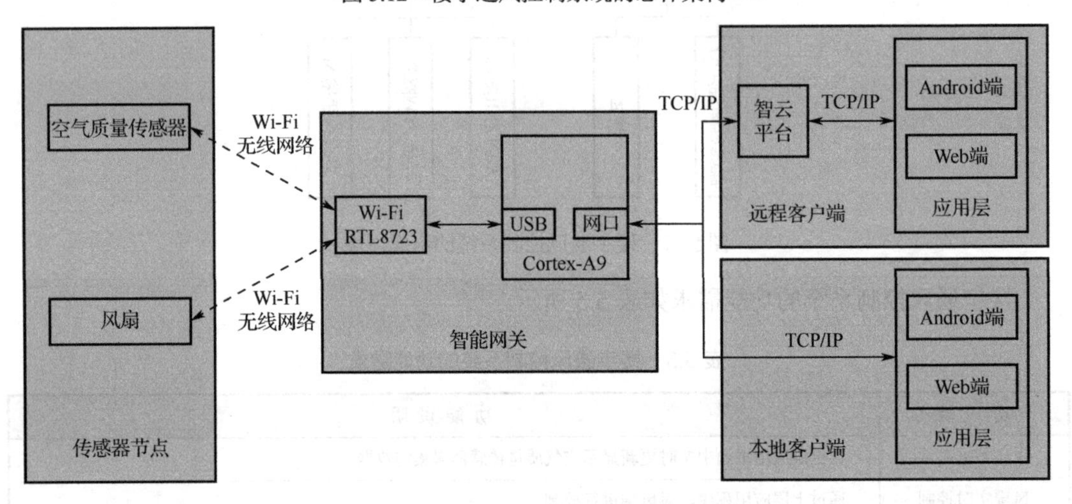

图 5.13　楼宇通风控制系统的数据传输

5.2.3　系统的软硬件开发：楼宇通风控制系统

1．系统底层软硬件设计

1）*感知层硬件设计*

楼宇通风控制系统的底层硬件主要包括 xLab 未来开发平台的智能网关、经典型无线节点 ZXBeeLiteB、采集类开发平台 Sensor-A 和控制类开发平台 Sensor-B。其中，智能网关负责汇集传感器采集的数据；Wi-Fi 无线节点（由经典型无线节点 ZXBeeLiteB 实现）通过无线通信的方式向智能网关发送传感器数据，接收智能网关发送的命令；采集类开发平台 Sensor-A 和控制类开发平台 Sensor-B 连接到 Wi-Fi 无线节点，由其中的 CC3200 对相关设备进行控制。本系统中传感器包括空气质量传感器和风扇，空气质量传感器的硬件接口电路如图 3.7 所示，风扇的硬件接口电路如图 3.35 所示。

2）系统底层开发

楼宇通风控制系统是基于 Wi-Fi 无线网络开发的。

（1）Wi-Fi 智云开发框架。本系统采用的智云框架和楼宇消防控制系统采用的智云框架相同，详见 5.1.3 节。

（2）智云平台底层 API。智云平台底层 API 详见 5.1.3 节。

3）感知层传感器驱动设计

楼宇通风控制系统的底层硬件主要是采集类传感器和控制类传感器，采集类传感器的逻辑事件详见 3.1.3 节，控制类传感器的逻辑事件详见 3.2.3 节。

（1）数据通信协议的定义。本系统主要使用采集类开发平台 Sensor-A 和控制类开发平台 Sensor-B，其 ZXBee 数据通信协议如表 5.6 所示。

表 5.6 采集类开发平台和控制类开发平台的 ZXBee 数据通信协议

开发平台	属　性	参　数	权　限	说　明
Sensor-A（601）	空气质量	A3	R	空气质量值
	上报状态	D0（OD0/CD0）	R/W	D0 的 Bit0～Bit7 分别代表 A0～A7 传感器数据的上报
	数据上报时间间隔	V0	R/W	A0～A7 传感器数据的循环上报时间间隔
	数据上报时间间隔	V0	R/W	传感器数据上报的时间间隔（循环上报）
Sensor-B（602）	风扇	D1（OD1/CD1）	R/W	D1 的 Bit3 代表风扇的开关状态，0 表示关闭，1 表示打开

（2）驱动程序的开发。在智云框架下不仅可以很容易地实现传感器驱动程序的开发，还可以省略无线传感器节点的组网和用户任务的创建等烦琐过程。例如，调用 sensorInit()函数可以实现传感器的初始化；调用 ZXBeeInfRecv()函数可以处理节点接收到的无线数据包；设备状态的定时上报使用 MyEventProcess()作为 sensorUpdate()函数的定时进入接口来反馈设备状态信息。

在 sensor.c 中，需要在 sensorInit()函数中添加传感器初始化的内容，通过定义上报事件和报警事件来实现设备工作状态的定时反馈，部分代码如下：

```
/*********************************************************************************************
*名称：sensorInit()
*功能：传感器初始化
*********************************************************************************************/
void sensorInit(void)
{
    airgas_init();                                          //空气质量传感器初始化
    ......
}
```

空气质量传感器的初始化函数是 airgas_init()，代码如下：

```
/*********************************************************************************
*名称：airgas_init()
*功能：空气质量传感器初始化
*********************************************************************************/
void airgas_init(void)
{
    PinTypeADC(PIN_60,PIN_MODE_255);
    ADCTimerConfig(ADC_BASE, 2^17);                    //配置 ADC 内部定时器
    ADCTimerEnable(ADC_BASE);                          //使能内部定时器
    ADCChannelEnable(ADC_BASE,ADC_CH_3);
    ADCEnable(ADC_BASE);
}
/*********************************************************************************
*名称：unsigned int get_airgas_data(void)
*功能：获取空气质量传感器数据
*********************************************************************************/
unsigned int get_airgas_data(void)
{
    unsigned int    value;
    while(!ADCFIFOLvlGet(ADC_BASE, ADC_CH_3));
    value = ((ADCFIFORead(ADC_BASE,ADC_CH_3)>>2)&0xfff)>>4;
    return value;                                                      //返回有效值
}
```

风扇的初始化函数是 FAN_init()，代码如下：

```
/*********************************************************************************
*名称：FAN_init()
*功能：风扇初始化
*********************************************************************************/
void FAN_init(void)
{
    PRCMPeripheralClkEnable(PRCM_GPIOA1, PRCM_RUN_MODE_CLK);      //使能时钟
    PinTypeGPIO(PIN_01,PIN_MODE_0,0);                              //配置引脚为GPIO模式
    GPIODirModeSet(GPIOA1_BASE, G10_UCPINS, GPIO_DIR_MODE_OUT); //配置引脚为输出模式
    FAN_off(0x01);
}
/*********************************************************************************
*名称：FAN_on()
*功能：打开风扇
*返回：0 表示打开成功，-1 表示参数错误
*********************************************************************************/
signed char FAN_on(unsigned char fan)
{
    if(fan & 0x01){                                                   //打开风扇
        GPIOPinWrite(GPIOA1_BASE, G10_UCPINS, 0xff);
```

```
        return 0;
    }
    return -1;                                                      //参数错误，返回-1
}
/*********************************************************************************
*名称：FAN_off()
*功能：关闭风扇
*返回：0 表示关闭成功，-1 表示参数错误
*********************************************************************************/
signed char FAN_off(unsigned char fan)
{
    if(fan &0x01){                                                  //关闭风扇
        GPIOPinWrite(GPIOA1_BASE, G10_UCPINS, 0x00);
        return 0;
    }
    return -1;                                                      //参数错误，返回-1
}
```

2. Android 端应用设计

1）Android 工程设计框架

打开 Android Studio 开发环境，可以看到本系统的工程目录，如图 5.14 所示。楼宇通风控制系统的工程框架和楼宇消防控制系统的工程框架相同，见表 5.4。

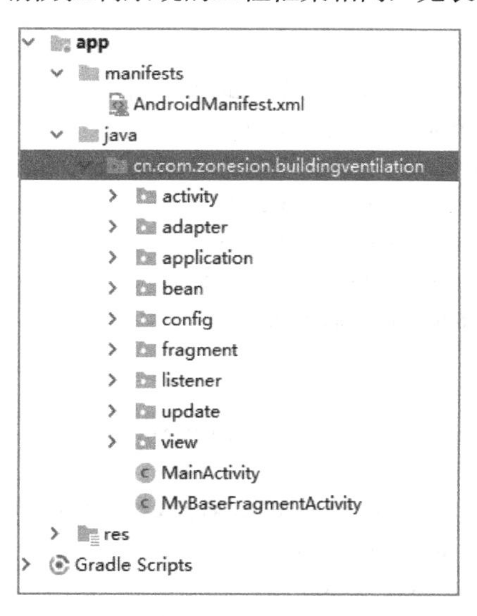

图 5.14　楼宇通风控制系统的工程目录

2）软件设计

根据智云 Android 端应用程序接口的定义，系统的应用设计主要采用实时数据 API 接口，实时数据 API 接口的流程见 3.1.3 节的图 3.11。

（1）LCApplication.java 程序代码剖析。楼宇通风控制系统中的 LCApplication.java 程序

代码和城市环境信息采集系统的 LCApplication.java 程序代码相同，详见 3.1.3 节的相关内容。

（2）HomepageFragment.java 程序代码剖析。下面的代码通过(LCApplication) getActivity().getApplication()获取 LCApplication 类中的 WSNRTConnect 对象。

```
private void initViewAndBindEvent() {
    preferences = getActivity().getSharedPreferences("user_info", Context.MODE_PRIVATE);
    lcApplication = (LCApplication) getActivity().getApplication();
    wsnrtConnect = lcApplication.getWSNRConnect();
    lcApplication.registerOnWSNDataListener(this);
    editor = preferences.edit();
}
```

下面的代码通过复写 onMessageArrive 方法来处理节点接收到的无线数据包，实现了设备的 MAC 地址获取，并在当前的页面显示设备的状态。

```
@Override
    public void onMessageArrive(String mac, String tag, String val) {
        if (sensorAMAC == null && sensorBMAC == null) {
            wsnrtConnect.sendMessage(mac, "{TYPE=?}".getBytes());
        }
        if ("TYPE".equals(tag) && "601".equals(val.substring(2, val.length()))) {
            sensorAMAC = mac;
        }
        if ("TYPE".equals(tag) && "602".equals(val.substring(2, val.length()))) {
            sensorBMAC = mac;
        }
        if (mac.equals(sensorAMAC) && "A3".equals(tag)) {
            textAtmosphereState.setText("在线");
            textAtmosphereState.setTextColor(getResources().getColor(R.color.line_text_color));
            atmosphereText.setText(val+"μg/m3");
            currentTemperature = Float.parseFloat(val);
            if(seekBarThreshold.getProgress() != 0) {
                limitofilluminationTooHigh();
            }
        }
        if (mac.equals(sensorBMAC) && "D1".equals(tag)) {
            textFanState.setText("在线");
            textFanState.setTextColor(getResources().getColor(R.color.line_text_color));
            int numResult = Integer.parseInt(val);
            if ((numResult & 0X8) == 0x8) {
                imageFanState.setImageDrawable(getResources().getDrawable(R.drawable.fan_on));
                openOrCloseLamp.setText("关闭");
                openOrCloseLamp.setBackground(getResources().getDrawable(R.drawable.close));
            }else {
                imageFanState.setImageDrawable(getResources().getDrawable(R.drawable.fan));
                openOrCloseLamp.setText("开启");
                openOrCloseLamp.setBackground(getResources().getDrawable(R.drawable.open));
```

```
            }
        }

    }
```

下面的代码实现了风扇的定时控制。

```
private void timerOpenOrCloseLED(){
    if(sensorBMAC!=null) {
        long time = System.currentTimeMillis();
        SimpleDateFormat sdf = new SimpleDateFormat("HH:mm");
        String format = sdf.format(new Date(time));
        Toast.makeText(lcApplication, format, Toast.LENGTH_SHORT).show();
        String[] result = format.split(":");
        //风扇
        String[] openTime = ledOpenTimeShow.getText().toString().split(":");
        String[] closeTime = ledCloseTimeShow.getText().toString().split(":");
        int resultHour = Integer.parseInt(result[0]);
        int resultMinute = Integer.parseInt(result[1]);
        //风扇开启的时间
        int openTimeHour = Integer.parseInt(openTime[0]);
        int openTimeMinute = Integer.parseInt(openTime[1]);
        //风扇关闭的时间
        int closeTimeHour = Integer.parseInt(closeTime[0]);
        int closeTimeMinute = Integer.parseInt(closeTime[1]);
        if (resultHour >= openTimeHour && resultMinute >= openTimeMinute && isSecurityMode == true) {
            wsnrtConnect.sendMessage(sensorBMAC, "{OD1=16，D1=?}".getBytes());
        }
        else if(resultHour <= openTimeHour && resultMinute <= closeTimeMinute && isSecurityMode ==
true){
            wsnrtConnect.sendMessage(sensorBMAC, "{CD1=16，D1=?}".getBytes());
        }
    }else{
        Toast.makeText(lcApplication, "请等待 MAC 地址上线", Toast.LENGTH_SHORT).show();
    }
}
```

3. Web 端应用设计

1）页面功能结构分析

楼宇通风控制系统的 Web 端默认显示的是“运营首页”页面，在“运营首页”页面上设计了 5 个模块，分别是空气质量显示模块、风扇控制模块、模式切换模块、定时器设置模块、阈值设置模块。楼宇通风控制系统 Web 端的“运营首页”页面如图 5.15 所示。

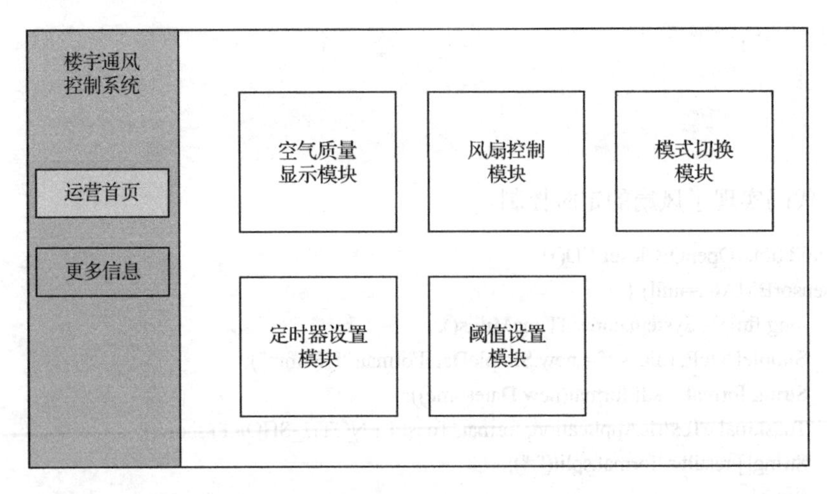

图 5.15 楼宇通风控制系统 Web 端的“运营首页”页面

“更多信息”页面的主要功能是进行智云服务器的连接配置，和城市环境信息采集系统中的 Web 端“更多信息”页面类似，参见图 3.14。

2）软件设计

楼宇通风控制系统 Web 端的 JS 开发逻辑与 Android 端的开发逻辑相似，首先通过配置 ID 和 KEY 与智云服务器进行连接，再通过实时监听数据的方法来获取相关传感器的数据并进行处理。JS 开发的部分代码如下。

在 getConnect()函数中定义了实时连接对象 rtc，连接成功回调函数是 rtc.onConnect，数据服务掉线回调函数是 rtc.onConnectLost，消息处理回调函数是 rtc.onmessageArrive。

```
function getConnect() {
    config["id"] = config["id"] ? config["id"] : $("#id").val();
    config["key"] = config["key"] ? config["key"] : $("#key").val();
    config["server"] = config["server"] ? config["server"] : $("#server").val();
    //创建数据连接服务对象
    rtc = new WSNRTConnect(config["id"], config["key"]);
    rtc.setServerAddr(config["server"] + ":28080");
    rtc.connect();
    rtc._connect = false;
    //连接成功回调函数
    rtc.onConnect = function() {
        $("#ConnectState").text("数据服务连接成功！");
        rtc._connect = 1;
        message_show("数据服务连接成功！");
        $("#idkeyInput").text("断开").addClass("btn-danger");
        $("#id,#key,#server").attr('disabled',true);
    };
    //数据服务掉线回调函数
    rtc.onConnectLost = function() {
        rtc._connect = 0;
        $("#ConnectState").text("数据服务连接掉线！");
```

```
                $("#idkeyInput").text("连接").removeClass("btn-danger");
                message_show("数据服务连接失败，检查网络或 ID、KEY");
                $("#RFIDLink").text("离线").css("color", "#e75d59");
                $("#doorLink").text("离线").css("color", "#e75d59");
                $("#id,#key,#server").removeAttr('disabled',true);
            };
            //消息处理回调函数
            rtc.onmessageArrive = function (mac, dat) {
                //console.log(mac+" >>> "+dat);
                if (dat[0]=='{' && dat[dat.length-1]=='}') {
                    dat = dat.substr(1, dat.length-2);
                    var its = dat.split(',');
                    for (var i=0; i<its.length; i++) {
                        var it = its[i].split('=');
                        if (it.length == 2) {
                            process_tag(mac, it[0], it[1]);
                        }
                    }
                    if (!mac2type[mac]) { //如果没有获取到 TYPE 值，主动去查询
                        rtc.sendMessage(mac,
"{TYPE=?,A0=?,A1=?,A2=?,A3=?,A4=?,A5=?,A6=?,A7=?,D1=?}");
                    }
                }
            }
        }
```

下述JS开发代码的功能是根据设备连接情况，在页面更新设备的状态，当风扇的状态是在线时，可根据当前空气质量与阈值的比较结果来控制风扇。

```
    var wsn_config = {
        "601" : {
                "online" : function() {
                    $(".online_601").text("在线").css("color", "#96ba5c");
                },
                "pro" : function (tag, val) {
                    if(tag=="A3"){
                        console.log(val);
                        soilTemper(val);
                        if(config["curMode"]=="auto-mode" && val>config["threshold"] && !state.fan){
                            rtc.sendMessage(config["mac_602"], "{OD1=8,D1=?}");
                            message_show("超出空气质量阈值，将打开风扇！");
                        }
                    }
                }
            },
            "602" : {
                "online" : function() {
```

```
                $(".online_602").text("在线").css("color", "#96ba5c");
            },
            "pro" : function (tag, val) {
                if(tag=="D1"){
                    if(val & 8){
                        $("#fanStatus").text("关闭");
                        $("#fanImg").attr("src", "img/fan-on.png").addClass("fan-active");
                        state.fan = true;
                    }else{
                        $("#fanStatus").text("打开");
                        $("#fanImg").attr("src", "img/fan.png").removeClass("fan-active");
                        state.fan = false;
                    }
                }
            }
        }
    }
```

下面的代码实现了风扇的打开和关闭操作。

```
//风扇控制开关
$("#fanStatus").on("click", function() {
    if (page.checkOnline() && page.checkMac("mac_602")){
        var state = $(this).text()=="关闭", cmd;
        if(state){
            cmd = "{CD1=8,D1=?}";
        }else{
            cmd = "{OD1=8,D1=?}";
        }
        rtc.sendMessage(config["mac_602"], cmd);
    }
});
```

下面的代码实现了风扇的定时器控制。

```
//初始化时间修改窗口
$("#dtBox").DateTimePicker({
    init : function() {
        //console.log("time picker init");
    },
    settingValueOfElement : function() {
        console.log("time picker click");
        config["open_time"] = $("#open_time").text();
        config["close_time"] = $("#close_time").text();
        storeStorage();
        console.log("更新开启时间："+config["open_time"]+"更新关闭时间："+config["close_time"]);
    }
});
```

```
    function getTime(){
        var nowdate = new Date();
        //获取年、月、日、时、分、秒
        var hours = nowdate.getHours(),
        minutes = nowdate.getMinutes(),
        date = nowdate.getDate();
        var hour_info = hours >=10 ? hours : "0"+hours;
        var minute_info = minutes >=10 ? minutes : "0"+minutes;
        var cur_time = hour_info + ":" + minute_info;
        //保存一个全局变量来缓存上一次的时间字符串，当最新的时间字符串与已保存的时间字符串不相
同时，则更新已保存的时间字符串，并执行相应的动作
        //console.log("每秒更新："+cur_time);
        if(last_time != cur_time){
            last_time = cur_time;
            console.log("每分钟更新：当前时间："+cur_time);
            if(config["curMode"]=="auto-mode" && (cur_time==config["open_time"] || cur_time==config
["close_time"])){
                if(page.checkOnline() && page.checkMac("mac_602")){
                    var cmd;
                    if(cur_time==config["open_time"]){
                        message_show("定时器执行风扇开启！");
                        cmd = "{OD1=8,D1=?}";
                    }
                    else if(cur_time==config["close_time"]){
                        message_show("定时器执行风扇关闭！");
                        cmd = "{CD1=8,D1=?}";
                    }
                    rtc.sendMessage(config["mac_602"], cmd);
                }else{
                    setTimeout(function() {
                        message_show("定时器控制失败！");
                    },4000)
                }
            }
        }
        setTimeout(getTime, 1000);
    }
```

楼宇通风控制系统 Web 端其他部分的代码请查看本书配套资料中的项目源文件。

5.2.4　开发验证

1．Web 端应用测试

在 Web 端打开楼宇通风控制系统后，在“运营首页”下可以看到 Web 端的主页，如图 5.16 所示。

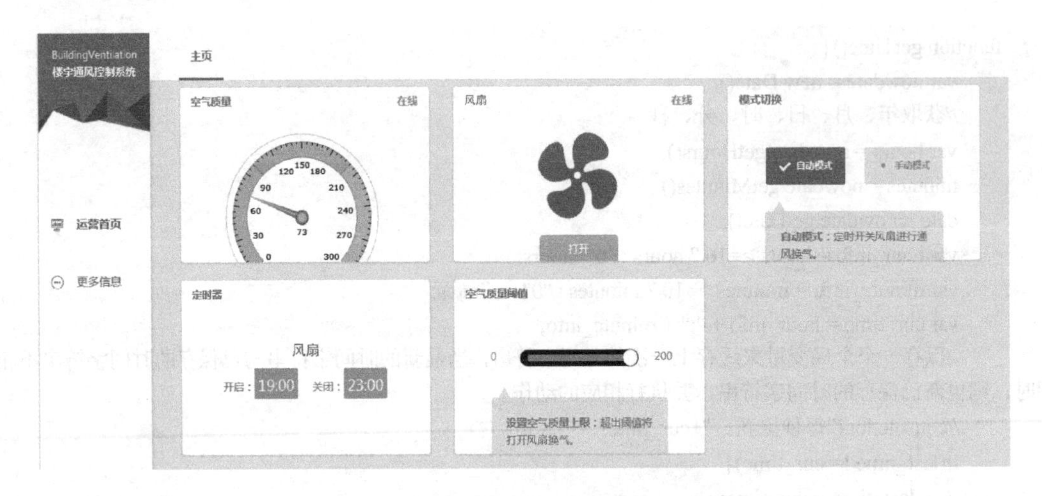

图 5.16 楼宇通风控制系统的 Web 端主页

当风扇的状态为在线时，可以通过风扇中的“打开”或“关闭”按钮来控制风扇，如图 5.17 所示。

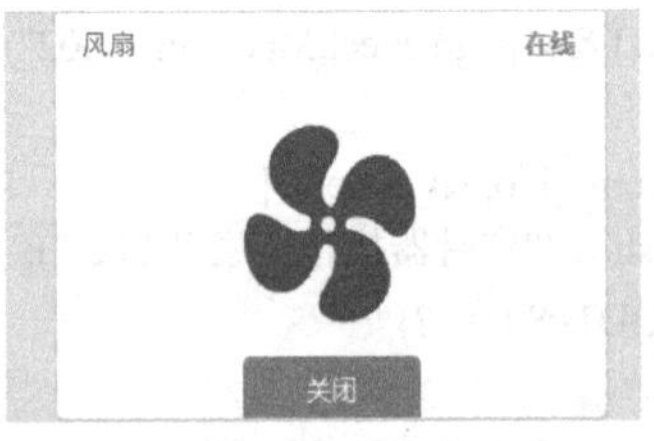

图 5.17 控制风扇

2. Android 端应用测试

Android 端应用测试同 Web 端应用测试流程基本一致，可参考本系统的 Web 端应用测试进行操作。楼宇通风控制系统的 Android 端“运营首页”页面如图 5.18 所示。

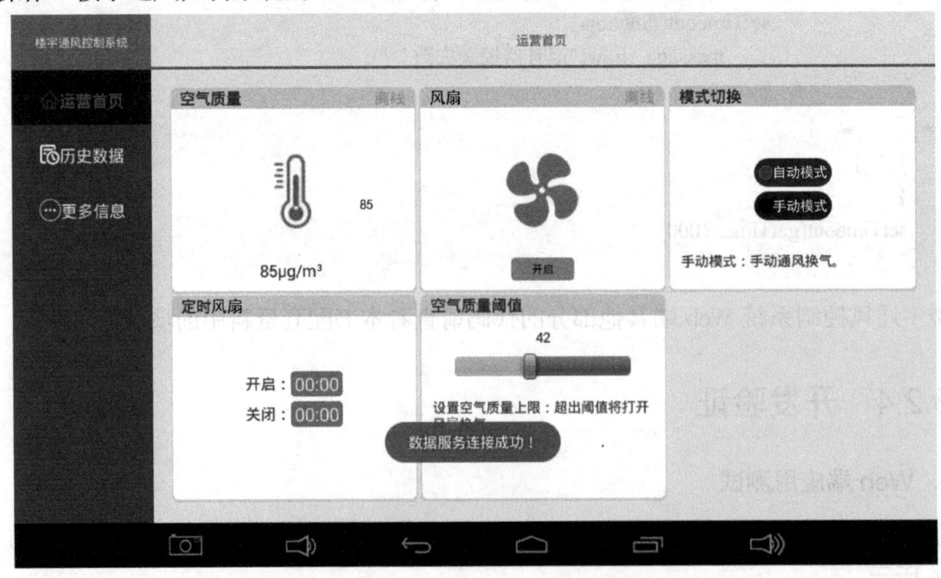

图 5.18 楼宇通风控制系统的 Android 端“运营首页”页面

5.2.5 总结与拓展

本节基于 CC3200 和 Wi-Fi 实现了空气质量传感器的数据采集和风扇的控制，通过 Android 和 HTML5 技术实现了 Android 端和 Web 端的应用设计，能根据空气质量传感器实时获取的数据来控制风扇，实现了基于 Wi-Fi 的楼宇通风控制系统。

第6章 LoRa高级应用开发

LoRa 是一种基于扩频技术的长距离无线通信技术，采用线性调频扩频调制技术，在保持低功耗的同时，明显增加了通信距离，消除了干扰，即使使用相同频率也不会产生相互干扰，使得其非常适合低功耗、远距离、大量连接以及定位跟踪等的物联网应用。更详尽的 LoRa 内容请参考《物联网长距离无线通信技术应用与开发》。

本章通过基于 LoRa 的农业土壤调节系统和农业光照度调节系统这两个贴近生活的开发案例，详细地介绍了基于 LoRa 的物联网系统架构和软硬件开发，实现了采集类传感器、控制类传感器和安防类传感器等驱动程序的开发，进行了 Android 端和 Web 端的应用开发。

6.1 基于 LoRa 的农业土壤调节系统

传统农业生态系统的物质能量主要依靠太阳能以及天然有机物质和矿物质的再利用，除了人的劳动，几乎没有其他补充投入，因而生产力极低。农业土壤调节系统是一个多传感器的采集、反馈与控制系统，通过温湿度传感器对土壤水分温度进行监测，并将监测到的数据上传到智云平台，通过客户端可以浏览这些信息，从而实时监测土壤的水分温度，判断土壤含水量。当含水量低于设定值时打开水泵对土壤进行补水，当含水量达到设定值时关闭水泵，停止补水。农业土壤调节系统如图 6.1 所示。

图 6.1　农业土壤调节系统

6.1.1 系统开发目标

（1）熟悉温湿度传感器和继电器等硬件原理和数据通信协议，实现基于 STM32F103 的温湿度传感器和继电器的驱动程序开发，通过对温湿度传感器的数据采集和继电器的控制，实现农业土壤调节系统的设计。

（2）实现农业土壤调节系统的 Android 端应用开发和 Web 端应用开发。

6.1.2 系统设计分析

1．系统的功能设计

从系统的功能角度出发，农业土壤调节系统可以分为两个模块：设备采集和控制模块以及系统设置模块，如图 6.2 所示。

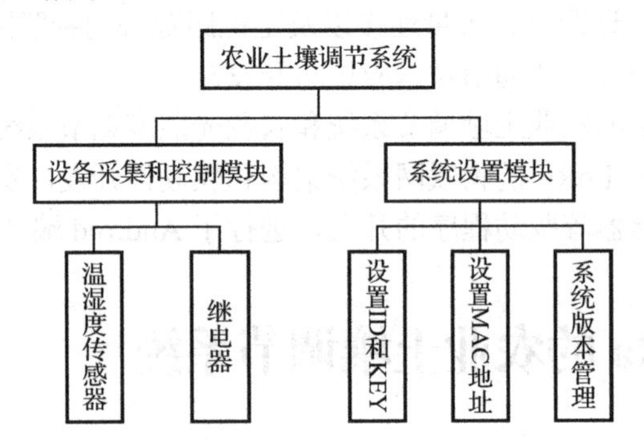

图 6.2　农业土壤调节系统的组成模块

农业土壤调节系统的功能需求如表 6.1 所示。

表 6.1　农业土壤调节系统的功能需求

功　能	功 能 说 明
采集数据显示	在上层应用页面中实时更新显示温湿度传感器采集的数据
水泵实时控制	通过上层应用程序，对水泵进行控制
模式设置	自动模式：通过设置温、湿度的阈值来控制水泵。手动模式：通过页面控制水泵
智云连接设置	设置智云服务器的参数和设备的 MAC 地址

2．系统的总体架构设计

农业土壤调节系统是基于物联网四层架构模型来设计的，其总体架构如图 6.3 所示。

感知层：通过采集类传感器（如温湿度传感器）采集温、湿度信息，控制类传感器（如继电器）由 STM32F103 微处理器控制。

网络层：感知层节点和智能网关之间的通信是通过 LoRa 来实现的，智能网关和智云服务器、上层应用设备间是通过局域网（互联网）进行数据传输的。

平台层：平台层提供物联网设备之间的基于互联网的存储、访问、控制。

应用层：应用层主要是物联网系统的人机交互接口，通过 Web 端和 Android 端提供页面友好、操作交互性强的应用。

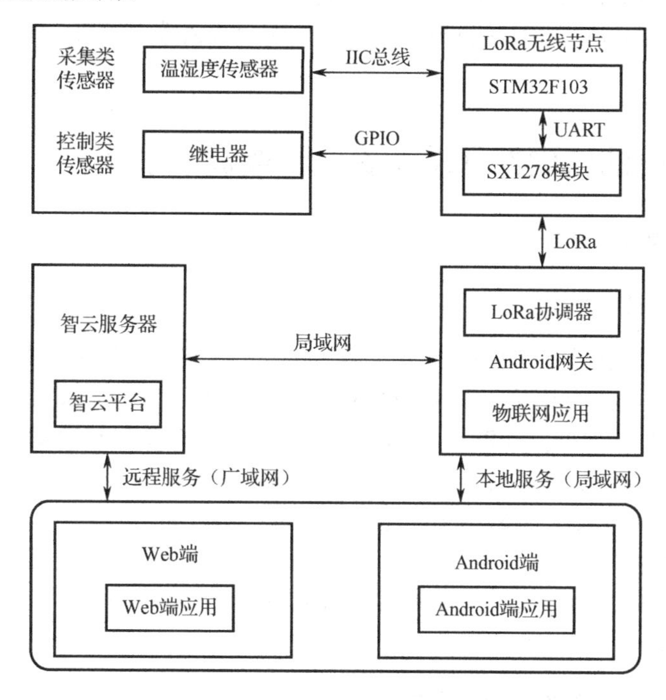

图 6.3　农业土壤调节系统的总体架构

3. 系统的数据传输

农业土壤调节系统的数据传输是在传感器节点、智能网关以及客户端（包括 Web 端和 Android 端）之间进行的，如图 6.4 所示。

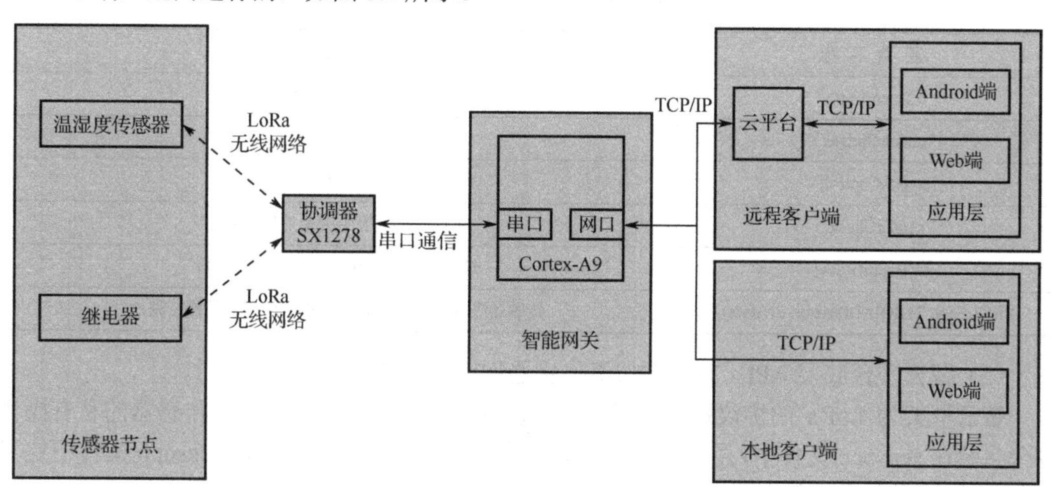

图 6.4　农业土壤调节系统的数据传输

（1）传感器节点通过 LoRa 无线网络与智能网关进行组网，协调器通过串口与智能网关进行数据通信。

（2）传感器采集的数据通过 LoRa 无线网络发送给协调器，协调器通过串口将数据转发给智能网关，然后将数据推送给所有连接到智能网关的客户端。

（3）客户端（Android 端和 Web 端）应用通过调用智云数据接口，实现实时数据采集等功能。

6.1.3 系统的软硬件开发：农业土壤调节系统

1. 系统底层软硬件设计

1）感知层硬件设计

农业土壤调节系统感知层的硬件主要包括 xLab 未来开发平台的智能网关、增强型无线节点 ZXBeePlusB、采集类开发平台 Sensor-A 和控制类开发平台 Sensor-B。其中，智能网关负责汇集传感器采集的数据；LoRa 无线节点（由增强型无线节点 ZXBeePlusB 实现）通过无线通信的方式向智能网关发送传感器采集的数据，接收智能网关发送的命令；采集类开发平台 Sensor-A 和控制类开发平台 Sensor-B 连接到 LoRa 无线节点，由其中的 STM32F103 微处理器对相关设备进行控制。本系统用到的传感器有温湿度传感器和继电器，温湿度传感器的硬件接口电路如图 3.5 所示，继电器的硬件接口电路如图 4.18 所示。

2）系统底层开发

本系统使用 LoRa 无线网络进行开发。

（1）LoRa 智云开发框架。智云框架是在传感器应用程序接口和 SAPI 框架的基础上搭建起来的，通过合理调用这些接口可以使 LoRa 的开发形成一套系统的开发逻辑。传感器应用程序接口函数是在 sensor.c 文件中实现的，如表 6.2 所示。

表 6.2　传感器应用程序接口函数

函数名称	函数说明
sensorInit()	传感器初始化
sensorUpdate()	传感器数据定时上报
sensorControl()	传感器控制函数
sensorCheck()	传感器预警监测及处理函数
ZXBeeInfRecv()	解析接收到的传感器控制命令函数
PROCESS_THREAD(sensor, ev, data)	传感器进程（处理传感器上报、传感器预警监测）

（2）智云平台底层 API。智云框架下传感器程序执行流程如图 6.5 所示。

智云框架为 LoRa 的协议栈的上层应用提供分层的软件设计结构，将传感器的私有操作部分封装在 sensor.c 文件中，用户任务中的处理事件和节点类型选择是在 sensor.h 文件中定义的。sensor.h 文件中事件宏定义如下：

```
#define NODE_NAME "601"
```

sensor.h 文件中声明了智云框架下的传感器应用文件 sensor.c 中的函数。传感器进程启动传感器任务及定时器任务，相关代码如下。

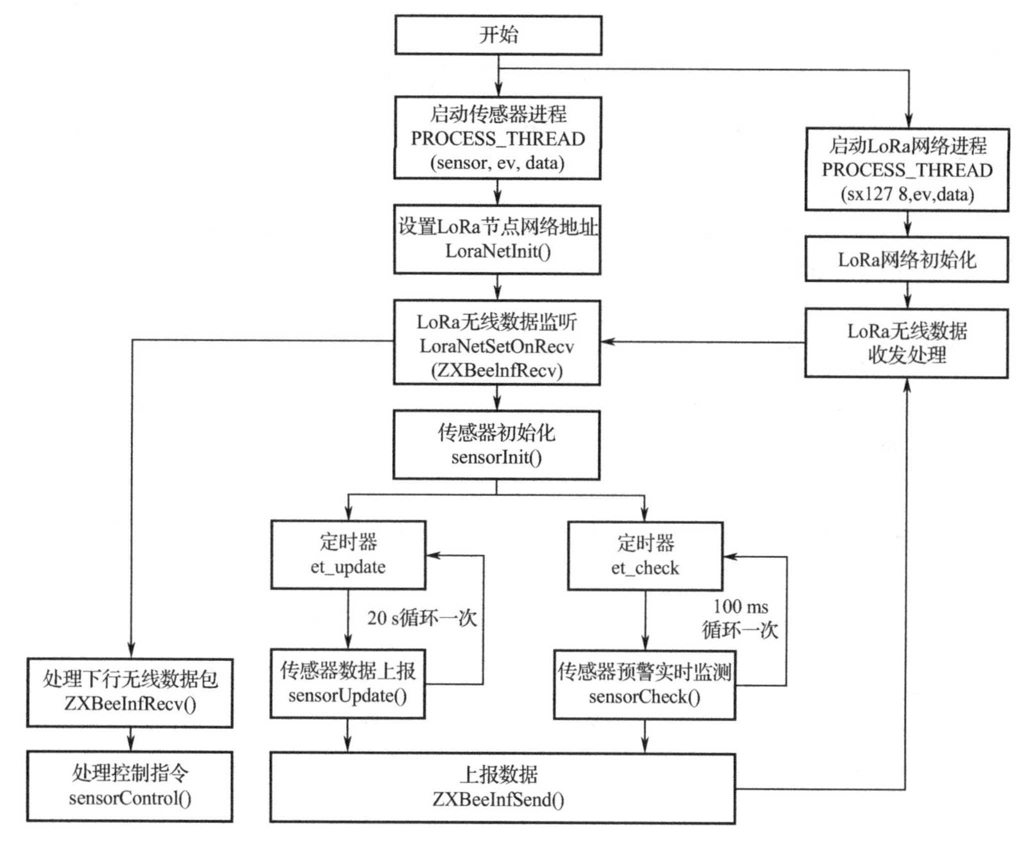

图 6.5　智云框架下传感器程序执行流程

```
/*********************************************************************************************
*名称：sensor()
*功能：传感器采集进程
*********************************************************************************************/
PROCESS_THREAD(sensor, ev, data)
{
    static struct etimer et_update;
    PROCESS_BEGIN();
    LoraNetInit();
    LoraNetSetOnRecv(ZXBeeInfRecv);
    sensorInit();
    etimer_set(&et_update, CLOCK_SECOND*10);
    while (1) {
        PROCESS_WAIT_EVENT_UNTIL(ev == PROCESS_EVENT_TIMER);
        if (etimer_expired(&et_update)) {
            printf("sensor->PROCESS_EVENT_TIMER: PROCESS_EVENT_TIMER trigger!\r\n");
            sensorUpdate();
            etimer_set(&et_update, CLOCK_SECOND*10);
        }
    }
    PROCESS_END();
```

```
}
```

sensorInit()函数用于对传感器进行初始化，相关代码如下：

```
/*********************************************************************************************
*名称：sensorInit()
*功能：传感器初始化
*********************************************************************************************/
void sensorInit(void)
{
    printf("sensor->sensorInit(): Sensor init!\r\n");
    …
    //继电器初始化
    relay_init();
}
```

sensorUpdate()函数用于对传感器采集的数据进行更新，并将更新后的数据打包上报，相关代码如下：

```
/*********************************************************************************************
*名称：sensorUpdate()
*功能：处理主动上报的数据
*********************************************************************************************/
void sensorUpdate(void)
{
    …
    if (pData != NULL) {
        ZXBeeInfSend(p, strlen(p));                              //将数据上传到智云平台
    }
    printf("sensor->sensorUpdate(): gas=%.1f\r\n", gas);
}
```

ZXBeeInfRecv()函数用于处理节点接收到的无线数据包，相关代码如下：

```
/*********************************************************************************************
*名称：ZXBeeInfRecv()
*功能：处理节点接收到的无线数据包
*参数：*pkg—接收到的无线数据包；len—无线数据包的长度
*********************************************************************************************/
void ZXBeeInfRecv(char *buf, int len)
{
    printf("sensor->ZXBeeInfRecv(): Receive LoRa Data!\r\n");
    ptag = buf;
    p = strchr(buf, '=');
    if (p != NULL) {
        *p++ = 0;
        pval = p;
    }
    val = atoi(pval);
    //控制命令解析
```

```
        if (0 == strcmp("cmd", ptag)){                    //对 D0 的位进行操作，CD0 表示位清零操作
            sensorControl(val);
        }
    }
```

sensorControl()函数用于控制传感器，相关代码如下：

```
/*********************************************************************************************
*名称：sensorControl()
*功能：传感器控制
*参数：cmd—控制命令
*********************************************************************************************/
void sensorControl(uint8_t cmd)
{
    //根据 cmd 参数执行对应的控制程序
}
```

通过在 sensor.c 文件中实现具体函数即可快速地完成 LoRa 项目开发。

3）感知层传感器驱动设计

农业土壤调节系统感知层使用采集类传感器和控制类传感器。采集类传感器主要用于实时采集并上报数据，控制类传感器主要用于对设备进行控制并反馈控制结果。

（1）数据通信协议的定义。本系统主要使用的是采集类开发平台 Sensor-A 和控制类开发平台 Sensor-B，其 ZXBee 数据通信协议如表 6.3 所示。

表 6.3　采集类开发平台和控制类开发平台的 ZXBee 数据通信协议

开发平台	属　性	参　数	权限	说　明
Sensor-A（601）	温度	A0	R	温度值，浮点型数据，精度为 0.1，范围为-40.0～105.0
	湿度	A1	R	湿度值，浮点型数据，精度为 0.1，范围为 0～100
	上报状态	D0（OD0/CD0）	R/W	D0 的 Bit0～Bit7 分别代表 A0～A7 传感器数据的上报
	数据上报时间间隔	V0	R/W	A0～A7 传感器数据的循环上报时间间隔
Sensor-B（602）	上报状态	D0（OD0/CD0）	R/W	D0 的 Bit0～Bit7 分别代表 A0～A7 传感器数据的上报
	继电器状态	D1（OD1/CD1）	R/W	D1 的 Bit6、Bit7 分别代表继电器 K1、K2 的状态，0 表示断开，1 表示吸合
	数据上报时间间隔	V0	R/W	A0～A7 传感器数据的循环上报时间间隔

（2）驱动程序的开发。在智云框架下不仅可以很容易地实现传感器驱动程序的开发，还可以省略无线传感器节点的组网和用户任务的创建等烦琐过程。例如，调用 sensorInit()函数可以实现传感器的初始化，调用 sensorUpdate()函数可以实现传感器数据的更新并打包上报。

在 sensor.c 中，需要在 sensorInit()函数中添加传感器初始化的内容，通过定义上报事件和报警事件来实现设备工作状态的定时反馈，部分代码如下：

```
/*********************************************************************************************
*名称：sensorInit()
*功能：传感器初始化
*********************************************************************************************/
```

```
void sensorInit(void)
{
    //初始化传感器代码
    htu21d_init();                                          //温湿度传感器初始化
    ……
}
```

温湿度传感器的初始化函数是 htu21d_init()，代码如下：

```
/*********************************************************************************************
*名称：htu21d_init()
*功能：HTU21D 型温湿度传感器初始化
*********************************************************************************************/
void htu21d_init(void)
{
    iic_init();                                             //IIC 总线初始化
    iic_start();                                            //开启 IIC 总线
    iic_write_byte(HTU21DADDR&0xfe);                        //写 HTU21D 型温湿度传感器的 IIC 总线地址
    iic_write_byte(0xfe);
    iic_stop();                                             //停止 IIC 总线
    delay(600);                                             //短延时
}
/*********************************************************************************************
*名称：htu21d_read_reg()
*功能：读取 HTU21D 型温湿度传感器寄存器的数据
*参数：cmd—寄存器地址
*返回：data—寄存器的数据
*********************************************************************************************/
unsigned char htu21d_read_reg(unsigned char cmd)
{
    unsigned char data = 0;
    iic_start();                                            //开启 IIC 总线
    if(iic_write_byte(HTU21DADDR & 0xfe) == 0){             //写 HTU21D 型温湿度传感器的 IIC 总线地址
        if(iic_write_byte(cmd) == 0){                       //写寄存器地址
            do{
                delay(30);                                  //延时 30 ms
                iic_start();                                //开启 IIC 总线
            }
            while(iic_write_byte(HTU21DADDR | 0x01) == 1);              //发送读信号
            data = iic_read_byte(0);                                    //读取一个字节数据
            iic_stop();                                                 //终止 IIC 总线
        }
    }
    return data;
}
/*********************************************************************************************
*名称：htu21d_get_data()
*功能：HTU21D 型温湿度传感器采集的数据
```

```
*参数：order—指令
*返回：temperature—温度值；humidity—湿度值
*********************************************************************************************/
int htu21d_get_data(unsigned char order)
{
    float temp = 0,TH = 0;
    unsigned char MSB,LSB;
    unsigned int humidity,temperature;
    iic_start();                                    //开启 IIC 总线
    if(iic_write_byte(HTU21DADDR & 0xfe) == 0){     //写 HTU21D 型温湿度传感器的 IIC 总线地址
        if(iic_write_byte(order) == 0){             //写寄存器地址
            do{
                delay(30);
                iic_start();
            }
            while(iic_write_byte(HTU21DADDR | 0x01) == 1);      //发送读信号
            MSB = iic_read_byte(0);                             //读取数据高 8 位
            delay(30);                                          //延时
            LSB = iic_read_byte(0);                             //读取数据低 8 位
            iic_read_byte(1);
            iic_stop();                                         //IIC 总线停止
            LSB &= 0xfc;                                        //取出数据有效位
            temp = MSB*256+LSB;                                 //数据合并
            if (order == 0xf3){                                 //触发开启温度监测
                TH=(175.72)*temp/65536-46.85;                   //温度：T= -46.85 + 175.72 *ST/2^16
                temperature =(unsigned int)(fabs(TH)*100);
                if(TH >= 0)
                    flag = 0;
                else
                    flag = 1;
                return temperature;
            }else{
                TH = (temp*125)/65536-6;
                humidity = (unsigned int)(fabs(TH)*100);        //湿度：RH%= -6 + 125 *SRH/2^16
                return humidity;
            }
        }
    }
    iic_stop();
    return 0;
}
```

继电器的初始化函数是 relay_init()，代码如下：

```
/*********************************************************************************************
*名称：relay_init()
*功能：继电器初始化
*********************************************************************************************/
```

```
void relay_init(void)
{
    GPIO_InitTypeDef    GPIO_InitStructure;
    RCC_APB2PeriphClockCmd(RCC_APB2Periph_GPIOA, ENABLE);
    GPIO_InitStructure.GPIO_Pin =    GPIO_Pin_5 | GPIO_Pin_4;
    GPIO_InitStructure.GPIO_Mode = GPIO_Mode_Out_PP;
    GPIO_InitStructure.GPIO_Speed = GPIO_Speed_2MHz;
    GPIO_Init(GPIOA, &GPIO_InitStructure);
    relay_control(0x00);
}
/*********************************************************************************************
*名称：relay_control()
*功能：继电器控制函数
*********************************************************************************************/
void relay_control(unsigned char cmd)
{
    if(cmd & 0x01){
        GPIO_ResetBits(GPIOA,    GPIO_Pin_5);
    } else{
        GPIO_SetBits(GPIOA,    GPIO_Pin_5);
    }
    if(cmd & 0x02){
        GPIO_ResetBits(GPIOA,    GPIO_Pin_4);
    }else{
        GPIO_SetBits(GPIOA,    GPIO_Pin_4);
    }
}
```

2. Android 端应用设计

1）Android 工程设计框架

打开 Android Studio 开发环境，可以看到农业土壤调节系统的工程目录，如图 6.6 所示，系统的工程框架如表 6.4 所示。

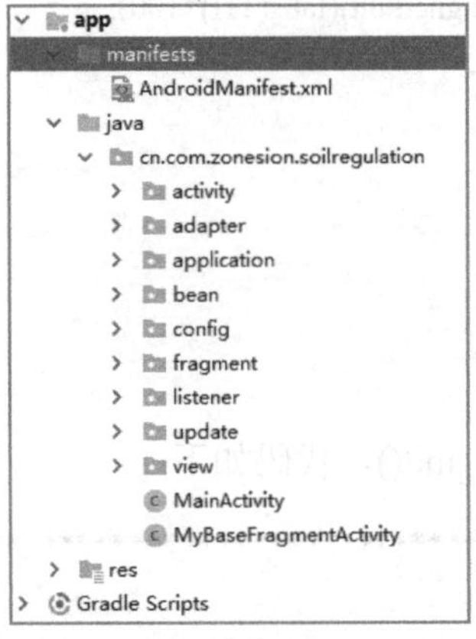

图 6.6　农业土壤调节系统的工程目录

表 6.4 农业土壤调节系统的工程框架

类 名	说 明
activity	
IdKeyShareActivity.java	在 IDKey 页面单击“分享”按钮时，可弹出 activity，用于分享二维码图片
TimePickerActivity.java	自动控制中的时间选择器
adapter	
HdArrayAdapter.java	历史数据显示适配器
application	
LCApplication.java	LCApplication 继承 application 类，使用单例模式（Singleton Pattern）创建 WSNRTConnect 对象
bean	
HistoricalData.java	历史数据的 bean 类，用于将从智云服务器获得的历史数据记录（JSON 形式）转换成该类对象
IdKeyBean.java	IdKeyBean 用来描述用户设备的 ID、KEY，以及智云服务器的地址 SERVER
config	
Config.java	config 用于修改用户的 ID、KEY，以及智云服务器的地址和 MAC 地址
fragment	
BaseFragment.java	页面基础 Fragment 定义类
HDFragment.java	历史数据页面
HistoricalDataFragment.java	历史数据显示页面
HomepageFragment.java	展示首页页面的 Fragment
IDKeyFragment.java	IDKey 选项的页面
MacSettingFragment.java	当用户设置被监测项的 MAC 地址时显示的页面
MoreInformationFragment.java	更多信息显示页面
RunHomePageFragment.java	运营首页显示页面
VersionInformationFragment.java	显示版本等相关信息的页面
listener	
IOnWSNDataListener.java	传感器数据监听器接口
update	
UpdateService.java	应用下载服务类
view	
APKVersionCodeUtils.java	获取当前本地 apk 的版本
CustomRadioButton.java	自定义按钮类
PagerSlidingTabStrip.java	自定义滑动控件类
MainActivity.java：主页面类	
MyBaseFragmentActivity.java：系统 Fragment 通信类	

2）软件设计

根据智云Android端应用程序接口的定义，系统的应用设计主要采用实时数据API接口，实时数据API接口的流程详见3.1.3节的图3.11。

（1）LCApplication.java程序代码剖析。农业土壤调节系统中的LCApplication.java程序代码和城市环境信息采集系统的LCApplication.java程序代码相同，详见3.1.3节的相关内容。

（2）HomepageFragment.java程序代码剖析。下面的代码通过(LCApplication) getActivity().getApplication()获取LCApplication类中的WSNRTConnect对象。

```
private void initViewAndBindEvent() {
    preferences = getActivity().getSharedPreferences("user_info", Context.MODE_PRIVATE);
    lcApplication = (LCApplication) getActivity().getApplication();
    wsnrtConnect = lcApplication.getWSNRConnect();
    lcApplication.registerOnWSNDataListener(this);
    editor = preferences.edit();
}
```

下面的代码通过复写onMessageArrive方法来处理节点接收到的无线数据包，实现了设备的MAC地址获取，并在当前的页面显示设备的状态。

```
@Override
    public void onMessageArrive(String mac, String tag, String val) {
        if (sensorAMac == null) {
            wsnrtConnect.sendMessage(mac, "{TYPE=?}".getBytes());
        }
        if ("TYPE".equals(tag) && "601".equals(val.substring(2, val.length()))) {
            sensorAMac = mac;
        }
        if (tag.equalsIgnoreCase("A0") && mac.equalsIgnoreCase(sensorAMac)) {
            textTemperatureState.setText("在线");
            textTemperatureState.setTextColor(getResources().getColor(R.color.line_text_color));
            temperatureText.setText(val + "℃");
            currentTemperature = Float.parseFloat(val);
        }
        if (tag.equalsIgnoreCase("A1") && mac.equalsIgnoreCase(sensorAMac)) {
            textHumidityState.setText("在线");
            textHumidityState.setTextColor(getResources().getColor(R.color.line_text_color));
            humidityText.setText(val + "%");
        }
        if (tag.equalsIgnoreCase("D1") && mac.equalsIgnoreCase(sensorAMac)) {
            textDehumidifierState.setText("在线");
            textDehumidifierState.setTextColor(getResources().getColor(R.color.line_text_color));
            int numResult = Integer.parseInt(val);
            if ((numResult & 0X40) == 0x40) {
                imageDehumidifierState.setImageDrawable(getResources().getDrawable(R.drawable.
waterpump_on));
                openOrCloseLamp.setText("关闭");
```

```
                openOrCloseLamp.setBackground(getResources().getDrawable(R.drawable.close));
            }else {
                imageDehumidifierState.setImageDrawable(getResources().getDrawable(R.drawable.
waterwump_off));
                openOrCloseLamp.setText("开启");
            openOrCloseLamp.setBackground(getResources().getDrawable(R.drawable.open));
            }
        }
    }
```

3. Web 端应用设计

1）页面功能结构分析

农业土壤调节系统的 Web 端默认显示的是“运营首页”页面，在“运营首页”页面上设计了 5 个模块，分别是土壤温度数据显示模块、土壤湿度数据显示模块、水泵控制模块、模式切换模块、湿度阈值设置模块。农业土壤调节系统 Web 端的“运营首页”页面如图 6.7 所示。

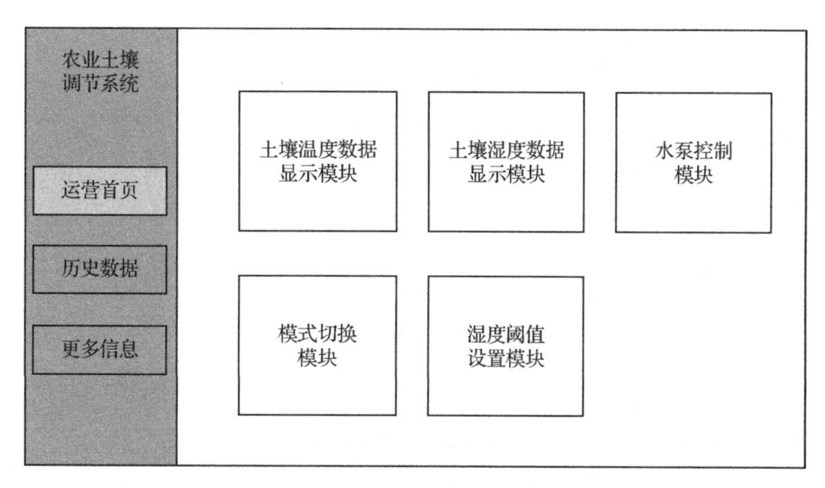

图 6.7　农业土壤调节系统 Web 端的“运营首页”页面

“历史数据”页面包括土壤温度历史数据查询模块和土壤湿度历史数据查询模块，如图 6.8 所示。

“更多信息”页面的主要功能是进行智云服务器的连接配置，和城市环境信息采集系统中的 Web 端“更多信息”页面类似，参见图 3.14。

2）软件设计

农业土壤调节系统 Web 端的 JS 开发逻辑与 Android 端的开发逻辑相似，首先通过配置 ID 和 KEY 与智云服务器进行连接，再通过实时监听数据的方法来获取相关传感器的数据并进行处理。JS 开发的部分代码如下。

在 getConnect()函数中定义了实时连接对象 rtc，连接成功回调函数是 rtc.onConnect，数据服务掉线回调函数是 rtc.onConnectLost，消息处理回调函数是 rtc.onmessageArrive。

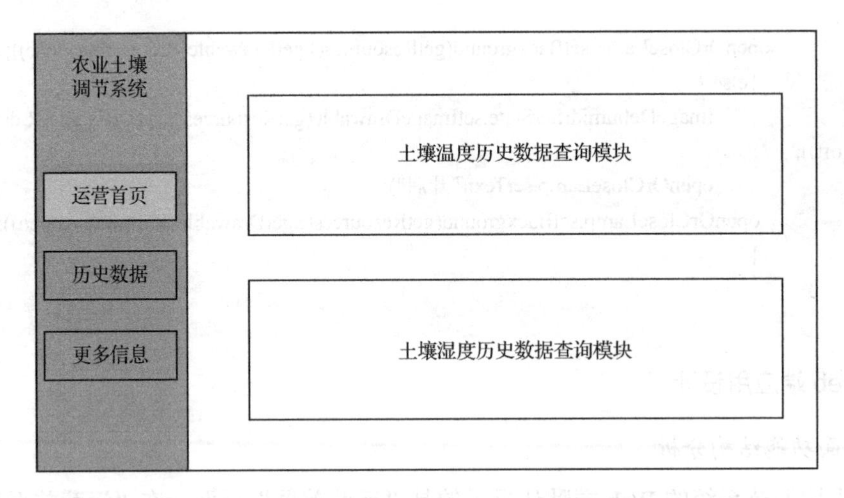

图 6.8　农业土壤调节系统 Web 端的“历史数据”页面

```
function getConnect() {
    config["id"] = config["id"] ? config["id"] : $("#id").val();
    config["key"] = config["key"] ? config["key"] : $("#key").val();
    config["server"] = config["server"] ? config["server"] : $("#server").val();
    //创建数据连接服务对象
    rtc = new WSNRTConnect(config["id"], config["key"]);
    rtc.setServerAddr(config["server"] + ":28080");
    rtc.connect();
    rtc._connect = false;
    //连接成功回调函数
    rtc.onConnect = function() {
        $("#ConnectState").text("数据服务连接成功！");
        rtc._connect = 1;
        message_show("数据服务连接成功！");
        $("#idkeyInput").text("断开").addClass("btn-danger");
        $("#id,#key,#server").attr('disabled',true);
    };
    //数据服务掉线回调函数
    rtc.onConnectLost = function() {
        rtc._connect = 0;
        $("#ConnectState").text("数据服务连接掉线！");
        $("#idkeyInput").text("连接").removeClass("btn-danger");
        message_show("数据服务连接失败，检查网络或 ID、KEY");
        $("#RFIDLink").text("离线").css("color", "#e75d59");
        $("#doorLink").text("离线").css("color", "#e75d59");
        $("#id,#key,#server").removeAttr('disabled',true);
    };
    //消息处理回调函数
    rtc.onmessageArrive = function (mac, dat) {
        //console.log(mac+" >>> "+dat);
        if (dat[0]=='{' && dat[dat.length-1]=='}') {
```

```
                dat = dat.substr(1, dat.length-2);
                var its = dat.split(',');
                for (var i=0; i<its.length; i++) {
                    var it = its[i].split('=');
                    if (it.length == 2) {
                        process_tag(mac, it[0], it[1]);
                    }
                }
                if (!mac2type[mac]) { //如果没有获取到 TYPE 值，主动去查询
                    rtc.sendMessage(mac,
"{TYPE=?,A0=?,A1=?,A2=?,A3=?,A4=?,A5=?,A6=?,A7=?,D1=?}");
                }
            }
        }
    }
```

下述JS开发代码的功能是根据设备连接情况，在页面显示设备是否在线，如果设备在线，则更新设备状态，并根据设置的阈值来控制水泵。

```
    var wsn_config = {
        "601" : {
            "online" : function() {
                $(".online_601").text("在线").css("color", "#96ba5c");
            },
            "pro" : function (tag, val) {
                if(tag == 'A0') {
                    thermometer('temp',"℃","#ff2400", -20, 80, val);
                } else if(tag == 'A1') {
                    thermometer('humi','%','#27A9E3', 0, 100, val);
                }
                if(config["curMode"]=="auto-mode" && tag=="A1" && config["mac_602"]!=""){
                    //超过上限阈值，关闭水泵
                    if(val>config["threshold"][1] && !state.pump){
                        rtc.sendMessage(config["mac_602"],"{CD1=1,D1=?}");
                        message_show("超出最大湿度阈值，将自动关闭水泵");
                    }
                    else if(val>config["threshold"][0] && state.pump){
                        rtc.sendMessage(config["mac_602"],"{OD1=1,D1=?}");
                        message_show("低于最小湿度阈值，将自动打开水泵");
                    }
                }
            }
        },
        "602" : {
            "online" : function() {
                $(".online_602").text("在线").css("color", "#96ba5c");
            },
```

```
            "pro" : function (tag, val) {
                if(tag=="D1"){
                    if(val & 0x20){
                        $("#pumpBtn").text("关闭");
                        $("#pumpStatus").attr("src", "img/WaterPump-on.png");
                        state.pump = true;
                    }else{
                        $("#pumpBtn").text("打开");
                        $("#pumpStatus").attr("src", "img/WaterPump-off.png");
                        state.pump = false;
                    }
                }
            }
        }
};
```

土壤温、湿度历史数据查询功能的实现代码如下：

```
//土壤温度历史数据
$("#airTempHistoryDisplay").click(function(){
    //初始化 API，实例化历史数据
    var myHisData = new WSNHistory(config["id"], config["key"]);
    //服务器接口查询
    myHisData.setServerAddr(config.server+":8080");
    //设置时间
    var time = $("#airTempSet").val();
    console.log(time);
    //设置数据通道
    var channel = $("#mac_601").val()+"_A0";
    console.log(channel);
    myHisData[time](channel, function(dat){
        if(dat.datapoints.length >0) {
            var data = DataAnalysis(dat);
            showChart('#her_air_temp', 'spline', '', false, eval(data));
        } else {
            message_show("该时间段没有数据");
        }
    });
});
//土壤湿度历史数据
$("#airHumiHistoryDisplay").click(function(){
    //初始化 API，实例化历史数据
    var myHisData = new WSNHistory(config["id"], config["key"]);
    //服务器接口查询
    myHisData.setServerAddr(config.server+":8080");
    //设置时间
    var time = $("#airHumiSet").val();
    console.log(time);
```

```
        //设置数据通道
        var channel = $("#mac_601").val()+"_A1";
        console.log(channel);
        myHisData[time](channel, function(dat){
            if(dat.datapoints.length >0) {
                var data = DataAnalysis(dat);
                showChart('#her_air_humi', 'spline', '', false, eval(data));
            } else {
                message_show("该时间段没有数据");
            }
        });
    });
```

农业土壤调节系统 Web 端其他部分的代码请查看本书配套资料中的项目源文件。

6.1.4 开发验证

1. Web 端应用测试

在 Web 端打开农业土壤调节系统后，在“运营首页”下可以看到 Web 端的主页，如图 6.9 所示。

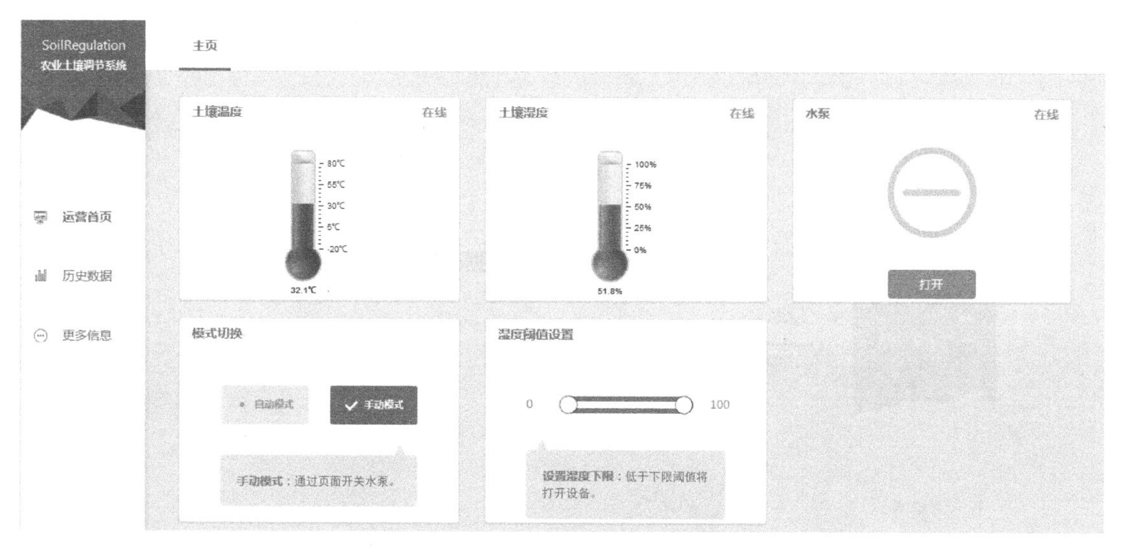

图 6.9 农业土壤调节系统的 Web 端主页

当水泵在线并且选择手动模式后，可以单击页面中“水泵”“打开”或“关闭”按钮来手动控制水泵，如图 6.10 所示。

土壤温度历史数据查询示例如图 6.11 所示。

土壤湿度历史数据查询示例如图 6.12 所示。

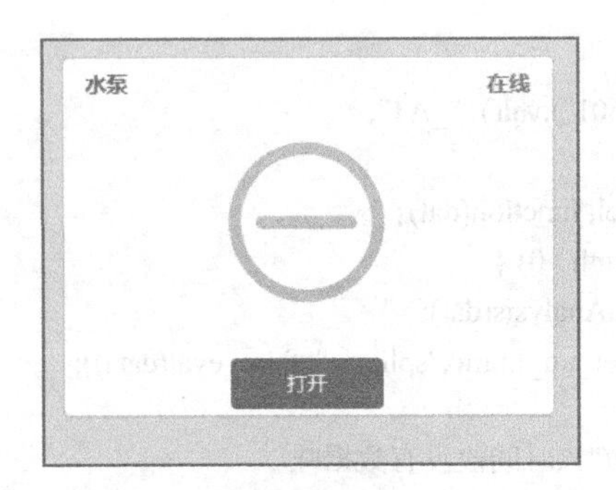

图 6.10　手动控制水泵

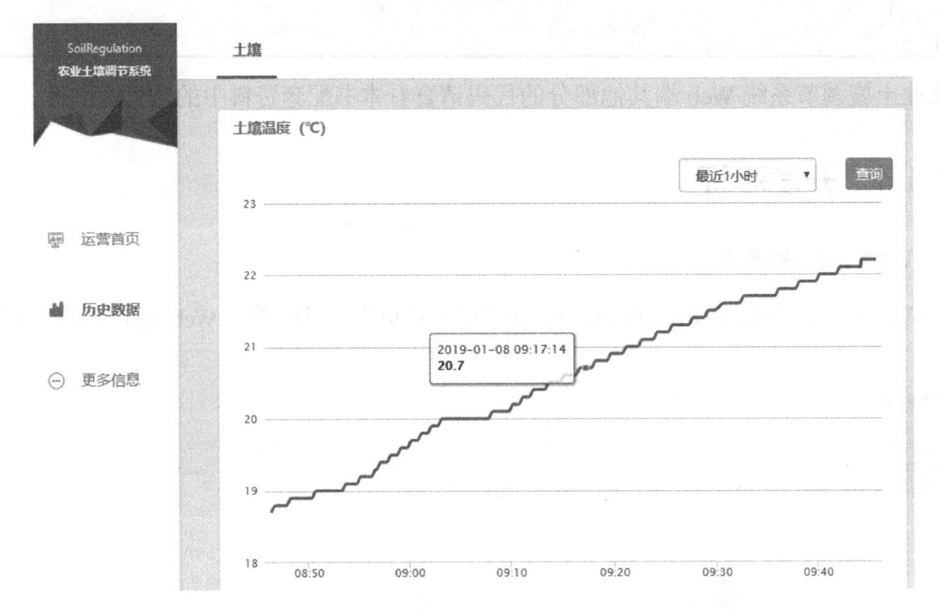

图 6.11　土壤温度历史数据查询示例

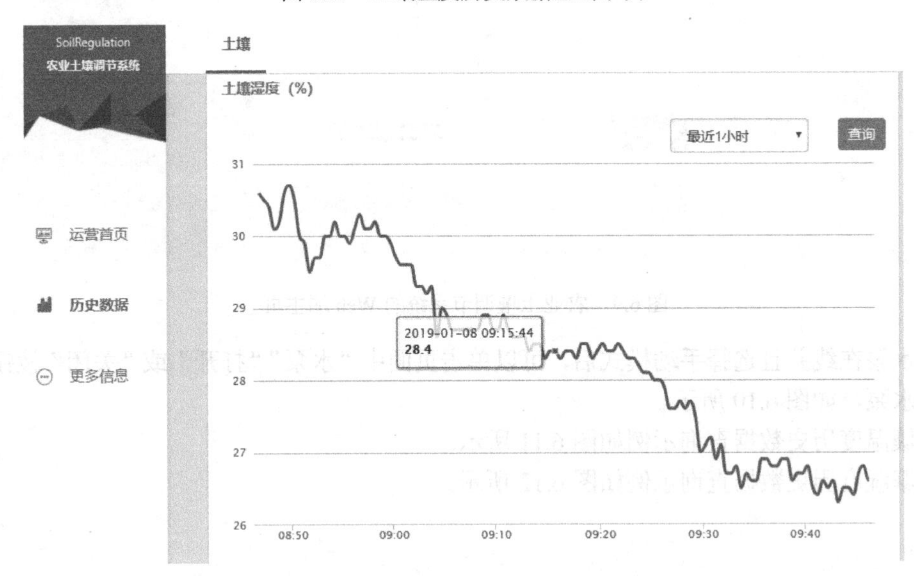

图 6.12　土壤湿度历史数据查询示例

3. Android 端应用测试

Android 端应用测试同 Web 端应用测试流程基本一致，可参考本系统的 Web 端应用测试进行操作。农业土壤调节系统 Android 端“运营首页”页面如图 6.13 所示。

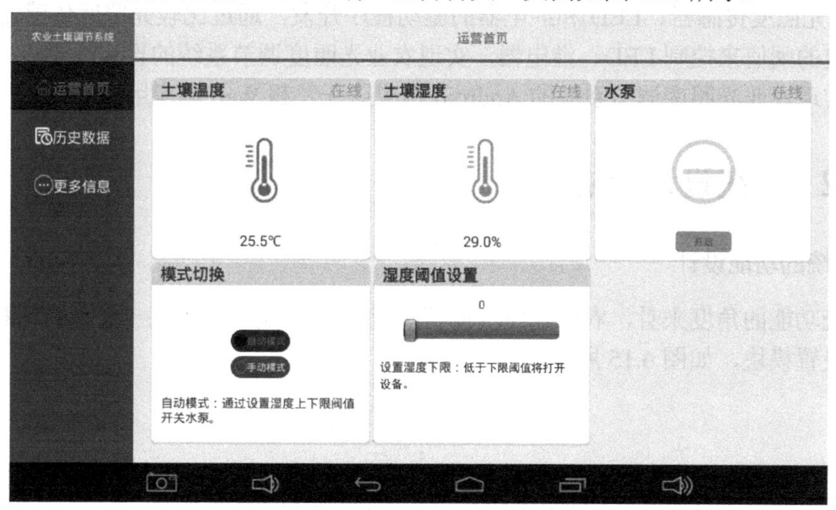

图 6.13　农业土壤调节系统 Android 端“运营首页”页面

6.1.5　总结与拓展

本节基于 LoRa 实现了温湿度传感器和继电器的驱动开发，通过 Android 和 HTML5 技术实现了 Android 端和 Web 端的应用设计，能根据传感器实时获取的数据来控制继电器，实现了基于 LoRa 的农业土壤调节系统。

6.2　基于 LoRa 的农业光照度调节系统

农业光照度调节系统（见图 6.14）是一个多传感器的采集、反馈与控制系统，光照度传感器采集到大棚内的光照度数据后将其上传到智云平台，在客户端中可以监测相关的信息。当光照度数据低于设定的阈值时，可以打开 LED 进行补光；当光照度数据高于设定的阈值时，可以打开遮阳电机，从而减少大棚的光照度。

图 6.14　农业光照度调节系统

6.2.1 系统开发目标

（1）熟悉光照度传感器、LED 和继电器等硬件原理和数据通信协议，实现基于 STM32F103 和 LoRa 的光照度传感器、LED 和继电器的驱动程序开发，通过比较光照度传感器采集到的数据和设定的阈值来控制 LED、继电器，实现农业光照度调节系统的设计。

（2）实现农业光照度调节系统的 Android 端应用开发和 Web 端应用开发。

6.2.2 系统设计分析

1. 系统的功能设计

从系统功能的角度来看，农业光照度调节系统可以分为两个模块：设备采集和控制模块以及系统设置模块，如图 6.15 所示。

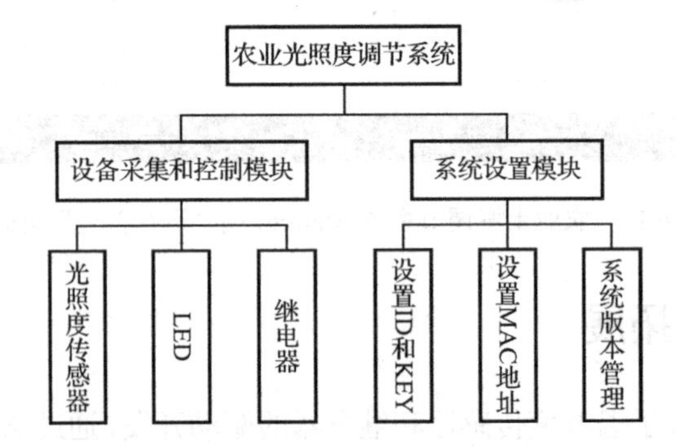

图 6.15 农业光照度调节系统的组成模块

农业光照度调节系统的功能需求如表 6.5 所示。

表 6.5 农业光照度系统的功能需求

功 能	功能说明
采集数据显示	在上层应用页面中实时更新显示光照度传感器采集的数据
遮阳帘实时控制	通过上层应用程序，对遮阳帘（遮阳电机）进行控制
补光灯实时控制	通过上层应用程序，对补光灯（LED）进行控制
模式设置	自动模式：通过设置光照度的阈值来控制遮阳帘和补光灯。手动模式：通过页面来控制遮阳帘和补光灯开关
智云连接设置	设置智云服务器的参数和设备的 MAC 地址

2. 系统的总体架构设计

农业光照度调节系统是基于物联网四层架构模型来设计的，其总体架构如图 6.16 所示。

3. 系统的数据传输

农业光照度调节系统的数据传输是在传感器节点、智能网关以及客户端（包括 Web 端和 Android 端）之间进行的，如图 6.17 所示。

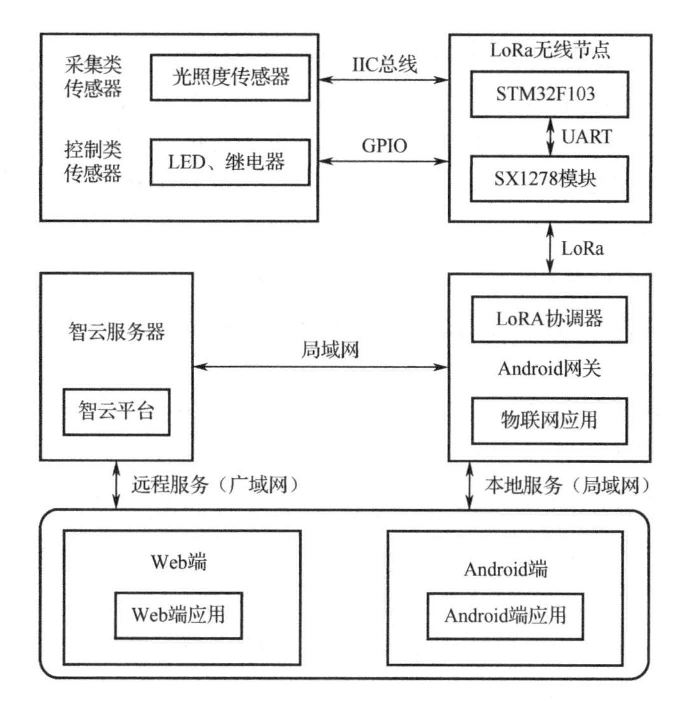

图 6.16　农业光照度调节系统的总体架构

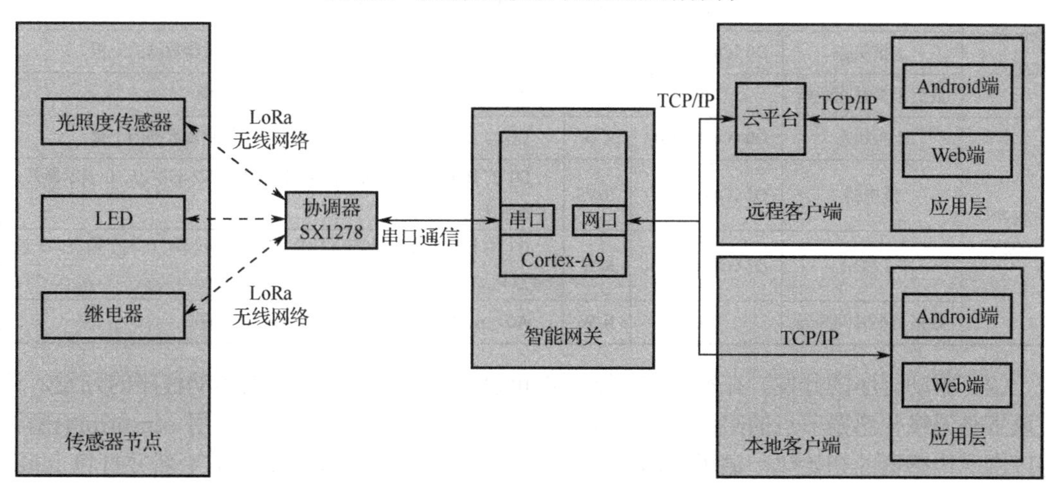

图 6.17　农业光照度调节系统的数据传输

6.2.3　系统的软硬件开发：农业光照度调节系统

1．系统底层软硬件设计

1）感知层硬件设计

农业光照度调节系统感知层的硬件包括 xLab 未来开发平台的智能网关、增强型无线节点 ZXBeePlusB、采集类开发平台 Sensor-A 和控制类开发平台 Sensor-B。其中，智能网关负责汇集传感器采集的数据；LoRa 无线节点（由增强型无线节点 ZXBeePlusB 实现）通过无线通信的方式向智能网关发送传感器数据，接收智能网关发送的命令；采集类开发平台 Sensor-A 和控制类开发平台 Sensor-B 连接到 LoRa 无线节点，由其中的 STM32F103 对相关设备进行控制。本系统使用光照度传感器、LED 和继电器，光照度传感器的硬件接口电路如图 3.6 所示，继电器的硬件接口电路如图 4.18 所示，LED 的硬件接口电路如图 3.24 所示。

2）系统底层开发

农业光照度调节系统是基于 LoRa 无线网络开发的。

（1）LoRa 智云开发框架。本系统采用的智云框架和农业土壤调节系统采用的智云框架相同，详见 6.1.3 节。

（2）智云平台底层 API。智云平台底层 API 详见 6.1.3 节。

3）感知层传感器驱动设计

农业光照度调节系统感知层的传感器主要是采集类传感器和控制类传感器。

（1）数据通信协议的定义。农业光照度调节系统使用的主要是采集类开发平台 Sensor-A 和控制类开发平台 Sensor-B，其 ZXBee 数据通信协议如表 6.6 所示。

表 6.6　采集类开发平台和控制类开发平台的 ZXBee 数据通信协议

传感器	属　性	参　数	权限	说　明
Sensor-A（601）	光照度	A2	R	光照度值，浮点型数据，精度为 0.1，范围为 0～65535
	上报状态	D0（OD0/CD0）	R/W	D0 的 Bit0～Bit7 分别代表 A0～A7 传感器数据的上报
	数据上报时间间隔	V0	R/W	A0～A7 传感器数据的循环上报时间间隔
Sensor-B（602）	上报状态	D0（OD0/CD0）	R/W	D0 的 Bit0～Bit7 分别代表 A0～A7 传感器数据的上报
	继电器	D1（OD1/CD1）	R/W	D1 的 Bit6、Bit7 分别代表继电器 K1、K2 的状态，0 表示断开，1 表示吸合
	LED	D1（OD1/CD1）	R/W	D1 的 Bit4 和 Bit5 代表 LED1 和 LED2 状态，0 表示熄灭，1 表示点亮
	数据上报时间间隔	V0	R/W	A0～A7 传感器数据的循环上报时间间隔

（2）驱动程序的开发。在智云框架下不仅可以很容易地实现传感器驱动程序的开发，还可以省略无线传感器节点的组网和用户任务的创建等烦琐过程。例如，调用 sensorInit()函数可以实现传感器的初始化，调用 sensorUpdate()函数可以实现传感器数据的更新并打包上报。

在 sensor.c 中，需要在 sensorInit()函数中添加传感器初始化的内容，通过定义上报事件和报警事件实现设备工作状态的定时反馈，部分代码如下：

```
/*********************************************************************************************
*名称：sensorInit()
*功能：传感器初始化
*********************************************************************************************/
void sensorInit(void)
{
    //初始化传感器代码
    bh1750_init();                                          //光照度传感器初始化
    ......
}
```

光照度传感器的驱动程序代码如下：

```
/*****************************************************************************************
*名称：bh1750_send_byte()
*功能：发送字节数据函数
*返回：如果返回 1 表示操作成功，否则操作有误
*****************************************************************************************/
uchar bh1750_send_byte(uchar sla,uchar c)
{
    iic_start();                                    //启动 IIC 总线
    if(iic_write_byte(sla) == 0){                   //发送器件地址
        if(iic_write_byte(c) == 0){                 //发送数据
        }
    }
    iic_stop();                                     //停止 IIC 总线
    return(1);
}
/*****************************************************************************************
*名称：bh1750_read_nbyte()
*功能：连续读出光照度传感器内部数据
*返回：应答或非应答信号
*****************************************************************************************/
uchar bh1750_read_nbyte(uchar sla,uchar *s,uchar no)
{
    uchar i;
    iic_start();                                    //起始信号
    if(iic_write_byte(sla+1) == 0){                 //发送设备地址+读信号
        for (i=0; i<no-1; i++){                     //连续读取 6 个地址数据，存储在 buf 中
            *s=iic_read_byte(0);
            s++;
        }
        *s=iic_read_byte(1);
    }
    iic_stop();
    return(1);
}
/*****************************************************************************************
*名称：bh1750_init()
*功能：光照度传感器初始化
*****************************************************************************************/
void bh1750_init()
{
    iic_init();
}
/*****************************************************************************************
*名称：bh1750_get_data()
*功能：光照度传感器数据处理函数
```

```
***********************************************************************************/
float bh1750_get_data(void)
{
    uchar *p=buf;
    bh1750_init();                                          //初始化光照度传感器
    bh1750_send_byte(0x46,0x01);                            //power on
    bh1750_send_byte(0x46,0X20);                            //高分辨率模式
    delay_ms(180);                                          //延时 180 ms
    bh1750_read_nbyte(0x46,p,2);                            //连续读出数据，存储在 buf 中
    unsigned short x = buf[0]<<8 | buf[1];
    return x/1.2;
}
```

LED 的初始化函数是 led_init()，代码如下：

```
/**********************************************************************************
*名称：led_init()
*功能：LED 控制引脚初始化
***********************************************************************************/
void led_init(void)
{
    GPIO_InitTypeDef   GPIO_InitStructure;
    RCC_APB2PeriphClockCmd(RCC_APB2Periph_GPIOA, ENABLE);  //使能 PA 端口时钟
    GPIO_InitStructure.GPIO_Pin = GPIO_Pin_7 | GPIO_Pin_6;
    GPIO_InitStructure.GPIO_Speed = GPIO_Speed_2MHz;
    GPIO_InitStructure.GPIO_Mode = GPIO_Mode_Out_PP;
    GPIO_Init(GPIOA, &GPIO_InitStructure);
    led_off(0x01);                                          //初始状态为关闭
    led_off(0x02);
}
/**********************************************************************************
*名称：led_on()
*功能：打开 LED
*参数：led—LED 号，在 led.h 中宏定义为 LED1、LED2
***********************************************************************************/
signed char led_on(unsigned char led)
{
    if(led & 0x01){                                         //打开 LED1
        GPIO_ResetBits(LED_port,LED1);
        return 0;
    }
    if(led & 0x02){                                         //打开 LED2
        GPIO_ResetBits(LED_port,LED2);
        return 0;
    }
    return -1;                                              //参数错误，返回-1
}
/**********************************************************************************
```

```
*名称：led_off()
*功能：关闭 LED
*参数：led—LED 号，在 led.h 中宏定义为 LED1、LED2
**********************************************************************************/
signed char led_off(unsigned char led)
{
    if(led &0x01){                                              //关闭 LED1
        GPIO_SetBits(LED_port,LED1);
        return 0;
    }
    if(led &0x2){                                               //关闭 LED2
        GPIO_SetBits(LED_port,LED2);
        return 0;
    }
    return -1;                                                  //参数错误，返回-1
}
```

继电器的初始化函数是 relay_init()，代码如下：

```
/*********************************************************************************
*名称：relay_init()
*功能：继电器初始化
**********************************************************************************/
void relay_init(void)
{
    GPIO_InitTypeDef    GPIO_InitStructure;
    RCC_APB2PeriphClockCmd(RCC_APB2Periph_GPIOA, ENABLE);
    GPIO_InitStructure.GPIO_Pin =    GPIO_Pin_5 | GPIO_Pin_4;
    GPIO_InitStructure.GPIO_Mode = GPIO_Mode_Out_PP;
    GPIO_InitStructure.GPIO_Speed = GPIO_Speed_2MHz;
    GPIO_Init(GPIOA, &GPIO_InitStructure);
    relay_control(0x00);
}
/*********************************************************************************
*名称：relay_control()
*功能：继电器控制函数
**********************************************************************************/
void relay_control(unsigned char cmd)
{
    if(cmd & 0x01){
        GPIO_ResetBits(GPIOA,    GPIO_Pin_5);
    }else{
        GPIO_SetBits(GPIOA,    GPIO_Pin_5);
    }
    if(cmd & 0x02){
        GPIO_ResetBits(GPIOA,    GPIO_Pin_4);
    }else{
        GPIO_SetBits(GPIOA,    GPIO_Pin_4);
```

```
    }
}
```

2. Android 端应用设计

1）Android 工程设计框架

打开 Android Studio 开发环境，可以看到农业光照度调节系统的工程目录，如图 6.18 所示，系统的工程框架如表 6.7 所示。

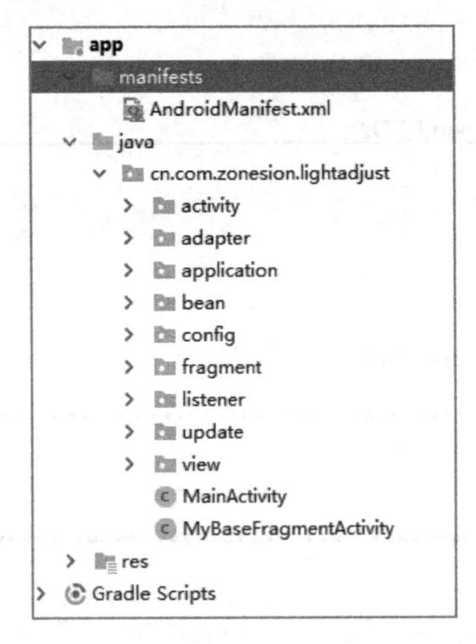

图 6.18　农业光照度调节系统的工程目录

表 6.7　农业光照度调节系统的工程框架

类　名	说　明
activity	
IdKeyShareActivity.java	在 IDKey 页面单击“分享”按钮时，可弹出 activity，用于分享二维码图片
Adapter	
HdArrayAdapter.java	历史数据显示适配器
Application	
LCApplication.java	LCApplication 继承 application 类，使用单例模式（Singleton Pattern）创建 WSNRTConnect 对象
bean	
HistoricalData.java	历史数据的 bean 类，用于将从智云服务器获得的历史数据记录（JSON 形式）转换成该类对象
IdKeyBean.java	IdKeyBean 用来描述用户设备的 ID、KEY，以及智云服务器的地址 SERVER
Config	
Config.java	config 用于修改用户的 ID、KEY，以及智云服务器的地址和 MAC 地址
Fragment	
BaseFragment.java	页面基础 Fragment 定义类

续表

类　名	说　明
HDFragment.java	历史数据页面
HistoricalDataFragment.java	历史数据显示页面
HomepageFragment.java	展示首页页面的 Fragment
IDKeyFragment.java	IDKey 选项的页面
MacSettingFragment.java	当用户设置被监测项的 MAC 地址时显示的页面
MoreInformationFragment.java	更多信息显示页面
RunHomePageFragment.java	运营首页显示页面
VersionInformationFragment.java	显示版本等相关信息的页面
listener	
IOnWSNDataListener.java	传感器数据监听器接口
Update	
UpdateService.java	应用下载服务类
View	
APKVersionCodeUtils.java	获取当前本地 apk 的版本
CustomRadioButton.java	自定义按钮类
PagerSlidingTabStrip.java	自定义滑动控件类
MainActivity.java：主页面类	
MyBaseFragmentActivity.java：系统 Fragment 通信类	

2）软件设计

根据智云 Android 端应用程序接口的定义，系统的应用设计主要采用实时数据 API 接口，实时数据 API 接口的流程见 3.1.3 节的图 3.11。

（1）LCApplication.java 程序代码剖析。农业光照度调节系统中的 LCApplication.java 程序代码和城市环境信息采集系统的 LCApplication.java 程序代码相同，详见 3.1.3 节的相关内容。

（2）HomepageFragment.java 程序代码剖析。下面的代码通过(LCApplication) getActivity().getApplication()获取 LCApplication 类中的 WSNRTConnect 对象。

```
private void initViewAndBindEvent() {
    preferences = getActivity().getSharedPreferences("user_info", Context.MODE_PRIVATE);
    lcApplication = (LCApplication) getActivity().getApplication();
    wsnrtConnect = lcApplication.getWSNRConnect();
    lcApplication.registerOnWSNDataListener(this);
    editor = preferences.edit();
}
```

下面的代码通过复写 onMessageArrive 方法来处理节点接收到的无线数据包，实现了设备的 MAC 地址获取，并在当前的页面显示设备的状态。

```
@Override
    public void onMessageArrive(String mac, String tag, String val) {
        if (sensorAMac == null && sensorBMac == null) {
            wsnrtConnect.sendMessage(mac, "{TYPE=?}".getBytes());
        }
        if ("TYPE".equals(tag) && "601".equals(val.substring(2, val.length()))) {
            sensorAMac = mac;
        }
        if ("TYPE".equals(tag) && "602".equals(val.substring(2, val.length()))) {
            sensorBMac = mac;
        }
        if (tag.equalsIgnoreCase("A2") && mac.equalsIgnoreCase(sensorAMac)) {
            textIlluminationState.setText("在线");
            textIlluminationState.setTextColor(getResources().getColor(R.color.line_text_color));
            currentTemperature = Float.parseFloat(val);
            dialChartIlluminationView.setCurrentStatus(currentTemperature);
            dialChartIlluminationView.invalidate();
        }
        if (tag.equalsIgnoreCase("D1") && mac.equalsIgnoreCase(sensorBMac)) {
            textLightState.setText("在线");
            textLightState.setTextColor(getResources().getColor(R.color.line_text_color));
            textSunshadeState.setText("在线");
            textSunshadeState.setTextColor(getResources().getColor(R.color.line_text_color));
            int numResult = Integer.parseInt(val);
            if ((numResult & 0X10) == 0x10) {
                imageLightState.setImageDrawable(getResources().getDrawable(R.drawable.
open_lamp));
                openOrCloseLight.setText("关闭");
                openOrCloseLight.setBackground(getResources().getDrawable(R.drawable.close));
            }else {
                imageLightState.setImageDrawable(getResources().getDrawable(R.drawable.
close_lamp));
                openOrCloseLight.setText("开启");
                openOrCloseLight.setBackground(getResources().getDrawable(R.drawable.open));
            }
            numResult1 = Integer.parseInt(val);
            if ((numResult1 & 0X04) == 0x04) {
                imageSunshadeState.setImageDrawable(getResources().getDrawable(R.drawable.
curtains_on));
                openOrCloseSunshade.setText("关闭");
                openOrCloseSunshade.setBackground(getResources().getDrawable(R.drawable.close));
            }else {
                imageSunshadeState.setImageDrawable(getResources().getDrawable(R.drawable.
curtains_off));
                openOrCloseSunshade.setText("开启");
                openOrCloseSunshade.setBackground(getResources().getDrawable(R.drawable.open));
            }
```

```
        }
    }
```

3. Web端应用设计

1）页面功能结构分析

农业光照度调节系统的Web端默认显示的是“运营首页”页面，在“运营首页”页面上设计了5个模块，分别是光照度数据显示模块、补光灯控制模块、遮阳帘控制模块、模式切换模块、光照度阈值设备模块。农业光照度调节系统Web端的“运营首页”页面如图6.19所示。

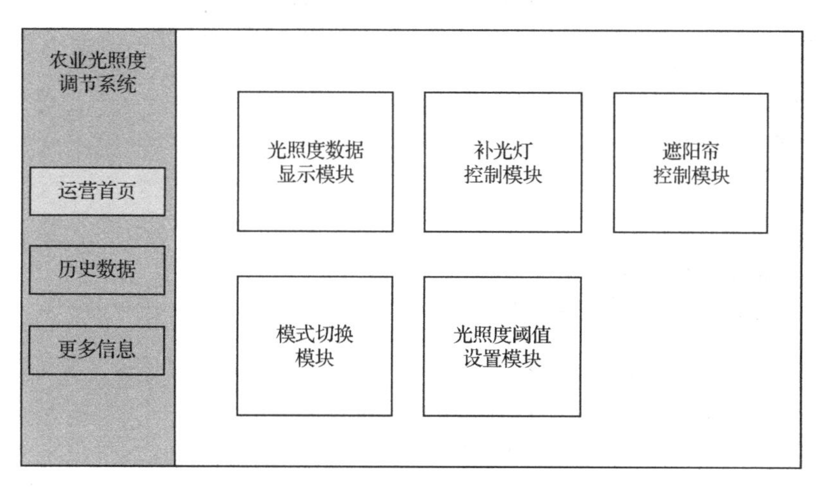

图6.19　农业光照度调节系统Web端的“运营首页”页面

“历史数据”页面可以查询显示光照度的历史数据，如图6.20所示。

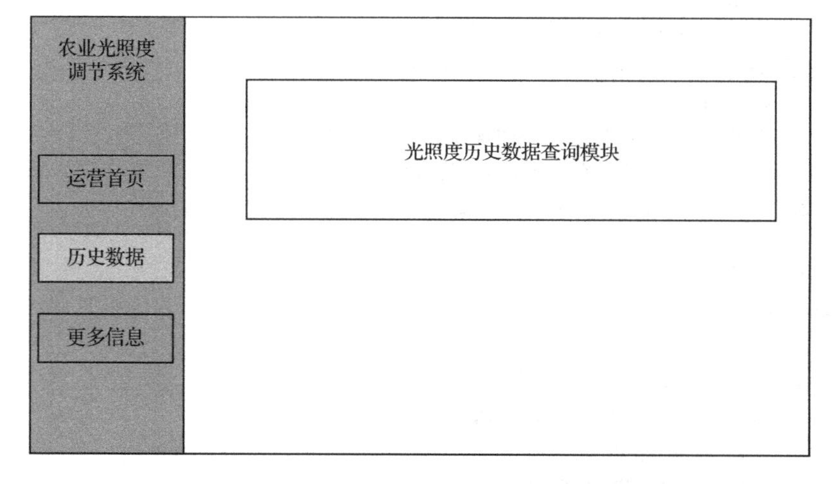

图6.20　农业光照度调节系统Web端的“历史数据”页面

“更多信息”页面的主要功能是进行智云服务器的连接配置，和城市环境信息采集系统中的Web端“更多信息”页面类似，参见图3.14。

2）软件设计

农业光照度调节系统 Web 端的 JS 开发逻辑与 Android 端的开发逻辑相似，首先通过配置 ID 和 KEY 与智云服务器进行连接，再通过实时监听数据的方法来获取相关传感器的数据并进行处理。JS 开发的部分代码如下。

在 getConnect()函数中定义了实时连接对象 rtc，连接成功回调函数是 rtc.onConnect，数据服务掉线回调函数是 rtc.onConnectLost，消息处理回调函数是 rtc.onmessageArrive。

```
function getConnect() {
    config["id"] = config["id"] ? config["id"] : $("#id").val();
    config["key"] = config["key"] ? config["key"] : $("#key").val();
    config["server"] = config["server"] ? config["server"] : $("#server").val();
    //创建数据连接服务对象
    rtc = new WSNRTConnect(config["id"], config["key"]);
    rtc.setServerAddr(config["server"] + ":28080");
    rtc.connect();
    rtc._connect = false;
    //连接成功回调函数
    rtc.onConnect = function() {
        $("#ConnectState").text("数据服务连接成功！");
        rtc._connect = 1;
        message_show("数据服务连接成功！");
        $("#idkeyInput").text("断开").addClass("btn-danger");
        $("#id,#key,#server").attr('disabled',true);
    };
    //数据服务掉线回调函数
    rtc.onConnectLost = function() {
        rtc._connect = 0;
        $("#ConnectState").text("数据服务连接掉线！");
        $("#idkeyInput").text("连接").removeClass("btn-danger");
        message_show("数据服务连接失败，检查网络或 ID、KEY");
        $("#RFIDLink").text("离线").css("color", "#e75d59");
        $("#doorLink").text("离线").css("color", "#e75d59");
        $("#id,#key,#server").removeAttr('disabled',true);
    };
    //消息处理回调函数
    rtc.onmessageArrive = function (mac, dat) {
        //console.log(mac+" >>> "+dat);
        if (dat[0]=='{' && dat[dat.length-1]=='}') {
            dat = dat.substr(1, dat.length-2);
            var its = dat.split(',');
            for (var i=0; i<its.length; i++) {
                var it = its[i].split('=');
                if (it.length == 2) {
                    process_tag(mac, it[0], it[1]);
                }
            }
```

```
                if (!mac2type[mac]) { //如果没有获取到 TYPE 值，主动去查询
                    rtc.sendMessage(mac, "{TYPE=?,A0=?,A1=?,A2=?,A3=?,A4=?,A5=?,A6=?,A7=?,D1=?}");
                }
            }
        }
    }
```

下述 JS 开发代码的功能是根据设备连接情况，在页面上更新设备的状态，当设备在线后，可以显示补光灯与遮阳帘的开关状态，并根据当前设置的光照度阈值，开关联动设备。

```
var wsn_config = {
    "601" : {
        "online" : function() {
            $(".online_601").text("在线").css("color", "#96ba5c");
        },
        "pro" : function (tag, val) {
            if(tag=="A2"){
                dial('illum','Lx', val);
                if(config["curMode"]=="auto-mode" && config["mac_602"]!=""){
                    //当超过光照度上限阈值时，关闭补光灯，打开遮阳帘
                    if(val>config["threshold"][1]){
                        if(!state.light){
                            console.log("{CD1=16,D1=?}");
                            rtc.sendMessage(config["mac_602"], "{CD1=16,D1=?}");
                        }
                        if(!state.curtain){
                            rtc.sendMessage(config["mac_602"], sensorBcmd.curtainsStatus.on);
                        }
                        message_show("超出最大光照度阈值，将自动关闭补光灯，打开遮阳帘");
                    }
                    //当低于光照度下限阈值时，打开补光灯，关闭遮阳帘
                    else if(val<config["threshold"][0]){
                        if(state.light){
                            rtc.sendMessage(config["mac_602"], "{OD1=16,D1=?}");
                        }
                        if(!state.curtain){
                            console.log(sensorBcmd.curtainsStatus.off);
                            rtc.sendMessage(config["mac_602"], sensorBcmd.curtainsStatus.off);
                        }
                        message_show("低于最小光照度阈值，将自动打开补光灯，关闭遮阳帘");
                    }
                }
            }
        }
    },
    "602" : {
        "online" : function() {
```

```
                $(".online_602").text("在线").css("color", "#96ba5c");
            },
            "pro" : function (tag, val) {
                if(tag=="D1"){
                    if(val & 16) {
                        $("#LEDStatus").text("关闭");
                        $("#LEDImg").attr("src", "img/LED-on.png");
                        state.pump = true;
                    }else{
                        $("#LEDStatus").text("打开");
                        $("#LEDImg").attr("src", "img/LED-off.png");
                    }
                    if(val & 4){
                        $("#curtainsStatus").text("关闭");
                        $("#curtainsImg").attr("src", "img/curtains-on.png");
                        state.pump = true;
                    }else{
                        $("#curtainsStatus").text("打开");
                        $("#curtainsImg").attr("src", "img/curtains-off.png");
                        state.pump = false;
                    }
                }
            }
        }
}
```

下面的代码通过智云平台的历史数据接口 WSNHistory 来查询并显示光照度历史数据。

```
//光照度历史数据
$("#airTempHistoryDisplay").click(function(){
    //初始化 API，实例化历史数据
    var myHisData = new WSNHistory(config["id"], config["key"]);
    //服务器接口查询
    myHisData.setServerAddr(config.server+":8080");
    //设置时间
    var time = $("#airTempSet").val();
    console.log(time);
    //设置数据通道
    var channel = $('#mac_601').val()+"_A2";
    console.log(channel);
    myHisData[time](channel, function(dat){
        if(dat.datapoints.length >0) {
            var data = DataAnalysis(dat);
            showChart('#her_air_temp', 'spline', '', false, eval(data));
        } else {
            message_show("该时间段没有数据");
        }
    });
});
```

农业光照度调节系统 Web 端其他部分的代码请查看本书配套资料中的项目源文件。

6.2.4　开发验证

1. Web 端应用测试

在 Web 端打开农业光照度调节系统后，在“运营首页”下可以看到 Web 端的主页，如图 6.21 所示。

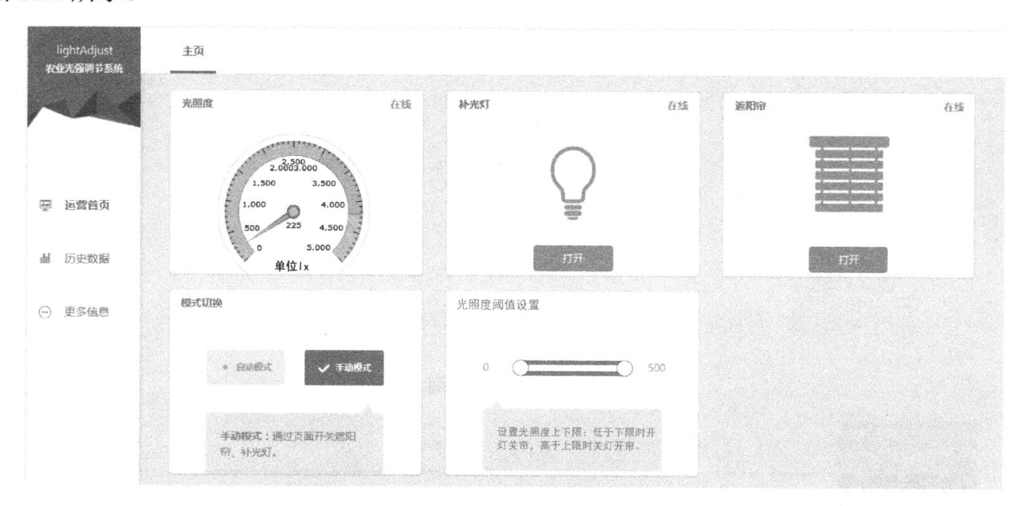

图 6.21　农业光照度调节系统的 Web 端主页

当设备（如补光灯、遮阳帘）在线并且选择手动模式后，可以通过单击“打开”或“关闭”按钮来手动地控制设备，如图 6.22 所示。

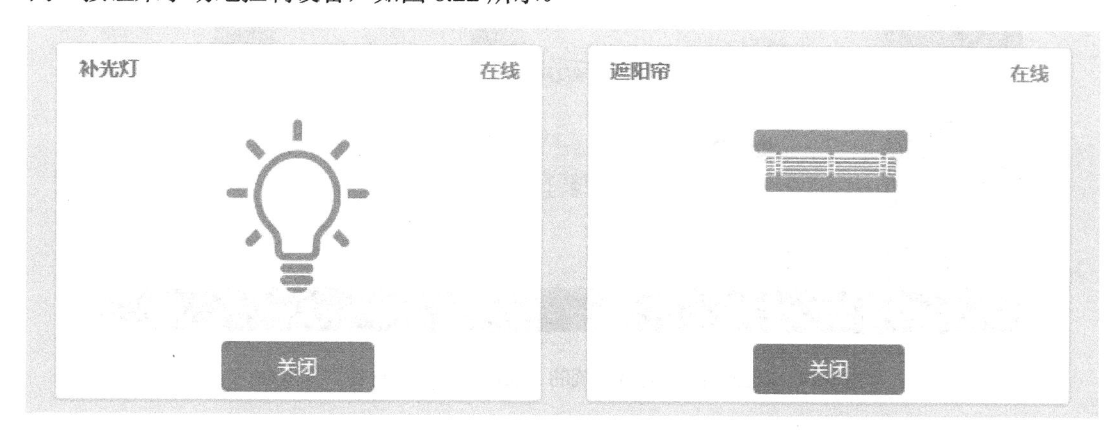

图 6.22　手动控制设备

进入“历史数据”页面，选择要查询的时间段后，单击“查询”按钮可以看到要查询的历史数据，如图 6.23 所示。

2. Android 端应用测试

Android 端应用测试同 Web 端应用测试流程基本一致，可参考本系统的 Web 端应用测试进行操作。农业光照度调节系统的 Android 端“运营首页”页面如图 6.24 所示。

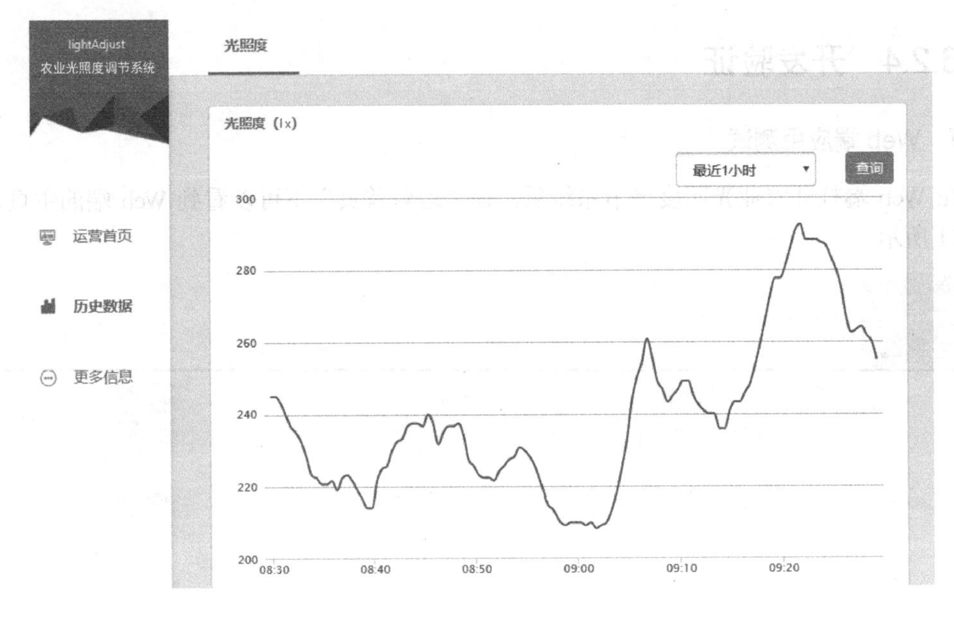

图 6.23　历史数据查询示例

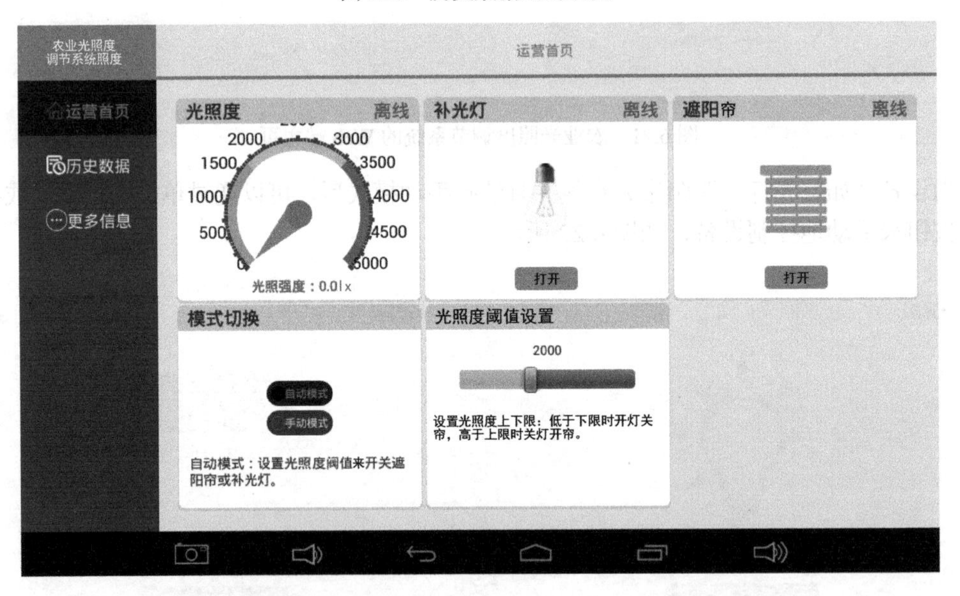

图 6.24　农业光照度调节系统的 Android 端“运营首页”页面

6.2.5　总结与拓展

本节基于 LoRa 实现了光照度传感器的数据采集以及对 LED 和继电器的控制，通过 Android 和 HTML5 技术实现了 Android 端和 Web 端的应用设计，能根据实时获取光照度传感器数据来控制 LED、继电器，实现了基于 LoRa 的农业光照度调节系统。

第7章 NB-IoT高级应用开发

窄带物联网（Narrow Band Internet of Things，NB-IoT）是在NB-CIOT和NB-LTE的基础上发展起来的。NB-IoT具有强连接、覆盖广、低功耗和低成本等特点，可支持大量新型物联网设备，广泛应用在交通行业、物流行业、卫生医疗、商品零售行业、智能抄表、公共设施、智能家居、智能农业、工业制造、企业能耗管理、企业安全防护等行业领域。有关更详尽的NB-IoT内容请参考《物联网长距离无线通信技术应用与开发》。

本章通过基于 NB-IoT 的停车收费管理系统和智能水表抄表系统这两个贴近生活的开发案例，详细地介绍了NB-IoT物联网系统的架构和软硬件开发，实现了采集类传感器、控制类传感器、安防类传感器和识别类传感器的驱动程序，进行了Android端和Web端的应用开发。

7.1 基于NB-IoT的停车收费管理系统

停车收费管理系统（见图7.1）是针对传统停车收费管理的局限性而研发出的一种基于视频监控与车牌自动识别技术的停车收费管理系统，可实现停车自助缴费、智能计时和多种交费方式等功能。

图7.1 停车收费管理系统

7.1.1 系统开发目标

（1）熟悉射频传感器的硬件原理和数据通信协议，基于STM32F103实现射频传感器的驱动程序开发，通过射频传感器进行信息采集，实现停车收费管理系统的设计。

（2）实现停车收费管理系统的Android端应用开发和Web端应用开发。

7.1.2 系统设计分析

1. 系统的功能设计

从系统功能的角度出发，停车收费管理系统可分为两个模块：车位和车辆管理模块以及系统设置模块，如图 7.2 所示。

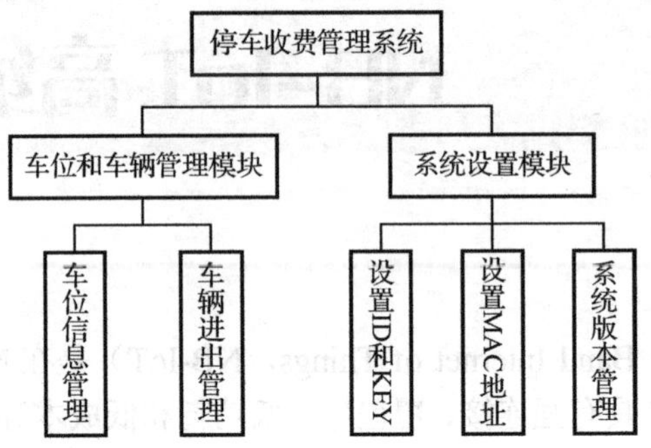

图 7.2 停车收费管理系统的组成模块

停车收费管理系统的功能需求如表 7.1 所示。

表 7.1 停车收费管理系统的功能需求

功 能	功 能 说 明
停车卡信息采集	在上层应用页面中实时更新显示车位的信息以及车辆出入场计费信息
智云连接设置	设置智云服务器的参数和设备的 MAC 地址

2. 系统的总体架构设计

停车收费管理系统是基于物联网四层架构模型来设计的，其总体架构如图 7.3 所示。

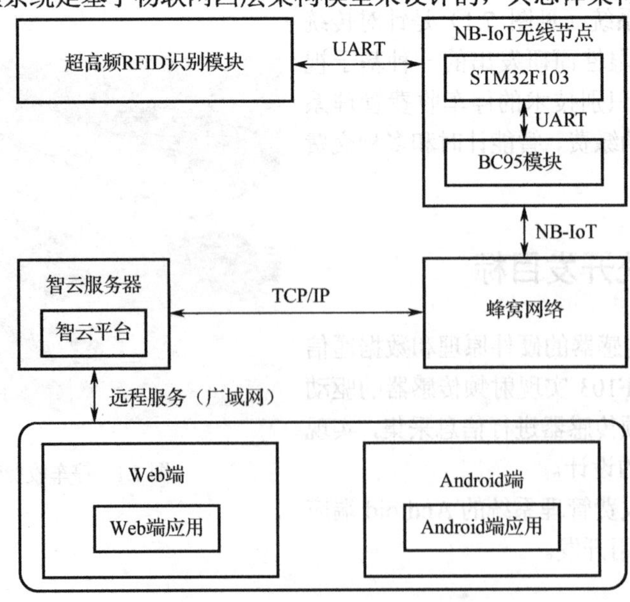

图 7.3 停车收费管理系统的总体架构

3. 系统的数据传输

停车收费管理系统的数据传输是在传感器节点、智云平台以及客户端（包括 Web 端和 Android 端）之间进行的，如图 7.4 所示。

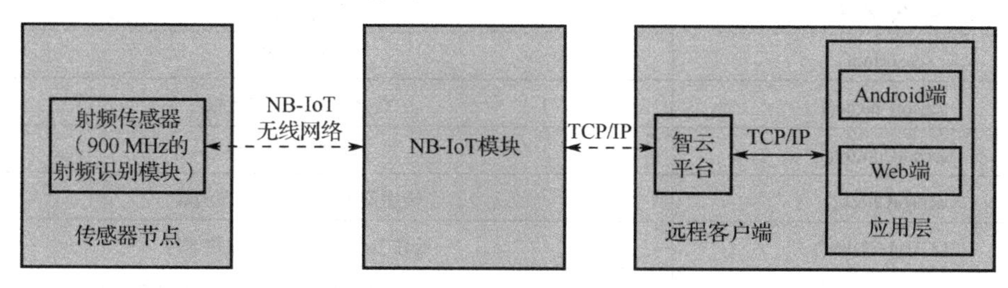

图 7.4 系统的数据传输

7.1.3 系统的软硬件开发：停车收费管理系统

1. 系统底层软硬件设计

1）感知层硬件设计

停车收费管理系统感知层硬件包括 xLab 未来开发平台的 NB-IoT 模块、增强型无线节点 ZXBeePlusB、识别类开发平台 Sensor-EH。其中，NB-IoT 模块负责汇集传感器采集的数据；NB-IoT 无线节点（由增强型无线节点 ZXBeePlusB 实现）通过无线通信的方式向 NB-IoT 模块发送数据；识别类传感器连接到 NB-IoT 无线节点，对相关设备进行信息采集。本系统使用的传感器是射频传感器，其硬件接口电路如图 7.5 所示。

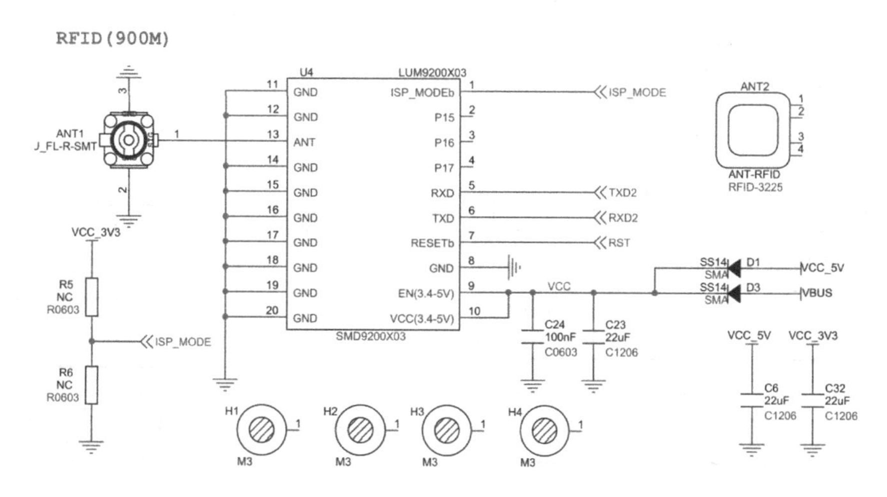

图 7.5 射频传感器的硬件接口电路

2）系统底层开发

本系统是基于 NB-IoT 无线网络开发的。

（1）NB-IoT 智云开发框架。智云框架是在传感器应用程序接口和 SAPI 框架的基础上搭建起来的，通过合理调用这些接口，可以使 NB-IoT 的开发形成一套系统的开发逻辑。传感器应用程序接口函数是在 sensor.c 文件中实现的，具体如表 7.2 所示。

表 7.2　传感器应用程序接口函数

函 数 名 称	函 数 说 明
sensorInit()	传感器初始化
sensorUpdate()	传感器数据定时上报
sensorControl()	传感器控制函数
sensorCheck()	传感器预警监测及处理函数
ZXBeeInfRecv()	处理节点接收到的无线数据包
PROCESS_THREAD(sensor, ev, data)	传感器进程（处理传感器上报、传感器预警监测）

（2）智云平台应用接口分析。NB-IoT 无线节点程序是基于智云框架开发的，详细的程序流程图如图 7.6 所示。

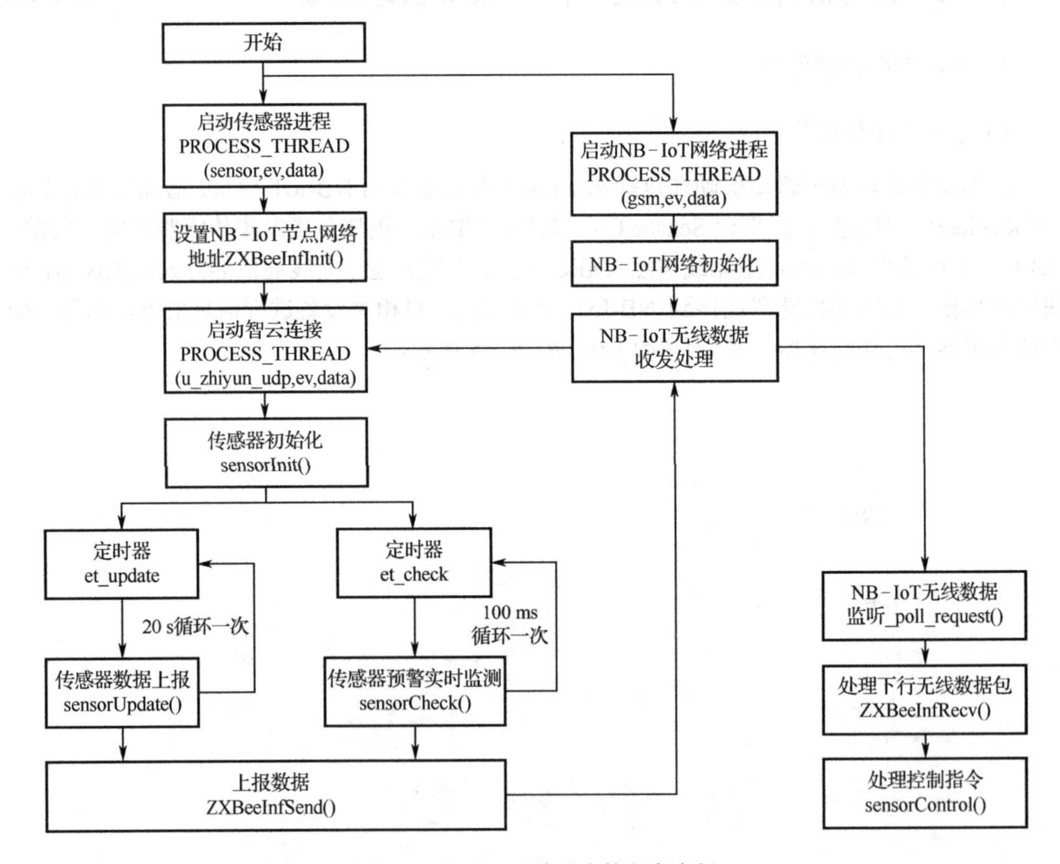

图 7.6　NB-IoT 无线节点的程序流程

图 7.6 中任务的调度主要是由智云框架完成的，用户只需要编写相应的传感器应用程序接口（API）函数即可，这些 API 函数都是在 sensor.c 文件中定义的，这样就和工程中的其他函数实现了隔离。用户在编写传感器 API 函数时，只需要修改在 sensor.c 文件中的代码即可，不需要修改其他文件。

智云框架为 NB-IoT 无线协议栈的上层应用提供分层的软件设计结构，将传感器的私有操作部分封装在 sensor.c 文件中，用户任务中的处理事件和节点类型选择在 sensor.h 文件中定义。sensor.h 文件中事件宏定义如下：

```
#define NODE_NAME "601"
```

sensor.h 文件中声明了智云框架下的传感器应用文件 sensor.c 中的函数，传感器进程用于启动传感器任务及定时器任务，相关代码如下：

```
PROCESS_THREAD(sensor, ev, data)
{
    static struct etimer et_update;
    PROCESS_BEGIN();
    ZXBeeInfInit();
    sensorInit();
    etimer_set(&et_update, CLOCK_SECOND*10);
    while (1) {
        PROCESS_WAIT_EVENT_UNTIL(ev == PROCESS_EVENT_TIMER);
        if (etimer_expired(&et_update)) {
            printf("sensor->PROCESS_EVENT_TIMER: PROCESS_EVENT_TIMER trigger!\r\n");
            sensorUpdate();
            etimer_set(&et_update, CLOCK_SECOND*10);
        }
    }
    PROCESS_END();
}
```

sensorInit()函数用于对传感器进行初始化，相关代码如下：

```
/*********************************************************************************************
*名称：sensorInit()
*功能：传感器初始化
*********************************************************************************************/
void sensorInit(void)
{
    printf("sensor->sensorInit(): Sensor init!\r\n");
    ……
}
```

sensorUpdate()函数用于更新传感器采集的数据，并将更新后的数据打包上报，相关代码如下：

```
/*********************************************************************************************
*名称：sensorUpdate()
*功能：处理主动上报的数据
*********************************************************************************************/
void sensorUpdate(void)
{
    ……
    //更新值并上报
```

```
    sprintf(p, "airGas=%.1f", airGas);
    if (pData != NULL) {
        ZXBeeInfSend(p, strlen(p));                               //将数据发送到智云平台
    }
    printf("sensor->sensorUpdate(): airGas=%.1f\r\n", airGas);
}
```

ZXBeeInfSend()函数用于发送数据，相关代码如下：

```
/*********************************************************************************************
*名称：ZXBeeInfSend()
*功能：节点将无线数据包发送到智云平台
*参数：*p—要发送的无线数据包；len—无线数据包的长度
*********************************************************************************************/
void ZXBeeInfSend(char *p, int len)
{
    leds_on(1);
    clock_delay_ms(50);
    if (nbConfig.mode == COAP) {
        zhiyun_send_coap(p);
    } else {
        zhiyun_send_udp(p);
    }
    leds_off(1);
}
```

ZXBeeInfRecv()函数用于处理节点接收到的无线数据包，相关代码如下：

```
/*********************************************************************************************
*名称：ZXBeeInfRecv()
*功能：处理节点接收到的无线数据包
*参数：*pkg—收到的无线数据包；len—无线数据包的长度
*********************************************************************************************/
void ZXBeeInfRecv(char *buf, int len)
{
    ……
    val = atoi(pval);
    //控制命令解析
    if (0 == strcmp("cmd", ptag)){                          //对D0的位进行操作，CD0表示位清零操作
        sensorControl(val);
    }
    leds_off(1);
}
```

sensorControl()函数用于控制传感器，相关代码如下：

```
/*********************************************************************************************
*名称：sensorControl()
*功能：传感器控制
```

```
*参数：cmd—控制命令
*********************************************************************************/
void sensorControl(uint8_t cmd)
{
    //根据cmd参数执行对应的控制程序
}
```

3）传感器驱动设计

（1）数据通信协议的定义。本系统主要使用的是识别类开发平台 Sensor-EH，其 ZXBee 数据通信协议如表 7.3 所示。

表 7.3 识别类开发平台 ZXBee 数据通信协议

开发平台	属　性	参　数	权限	说　明
Sensor-EH（606）	卡号	A0	—	字符串（主动上报，不可查询）
	卡余额	A2	R	整型数据，范围为0～800000，手动查询
	ETC杆开关	D1(OD1/CD1)	R/W	D1的Bit0表示ETC杆的状态，0表示落下，1表示抬起一次，3 s后自动落下，同时将Bit0置0
	充值金额	V1	R/W	返回充值状态，0表示操作未成功，1表示操作成功
	扣款金额	V2	R/W	返回扣款状态，0表示操作未成功，1表示操作成功

（2）驱动程序的开发。在智云框架下不仅可以很容易地实现传感器驱动程序的开发，还可以省略无线传感器节点的组网和用户任务的创建等烦琐过程。例如，直接调用 sensorInit()函数可以实现节点传感器的初始化；调用 ZXBeeInfRecv()函数可以处理节点接收到的无线数据包；ZXBeeInfSend()函数用于将无线数据包发送到智云平台；ZXBeeUserProcess()函数用于解析接收到的控制命令；设备状态的定时上报需要使用 PROCESS_THREAD()作为 sensorUpdate()函数和 sensorCheck()函数的定时进入接口来反馈设备状态信息。

在 sensor.c 中，需要在 sensorInit()函数中添加传感器初始化的内容，并通过定义上报事件和报警事件实现设备工作状态的定时反馈，部分代码如下：

```
/*********************************************************************************
*名称：sensorInit()
*功能：传感器初始化
*********************************************************************************/
void sensorInit(void)
{
    motorInit();
    RFID900MInit();
}
```

本系统主要使用的是射频传感器（900 MHz 的射频模块），下面重点介绍射频传感器的初始化函数 RFID900MInit()，通过设置串口的引脚、波特率、数据格式来初始化串口。实现代码在 rfid900M.c 文件中，特别要注意读写命令的操作，一定按照模块数据手册编写。rfid900M.c 部分代码如下：

```
void RFID900MInit(void)
{
    uart2_init(115200);
    uart2_set_input(uartRecvCallBack);
}
void RFID900MCheckCardReq(void)
{
    /*read type c uii, response pc+epc*/
    char cmd[] =   {0xBB,0x00,0x22,0x00,0x00,0x7E,0x54,0x73};
    RFID900MReq(cmd, sizeof cmd);
    return;
}
int RFID900MCheckCardRsp(char *cid)
{
    if (rxPackage == 22 && u8Buff[2] == 0x22) {
        memcpy(cid, &u8Buff[7], 12);
        return 12;
    }
    return 0;
}
void RFID900MReadReq(char*ap, char*epc, int mb, int sa, int dl)
{
    /*read type c tag memory */
    char cmd[] = {0xBB,0x00,0x29,0x00,0x17,0x00,0x00,0x00,0x00,0x00,0x0C,0xE2, 0x00,0x30,0x32,0x76,0x13,
                0x01,0x52,0x21,0x40,0x34,0x6D,0x03,0x00,0x00,0x00,0x04,0x7E,0x49,0xF2};
    if (ap != NULL) {
        cmd[5] = ap[0], cmd[6] = ap[1], cmd[7] = ap[2], cmd[8] = ap[3];
    }
    memcpy(&cmd[11], epc, 12); //EPC
    cmd[23] = mb;
    cmd[24] = sa>>8;
    cmd[25] = sa & 0xff;
    cmd[26] = dl>>8;
    cmd[27] = dl & 0xff;
    unsigned int c = CRC16_CCITT_FALSE((unsigned char*)&cmd[1], 28);
    cmd[29] = c>>8;
    cmd[30] = c & 0xff;

    RFID900MReq(cmd, sizeof cmd);
    return;
}
int RFID900MReadRsp(char *out)
{
    int rlen = (u8Buff[3]<<8 | u8Buff[4]);
    if (rxPackage == (rlen+8) && u8Buff[2] == 0x29) {
        memcpy(out, &u8Buff[5], rlen);
        return rlen;
```

```
    }
    return 0;
}
void RFID900MWriteReq(char*ap, char*epc, int mb, int sa, int dl, char *in)
{
    static char cmd[96] = {0xBB,0x00,0x46,0x00,0x17,0x00,0x00,0x00,0x00,0x00,0x0C,0xE2,
    0x00,0x30,0x32,0x76,0x13,0x01,0x52,0x21,0x40,0x34,0x6D,0x03,0x00,0x00,0x00,
    0x04,0x7E,0x49,0xF2};
    if (dl > 32) return;
    if (ap != NULL) {
        cmd[5] = ap[0], cmd[6] = ap[1], cmd[7] = ap[2], cmd[8] = ap[3];
    }
    memcpy(&cmd[11], epc, 12); //len(epc)+epc
    cmd[23] = mb;
    cmd[24] = sa>>8;
    cmd[25] = sa & 0xff;
    cmd[26] = dl>>8;
    cmd[27] = dl & 0xff;
    memcpy(&cmd[28], in, dl*2);
    int wlen = 28 + dl*2;
    cmd[wlen++] = 0x7e;
    unsigned int c = CRC16_CCITT_FALSE((unsigned char*)&cmd[1], wlen-1);
    cmd[wlen++] = c>>8;
    cmd[wlen++] = c & 0xff;
    cmd[3] = (wlen-8)>>8;
    cmd[4] = (wlen-8)&0xff;
    RFID900MReq(cmd, wlen);
}
int RFID900MWriteRsp(void)
{
    if (rxPackage == 9 && u8Buff[2] == 0x46) {
        if (u8Buff[5] == 0) return 1;
        else return -2;
    }
    return 0;
}
```

监测射频传感器的函数是 sensor_check()，代码如下：

```
/*********************************************************************************************
*名称：sensorCheck()
*功能：监测传感器
*********************************************************************************************/
void sensorCheck(void)
{
    static char last_epc[12];
    static int new_tag = 0;                              //监测到卡片并读取余额成功标记
    static char epc[12];                                 //记录当前监测到卡片 EPC
```

```
static char write = 0;
static char status = 1;                                   //状态转换标识
if (status == 1) {
    if (RFID900MCheckCardRsp(epc) > 0) {
        if (memcmp(last_epc, epc, 12) != 0)    {
            RFID900MReadReq(NULL, epc, BLK_USER, 0, 2); //发送读卡请求
            status = 2;
        } else {
            new_tag = 8;
            if (V1 != 0) {
                long money = A2 + V1;
                RFID900MWriteReq(NULL, epc, BLK_USER, 0, 2, (char*)&money);
                write = 1;
                status = 3;
            } else if (V2 != 0) {
                long money = A2 - V2;
                RFID900MWriteReq(NULL, epc, BLK_USER, 0, 2, (char*)&money);
                write = 2;
                status = 3;
            } else {
                RFID900MCheckCardReq();
                status = 1;
            }
        }
    } else {
        //没有监测到卡片
        if (new_tag > 0) {
            new_tag -= 1;
            if (new_tag == 0) {
                memset(last_epc, 0, 12);
            }
        }
        RFID900MCheckCardReq();
        status = 1;
    }
} else if (status == 2) {
    int money;
    if (RFID900MReadRsp((char*)&money) > 0) {
        //读取到金额并上报
        for (int j=0; j<12; j++) sprintf(&A0[j*2], "%02X", epc[j]);
        char buf[16];
        if (money < 0) money = 0;
        ZXBeeBegin();
        ZXBeeAdd("A0", A0);
        A2 = money;
        sprintf(buf, "%ld.%ld", A2/100, A2%100);
        ZXBeeAdd("A2", buf);
```

```
                char *p = ZXBeeEnd();
                if (p != NULL) {
                    ZXBeeInfSend(p, strlen(p));
                }
                memcpy(last_epc, epc, 12);          //保存当前卡片的 ID
                new_tag = 8;
                V1 = 0;                             //初始化扣金额
                V2 = 0;
            }
            RFID900MCheckCardReq();
            status = 1;
        } else if (status == 3) {
            if (RFID900MWriteRsp() > 0) {
                char buf[16];
                ZXBeeBegin();
                if (write == 1) {
                    A2 = A2 + V1;
                    V1 = 0;
                    ZXBeeAdd("V1", "1");
                } else if (write == 2) {
                    A2 = A2 - V2;
                    V2 = 0;
                    ZXBeeAdd("V2", "1");
                }
                sprintf(buf, "%d.%d", A2/100, A2%100);
                ZXBeeAdd("A2", buf);
                char *p = ZXBeeEnd();
                if (p != NULL) {
                    ZXBeeInfSend(p, strlen(p));
                }
            }
            RFID900MCheckCardReq();
            status = 1;
        }
        /*电机状态监测*/
        static int tick = 0;
        if (tick == motor){
            motorSet(0);                    //关闭电机
        }else if (tick > motor) {
            motorSet(2);                    //开启闸机
            tick -= 1;
        } else if (tick < motor) {
            motorSet(1);                    //关闭闸机
            tick += 1;
        }
    }
```

2. Android 端应用设计

1）Android 工程设计框架

打开 Android Studio 开发环境，可以看到停车收费管理系统的工程目录，如图 7.7 所示，系统的工程框架如表 7.4 所示。

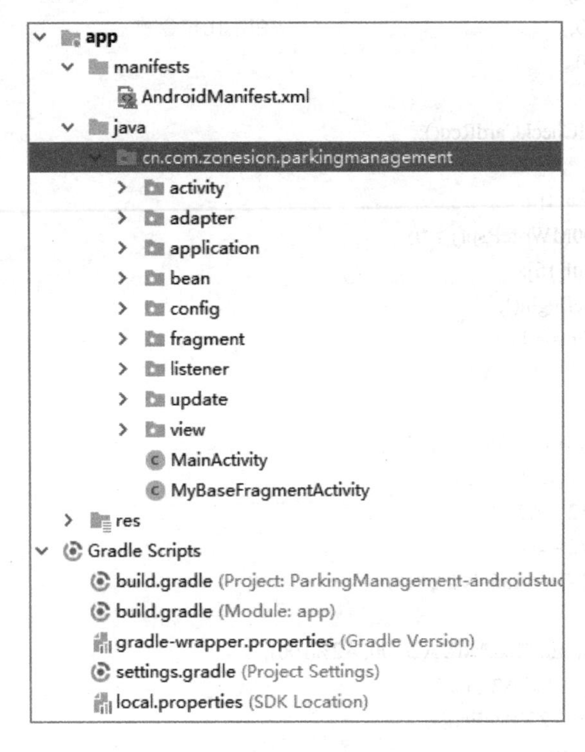

图 7.7　停车收费管理系统的工程目录

表 7.4　停车收费管理系统的工程框架

类　名	说　明
activity	
IdKeyShareActivity.java	在 IDKey 页面单击“分享”按钮时，可弹出 activity，用于分享二维码图片
adapter	
HdArrayAdapter.java	历史数据显示适配器
application	
LCApplication.java	LCApplication 继承 application 类，使用单例模式（Singleton Pattern）创建 WSNRTConnect 对象
bean	
HistoricalData.java	历史数据的 bean 类，用于将从智云服务器获得的历史数据记录（JSON 形式）转换成该类对象
IdKeyBean.java	IdKeyBean 用来描述用户设备的 ID、KEY，以及智云服务器的地址 SERVER
config	
Config.java	config 用于修改用户的 ID、KEY，以及智云服务器的地址和 MAC 地址

续表

类　名	说　明
fragment	
BaseFragment.java	页面基础 Fragment 定义类
HDFragment.java	历史数据页面
HistoricalDataFragment.java	历史数据显示页面
HomepageFragment.java	展示首页页面的 Fragment
IDKeyFragment.java	IDKey 选项的页面
MacSettingFragment.java	当用户设置被监测项的 MAC 地址时显示的页面
MoreInformationFragment.java	更多信息显示页面
RunHomePageFragment.java	运营首页显示页面
VersionInformationFragment.java	显示版本等相关信息的页面
listener	
IOnWSNDataListener.java	传感器数据监听器接口
update	
UpdateService.java	应用下载服务类
view	
APKVersionCodeUtils.java	获取当前本地 apk 的版本
CustomRadioButton.java	自定义按钮类
PagerSlidingTabStrip.java	自定义滑动控件类
MainActivity.java：主页面类	
MyBaseFragmentActivity.java：系统 Fragment 通信类	

2）软件设计

根据智云 Android 端应用程序接口的定义，系统的应用设计主要采用实时数据 API 接口，实时数据 API 接口的流程见 3.1.3 节的图 3.11。

（1）LCApplication.java 程序代码剖析。停车收费管理系统中的 LCApplication.java 程序代码和城市环境信息采集系统的 LCApplication.java 程序代码相同，详见 3.1.3 节的相关内容。

（2）HomepageFragment.java 程序代码剖析。下面的代码通过(LCApplication) getActivity().getApplication()获取 LCApplication 类中的 WSNRTConnect 对象。

```
private void initInstance(){
    config = Config.getConfig();
    lcApplication = (LCApplication) getActivity().getApplication();
    lcApplication.registerOnWSNDataListener(this);
    wsnrtConnect = lcApplication.getWSNRConnect();
    preferences = getActivity().getSharedPreferences("user_info", Context.MODE_PRIVATE);
    editor = preferences.edit();
    }
```

下面的代码通过复写 onMessageArrive 方法来处理接收到的无线数据包，实现了设备的

MAC 地址获取，并在当前的页面显示设备的状态。

```
@Override
    public void onMessageArrive(String mac, String tag, String val) {
        if ("TYPE".equals(tag) && "606".equals(val.substring(2, val.length()))) {
            totalMac = mac;
        }
        if (mac.equals(totalMac) && "A0".equals(tag)) {
            enterState.setText("在线");
            enterState.setTextColor(getResources().getColor(R.color.line_text_color));
            cardFlag = true;
            if (!textNumber.getText().toString().equals(val)) {
                currentCardNo = val;
                textNumber.setText(val);
            }else {
                hasShowedChargeInfo = false;
                hasShowedConsumeInfo = false;
            }
        }
        if (mac.equals(totalMac) && "A2".equals(tag)) {
            try{
                cardBalance = Float.parseFloat(val);
            }catch(Exception e){
                cardBalance = 0;
            }
            textBalance.setText(val);
            if (cardFlag) {
                cardFlag = false;
                if (!textNumber.getText().toString().equals("")) {
                    if (cardNos.containsKey(textNumber.getText().toString())) {
                        long t1 = cardNos.get(textNumber.getText().toString());
                        int v = (int)((java.lang.System.currentTimeMillis() - t1)/1000);
                        charge   = v;
                        if (cardBalance >= charge) {
                            String v1 = "{V2=" + String.valueOf(charge) + "}";
                            wsnrtConnect.sendMessage(totalMac, v1.getBytes());
                            textDeductions.setText(String.valueOf(charge)+"元");
                        }else {
                            wsnrtConnect.sendMessage(totalMac, "{V1=2000}".getBytes());
                        }
                    }else {
                        cardNos.put(textNumber.getText().toString(),
                        java.lang.System.currentTimeMillis());
                        textDeductions.setText("");
                        Date date = new Date(java.lang.System.currentTimeMillis());
                        textTime.setText(format.format(date));
                        Toast.makeText(getContext(), "欢迎光临", Toast.LENGTH_SHORT).show();
```

```
                        wsnrtConnect.sendMessage(totalMac, "{OD1=1,D1=?}".getBytes());
                    }
                }
            }
        }
        if (mac.equals(totalMac) && "D1".equals(tag)) {
            int numResult = Integer.parseInt(val);
        }
        if (mac.equals(totalMac) && "V2".equals(tag)) {
            int numResult = Integer.parseInt(val);
            if ((numResult & 0X01) == 0x01) {//扣款成功，抬杆
                if (!hasShowedConsumeInfo) {
                    float balance = Float.parseFloat(textBalance.getText().toString()) - (float)charge;
                    textBalance.setText(String.format("%.2f", balance));
                    wsnrtConnect.sendMessage(totalMac, "{A2=?,OD1=1,D1=?}".getBytes());
                    cardNos.remove(currentCardNo);
                    Toast.makeText(getContext(), "欢迎下次光临", Toast.LENGTH_SHORT).show();
                    hasShowedConsumeInfo = true;
                }
            }else {//扣款失败，充值
                Toast.makeText(getContext(), "扣款失败，请将卡片取下并重新放在机器上面进行扣款",
                                Toast.LENGTH_SHORT).show();
                dataBecomeNull();
            }
        }
        if (mac.equals(totalMac) && "V1".equals(tag)) {
            int numResult = Integer.parseInt(val);
            if ((numResult & 0X01) == 0x01) {
                if (!hasShowedChargeInfo) {
                    float balance = Float.parseFloat(textBalance.getText().toString()) + (float)2000;
                    textBalance.setText(String.format("%.2f", balance));
                    Toast.makeText(getContext(), "充值成功", Toast.LENGTH_SHORT).show();
                    textDeductions.setText(charge+"元");
                    hasShowedChargeInfo = true;
                }
            }else {
                Toast.makeText(getContext(), "充值失败，请将卡片重新放在机器上面进行充值",
                                Toast.LENGTH_SHORT).show();
                dataBecomeNull();
            }
        }
    }
```

3．Web 端应用设计

1）页面功能结构分析

停车收费管理系统的 Web 端默认显示的是“运营首页”页面，在“运营首页”页面上设

计了两个模块，分别是车位显示模块、收费管理模块。停车收费管理系统 Web 端的“运营首页”页面如图 7.8 所示。

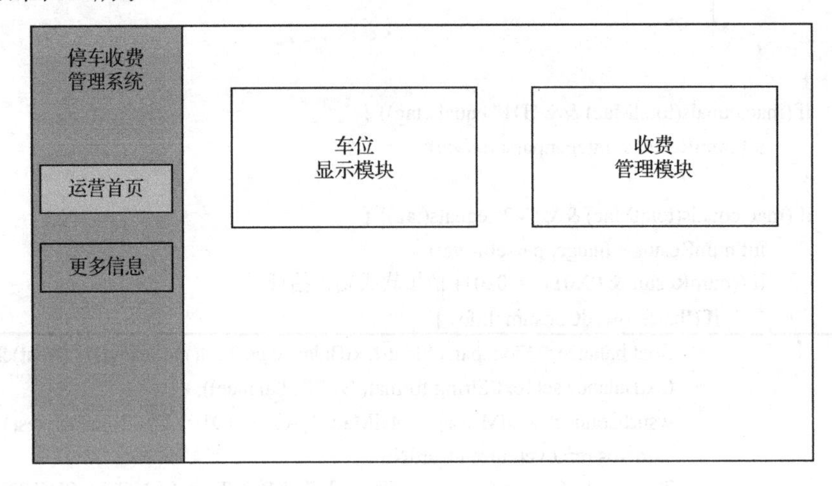

图 7.8　停车收费管理系统 Web 端的“运营首页”页面

“更多信息”页面的主要功能是进行智云服务器的连接配置，和城市环境信息采集系统中的 Web 端“更多信息”页面类似，参见图 3.14。

2）软件设计

停车收费管理系统 Web 端的 JS 开发逻辑与 Android 端的开发逻辑相似，首先通过配置 ID 和 KEY 与智云服务器进行连接，再通过实时监听数据的方法来获取相关传感器的数据并进行处理。JS 开发的部分代码如下。

在 getConnect()函数中定义了实时连接对象 rtc，连接成功回调函数是 rtc.onConnect，数据服务掉线回调函数是 rtc.onConnectLost，消息处理回调函数是 rtc.onmessageArrive。

```
function getConnect() {
    config["id"] = config["id"] ? config["id"] : $("#id").val();
    config["key"] = config["key"] ? config["key"] : $("#key").val();
    config["server"] = config["server"] ? config["server"] : $("#server").val();
    //创建数据连接服务对象
    rtc = new WSNRTConnect(config["id"], config["key"]);
    rtc.setServerAddr(config["server"] + ":28080");
    rtc.connect();
    rtc._connect = false;
    //连接成功回调函数
    rtc.onConnect = function() {
        $("#ConnectState").text("数据服务连接成功！");
        rtc._connect = 1;
        message_show("数据服务连接成功！");
        $("#idkeyInput").text("断开").addClass("btn-danger");
        $("#id,#key,#server").attr('disabled',true);
    };
    //数据服务掉线回调函数
```

```
        rtc.onConnectLost = function() {
            rtc._connect = 0;
            $("#ConnectState").text("数据服务连接掉线！");
            $("#idkeyInput").text("连接").removeClass("btn-danger");
            message_show("数据服务连接失败，检查网络或 ID、KEY");
            $("#RFIDLink").text("离线").css("color", "#e75d59");
            $("#doorLink").text("离线").css("color", "#e75d59");
            $("#id,#key,#server").removeAttr('disabled',true);
        };
        //消息处理回调函数
        rtc.onmessageArrive = function (mac, dat) {
            //console.log(mac+" >>> "+dat);
            if (dat[0]=='{' && dat[dat.length-1]=='}') {
                dat = dat.substr(1, dat.length-2);
                var its = dat.split(',');
                for (var i=0; i<its.length; i++) {
                    var it = its[i].split('=');
                    if (it.length == 2) {
                        process_tag(mac, it[0], it[1]);
                    }
                }
                if (!mac2type[mac]) { //如果没有获取到 TYPE 值，主动去查询
                    rtc.sendMessage(mac, "{TYPE=?,A0=?,A1=?,A2=?,A3=?,A4=?,A5=?,A6=?,A7=?,D1=?}");
                }
            }
        }
    }
```

下述 JS 开发代码的功能是根据设备连接情况来实现停车收费功能，并在页面更新显示车位的数量。

```
    var wsn_config = {
        "606" : {
            "online" : function() {
                $(".online_606").text("在线").css("color", "#96ba5c");
            },
            "pro" : function (tag, val) {
                if(tag=="A0"){
                    var msg ="";
                    //判断当前是第几次读卡，
                    //奇数次读卡开始计时，不扣费，剩余车位减 1，偶数次读卡时计算停车时间
                    //根据时间乘以单价进行扣费，剩余车位加 1，card 对象内移除当前卡号记录
                    if(!config.card.val){
                        config.card.val = {};
                        config.card.val.startTime = new Date().getTime();       //第一次刷卡时间
                        console.log(config.card.val.startTime);
                        config.parkNum --;
                        msg = "欢迎进场，卡号："+val;
```

```
            }else{
                //获取出场时间和入场时间的分钟差
                config.card.val.endTime = new Date().getTime();
                var time = parseInt( ( config.card.val.endTime - config.card.val.startTime ) / 1000 / 60 );
                //按每分钟两元的单价计算停车费用
                var money = time *2;
                config.parkNum ++;
                delete config["card"][val];
                msg = "谢谢光临，卡号："+val+"扣费："+money;
                rtc.sendMessage(config["mac_605"], "{V2="+money+",V2=?}");
                $("#parkMoney").text(money);
            }
            //显示当前卡号，并更新剩余车位
            $("#ruc_rfid").text(val);
            $("#parkNum").text(config.parkNum);
            message_show(msg);
        }
        else if(tag=="A4"){
            message_show("扣费成功！");
        }
    }
  },
};
```

停车收费管理管理系统 Web 端其他部分的代码请查看本书配套资料中的项目源文件。

7.1.4 开发验证

1. Web 端应用测试

在 Web 端打开停车收费管理系统后，在“运营首页”下可以看到 Web 端的主页，如图 7.9 所示。

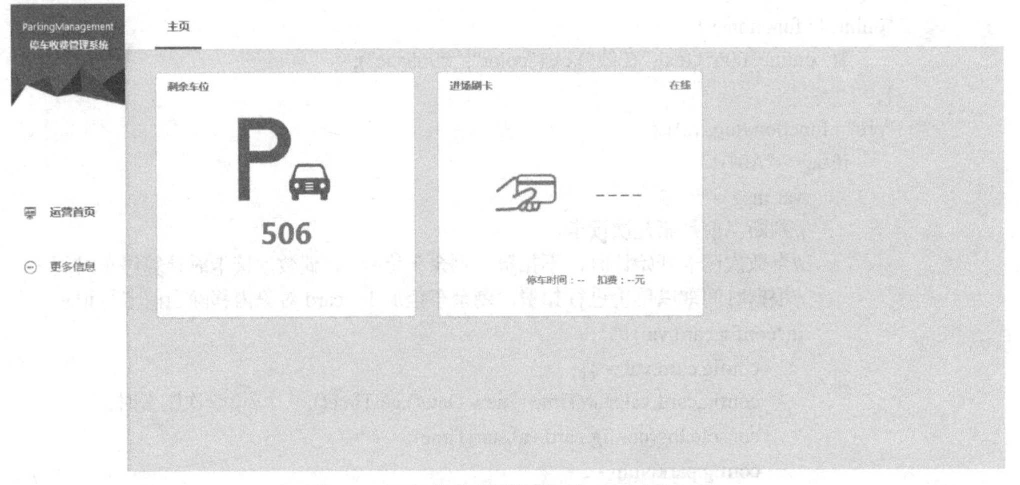

图 7.9　停车收费管理系统的 Web 端主页

设备在线后可以通过刷用户卡片来显示用户卡片信息，如卡片 ID（卡号）、扣费信息、剩余停车位，如图 7.10 所示。

图 7.10　显示的用户卡片信息

2. Android 端应用测试

Android 端应用测试同 Web 端应用测试流程基本一致，可参考本系统的 Web 端应用测试进行操作。停车收费管理系统的 Android 端“运营首页”页面如图 7.11 所示。

图 7.11　停车收费管理系统的 Android 端“运营首页”页面

7.1.5　总结与拓展

本节基于 NB-IoT 实现了射频传感器的识别，通过 Android 和 HTML5 技术实现了 Android 端和 Web 端的应用设计，能实时获取射频传感器数据，实现了基于 NB-IoT 的停车收费管理系统。

7.2 基于 NB-IoT 的智能水表抄表系统

在智能水表抄表系统中，每块水表（见图 7.12）都有 SIM 卡，既可以用来存储用户的用水信息，也可以实时监控每块水表的在线运行情况。智能水表抄表系统的优势是，只要有信号，就可以进行覆盖范围很广的数据传输，也可以用于一些分散式空旷的地区。

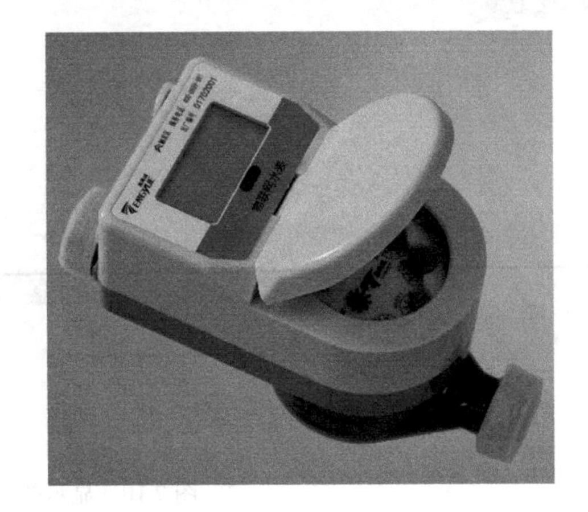

图 7.12　智能水表抄表系统中的水表

7.2.1　系统开发目标

（1）熟悉射频传感器的硬件原理和数据通信协议，基于 STM32F103 实现射频传感器的驱动程序开发，通过射频传感器采集的数据和 NB-IoT，实现智能水表抄表系统的设计。

（2）实现智能水表抄表系统的 Android 端应用开发和 Web 端应用开发。

7.2.2　系统设计分析

1．系统的功能设计

从系统功能的角度出发，智能水表抄表系统可以分为两个模块：智能水表管理模块和系统设置模块，如图 7.13 所示。

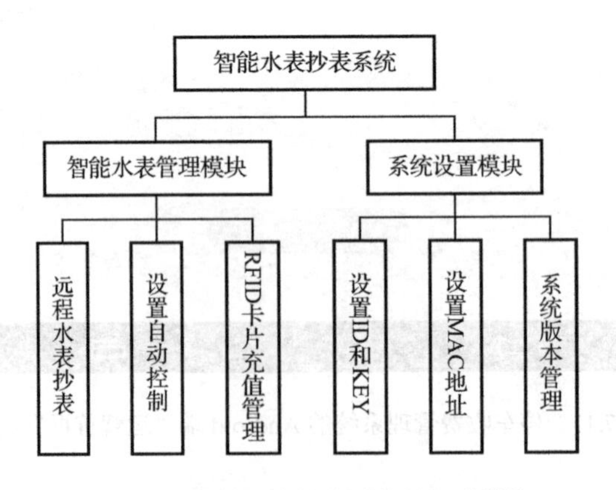

图 7.13　智能水表抄表系统的组成模块

智能水表抄表系统的功能需求如表 7.5 所示。

表7.5 智能水表抄表系统的功能需求分析

功 能	功 能 说 明
智能水表控制管理	远程设备数据读取、设备开关自动控制、用户卡片充值管理
智云连接设置	设置智云服务器的参数和设备的MAC地址

2. 系统的总体架构设计

智能水表抄表系统是基于物联网四层架构模型来设计的，其总体架构如图7.14所示。

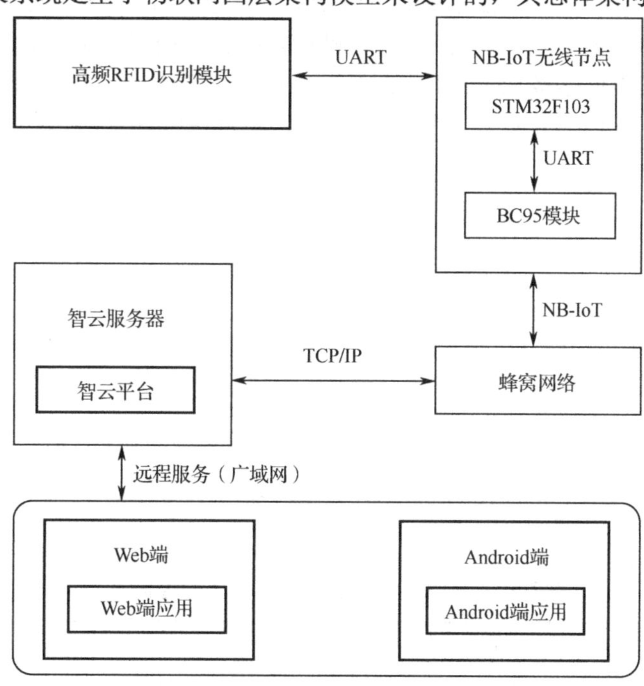

图7.14 智能水表抄表系统的总体架构

3. 系统的数据传输

智能水表抄表系统传输中的数据传输是在传感器节点、智云平台以及客户端（包括Web端和Android端）之间进行的，如图7.15所示。

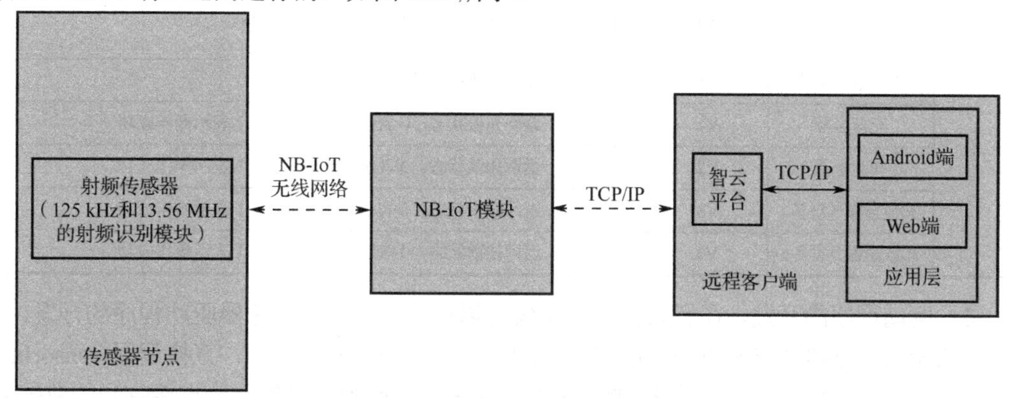

图7.15 智能水表抄表系统的数据传输

7.2.3 系统的软硬件开发：智能水表抄表系统

1. 系统底层软硬件设计

1）感知层硬件设计

智能水表抄表系统的感知层硬件主要包括 xLab 未来开发平台的 NB-IoT 模块、增强型无线节点 ZXBeePlusB、识别类传感器 Sensor-EL。其中传感器节点数据通过 NB-IoT 模块汇集；NB-IoT 无线节点（由增强型无线节点 ZXBeePlusB 实现）通过无线通信的方式向智云平台发送传感器数据；识别类传感器 Sensor-EL 连接到 NB-IoT 无线节点，对相关设备进行识别管理。传感器包括 125 kHz 和 13.56 MHz 的射频传感器，其硬件接口电路如图 4.17 所示。

2）系统底层开发

本系统使用的 NB-IoT 无线网络进行开发。

（1）NB-IoT 智云开发框架。本系统的智云开发框架和停车收费管理系统相同，详见 7.1.3 节。

（2）智云平台应用接口分析。本系统的智云平台应用接口和停车收费管理系统相同，详见 7.1.3 节。

3）传感器驱动设计

（1）数据通信协议的定义。本系统主要使用的是识别类开发平台 Sensor-EL，其 ZXBee 数据通信协议如表 7.6 所示。

表 7.6 识别类开发平台的 ZXBee 数据通信协议

传感器	属 性	参 数	权限	说 明
Sensor-EL（605）	卡号	A0	—	字符串（主动上报，不可查询）
	卡类型	A1	R	整型数据，0 表示 125K 型射频识别模块，1 表示 13.56M 型射频识别模块
	卡余额	A2	R	整型数据，范围 0～800000，手动查询
	设备余额	A3	R	浮点型数据，设备金额
	设备单次消费金额	A4	R	浮点型数据，设备本次消费扣款金额
	设备累计消费	A5	R	浮点型数据，设备累计扣款金额
	充值金额	V1	R/W	返回充值状态，0 表示操作未成功，1 表示操作成功
	扣款金额	V2	R/W	返回扣款状态，0 表示操作未成功，1 表示操作成功
	充值金额（设备）	V3	R/W	返回充值状态，0 表示操作未成功，1 表示操作成功
	扣款金额（设备）	V4	R/W	返回扣款状态，0 表示操作未成功，1 表示操作成功

（2）驱动程序的开发。在智云框架下不仅可以很容易地实现传感器驱动程序的开发，还可以省略无线传感器节点的组网和用户任务的创建等烦琐过程。例如，直接调用 sensorInit()函数可以实现节点传感器的初始化；调用 ZXBeeInfRecv()函数可以处理节点接收到的数据包；ZXBeeInfSend()函数用于将数据发送到智云平台；ZXBeeUserProcess()函数用于解析接收到的

控制命令；设备状态的定时上报需要使用 PROCESS_THREAD()作为 sensorUpdate()函数和 sensorCheck()函数的定时进入接口来反馈设备状态信息。

在 sensor.c 中，需要在 sensorInit()函数中添加传感器初始化的内容，并通过定义上报事件和报警事件来实现设备工作状态的定时反馈。部分代码如下：

```
/*********************************************************************************************
*名称：sensorInit()
*功能：传感器初始化
*********************************************************************************************/
void sensorInit(void)
{
    RFID7941Init();
    key_init();
    relay_init();
    relay_control(2);;     //开启继电器 2
    buzzer_init();
    OLED_Init();
    OLED_Clear();
    OLED_ShowCHinese(21,2,4);
    OLED_ShowCHinese(21+13,2,5);
    OLED_ShowCHinese(21+13+13,2,12);
    OLED_ShowCHinese(21,0,8);
    OLED_ShowCHinese(21+13,0,9);
    OLED_ShowCHinese(21+13+13,0,12);
    if(work_mod == 1){
        OLED_ShowCHinese(21+13*3,0,6);
        OLED_ShowCHinese(21+13*4,0,7);
    }else{
        OLED_ShowCHinese(21+13*3,0,10);
        OLED_ShowCHinese(21+13*4,0,11);
    }
    char buf[16];
    sprintf(buf, "%d.%d", A3/100, A3%100);
    OLED_ShowString(21+13*3,2,(unsigned char *)buf,12);
}
```

本系统使用射频传感器，下面重点介绍射频传感器的初始化函数 RFID7941Init()，代码如下：

```
void RFID7941Init(void)
{
    uart2_init(115200);
    uart2_set_input(uartRecvCallBack);
}
void RFID7941CheckCard1356MHzReq(void)
{
    char cmd[] =   {0xAB,0xBA,0x00,0x10,0x00,0x10};
```

```
        RFID7941WriteReq(cmd, sizeof cmd);
    }
    int RFID7941CheckCard1356MHzRsp(char *cid)
    {
        if (rxPackage == 10 && u8Buff[3] == 0x81) {
            memcpy(cid, &u8Buff[5], 4);
            return 4;
        }
        return 0;
    }
    void RFID7941ReadCard1356MHzReq(char blk, char*key)
    {
        char cmd[] = {0xAB,0xBA,0x00,0x12,0x09,0x00,0x01,0x0A,0xFF,0xFF,0xFF,0xFF,0xFF,0xFF,0x10};
        cmd[6] = blk;
        memcpy(&cmd[8], key, 6);
        cmd[14] = xor(&cmd[2], 12);
        RFID7941WriteReq(cmd, sizeof cmd);
    }
    int RFID7941ReadCard1356MHzRsp(char*buf)
    {
        if (rxPackage == 24 && u8Buff[3] == 0x81) {
            memcpy(buf, &u8Buff[7], 16);
            return 16;
        }
        return 0;
    }

    void RFID7941WriteCard1356MHzReq(char blk, char*key, char*buf)
    {
        char cmd[] = {0xAB,0xBA,0x00,0x13,0x19,0x00,0x01,0x0A,0xFF,0xFF,0xFF,0xFF,0xFF,0xFF,
                 0x00,0x00,0x00,0x00,0x00,0x00,0x00,0x00,0x00,0x00,0x00,0x00,0x00,0x00,0x00,0x00,0x00};
        cmd[6] = blk;
        memcpy(&cmd[8], key, 6);
        memcpy(&cmd[14], buf, 16);
        cmd[14] = xor(&cmd[2], sizeof cmd - 2);
        RFID7941WriteReq(cmd, sizeof cmd);
    }
    int RFID7941WriteCard1356MHzRsp(void)
    {
        if (rxPackage == 6 && u8Buff[3] == 0x81){
            return 1;
        }
        return 0;
    }
```

项目中模拟的计量数据更新是通过 sensorUpdate()函数实现的。

```
/*********************************************************************************************
*名称：sensorUpdate()
*功能：处理主动上报的数据
*********************************************************************************************/
void sensorUpdate(void)
{
    char pData[16];
    char *p = pData;
    if (A3 > 0) {
        A4 = 1 + rand()%500;                            //设备本次消费扣款金额，通过随机数来模拟
        if (A4 > A3) A4 = A3;
        A3 = (A3 - A4);                                 //设备余额
        A5 = (A5 + A4);                                 //设备累计消费金额
        ZXBeeBegin();
        sprintf(p, "%d.%d", A3/100, A3%100);            //水表余额
        ZXBeeAdd("A3", p);
        sprintf(p, "%d.%d", A4/100, A4%100);            //本次消费扣款金额
        ZXBeeAdd("A4", p);
        sprintf(p, "%d.%d", A5/100, A5%100);            //水表累积消费金额
        ZXBeeAdd("A5", p);
        p = ZXBeeEnd();                                 //智云数据帧格式包尾
        if (p != NULL) {
            //将需要上传的数据进行打包操作，并通过 zb_SendDataRequest()发送到协调器
            ZXBeeInfSend(p, strlen(p));
        }
        sprintf(p, "%d.%d      ", A3/100, A3%100);
        OLED_ShowString(21+13*3,2,(unsigned char *)p,12);          //OLED 显示水表余额
        if (A3 == 0) { //余额不足关闭继电器
            D1 &= ~2;
            relay_control(D1);
        }
    }
}
```

射频传感器的监测功能代码在 sensor_check()函数中实现，代码如下：

```
/*********************************************************************************************
*名称：sensorCheck()
*功能：传感器监测
*********************************************************************************************/
void sensorCheck(void)
{
    char buff[16];
    ZXBeeBegin();
    /*按键处理 */
    uint8_t k = get_key_status();
    static uint8_t last_key = 0;
```

```
if (k & 0x01 == 0x01) {
    if ((last_key & 0x01) == 0) {
        if (work_mod == 1) {
            OLED_ShowCHinese(21+13*3,0,10);
            OLED_ShowCHinese(21+13*4,0,11);
            work_mod = 2;
        } else {
            OLED_ShowCHinese(21+13*3,0,6);
            OLED_ShowCHinese(21+13*4,0,7);
            work_mod = 1;
        }
    }
} else if ((k & 0x02) == 0x02) {
    if ((last_key & 0x02) == 0) {
        //模拟消费
        if (A3 > 0) {
            A4 = rand()%500;   //设备本次消费扣款金额，通过产生的随机数来模拟
            if (A4 > A3) A4 = A3;
            A3 = A3 - A4;
            A5 = A5 + A4;
            sprintf(buff, "%d.%d", A3/100,A3%100);
            ZXBeeAdd("A3", buff);
            sprintf(buff, "%d.%d", A4/100,A4%100);
            ZXBeeAdd("A4", buff);
            sprintf(buff, "%d.%d", A5/100,A5%100);
            ZXBeeAdd("A5", buff);
            sprintf(buff, "%d.%d      ", A3/100,A3%100);
            OLED_ShowString(21+13*3,2,(unsigned char *)buff,12);
            if (A3 == 0) { //余额不足关闭继电器
                D1 &= ~2;
                relay_control(D1);
            }
        }
    }
}
last_key = k;
//按键处理完毕
//读卡处理
static char status = 1;
static char last_cid[7];
static char card_cnt = 0;                                   //卡片监测计数
static char cid[7];
static char write = 0;
if (status == 1) {                                          //监测 125K 型卡片的读取结果
    buzzer_off();
    if (RFID7941CheckCard125kHzRsp(cid) > 0) {
        if (memcmp(last_cid, cid, 5) != 0) {
```

```
                sprintf(A0, "%02X%02X%02X%02X%02X", cid[0],cid[1],cid[2],cid[3],cid[4]);
                ZXBeeAdd("A0", A0);
                A1 = 0;
                sprintf(buff, "%d", A1);
                ZXBeeAdd("A1", buff);
                memcpy(last_cid, cid, 5);
                card_cnt = 5;
                buzzer_on();
            } else { //同一张卡片
                card_cnt = 5;
            }
            RFID7941CheckCard125kHzReq();
        } else {
            if (card_cnt > 0) {
                card_cnt -= 1;
            }
            if (card_cnt == 0) {
                memset(last_cid, 0, sizeof last_cid);
                RFID7941CheckCard1356MHzReq();                  //监测 13.56M 型卡片
                status = 2;
            } else {
                RFID7941CheckCard125kHzReq();
            }
        }
    } else if (status == 2) {                                   //13.56M 型卡片监测处理
        buzzer_off();
        if (RFID7941CheckCard1356MHzRsp(cid) > 0) {
            if (memcmp(last_cid, cid, 4) != 0) {
                RFID7941ReadCard1356MHzReq(8, KEY);
                status = 3;
            } else {
                card_cnt = 5;                                   //刷新卡片
                if (work_mod == 1) {                            //充值设备
                    if (write == 3 && A2 > 0) {
                        memset(buff, 0, 16);
                        RFID7941WriteCard1356MHzReq(8, KEY, buff);
                        status = 4;
                    } else {
                        RFID7941CheckCard1356MHzReq();
                    }
                } else
                if (work_mod == 2) {                            //读卡模式
                    int money;
                    if (V1 != 0) {                              //充值
                        money = A2 + V1;
                        write = 1;
                    } else if (V2 != 0) {
```

```
                        money = A2 - V2;
                        write = 2;
                    }
                    if (V1 != 0 || V2 != 0) {
                        memset(buff, 0, 16);
                        buff[12] = (money>>24) & 0xff;
                        buff[13] = (money>>16) & 0xff;
                        buff[14] = (money>>8) & 0xff;
                        buff[15] = money&0xff;
                        RFID7941WriteCard1356MHzReq(8, KEY, buff);
                        status = 5;
                    } else {
                        RFID7941CheckCard1356MHzReq();
                    }
                } else {
                    RFID7941CheckCard1356MHzReq();
                }
            }
        } else {
            if (card_cnt > 0) {
                card_cnt -= 1;
            }
            if (card_cnt == 0) {
                memset(last_cid, 0, sizeof last_cid);
                RFID7941CheckCard125kHzReq();
                status = 1;
                A2 = 0;   //余额清零
            } else {
                RFID7941CheckCard1356MHzReq();
            }
        }
    } else if (status == 3) {                                          //处理卡片读取结果
        if (RFID7941ReadCard1356MHzRsp(buff) > 0) {                    //读取到余额，保存当前卡片的 ID
            memcpy(last_cid, cid, 4);
            card_cnt = 5;
            sprintf(A0, "%02X%02X%02X%02X", cid[0],cid[1],cid[2],cid[3]);
            int money = ((uint32_t)buff[12]<<24) | ((uint32_t)buff[13]<<16) | (buff[14]<<8) | (buff[15]);
            A2 = money;
            if (work_mod == 1) {                                       //设备充值模式
                if (money > 0) {
                    buff[12] = buff[13] = buff[14] = buff[15] = 0;
                    write = 3;                    //再次监测卡片后通过 write = 3 来写卡
                    RFID7941CheckCard1356MHzReq();
                    status = 2;
                } else {
                    buzzer_on();
                    RFID7941CheckCard1356MHzReq();
```

```
                status = 2;
            }
        } else {  //work_mod == 2;
            ZXBeeAdd("A0", A0);
            ZXBeeAdd("A1", "1");
            sprintf(buff, "%d.%d", A2/100,A2%100);
            ZXBeeAdd("A2", buff);
            V1 = 0;
            V2 = 0;
            buzzer_on();
            RFID7941CheckCard1356MHzReq();
            status = 2;
        }
    } else {
        RFID7941CheckCard1356MHzReq();                    //重新监测卡片
        status = 2;
    }
} else if (status == 4) {                                 //处理充值设备写卡结果
    if (RFID7941WriteCard1356MHzRsp() > 0) {
        //充值设备成功
        A3 += A2;
        A2 = 0;
        sprintf(buff, "%d.%d      ", A3/100,A3%100);
        OLED_ShowString(21+13*3,2,(unsigned char *)buff,12);
        write = 0;
        /*充值成功，开启继电器*/
        D1 |= 2;
        relay_control(D1);
        buzzer_on();
    } else { //写卡失败
        //memset(last_cid, 0, sizeof last_cid);
        //card_cnt = 0;
    }
    RFID7941CheckCard1356MHzReq();
    status = 2;
} else if (status == 5) {                                 //卡片充值扣费结果监测
    if (RFID7941WriteCard1356MHzRsp() > 0) {
        if (write == 1) {
            ZXBeeAdd("V1", "1");
            A2 += V1;
            V1 = 0;
            sprintf(buff, "%d.%d", A2/100,A2%100);
            ZXBeeAdd("A2", buff);
            write = 0;
        } else if (write == 2) {
            ZXBeeAdd("V2", "1");
            A2 -= V2;
```

```
                    V2 = 0;
                    sprintf(buff, "%d.%d", A2/100,A2%100);
                    ZXBeeAdd("A2", buff);
                    write   = 0;
                }
                buzzer_on();
            }
            RFID7941CheckCard1356MHzReq();
            status = 2;
        }
        char *p = ZXBeeEnd();
        if (p != NULL) {
            ZXBeeInfSend(p, strlen(p));
        }
}
```

2. Android 端应用设计

1）Android 工程设计框架

打开 Android Studio 开发环境，可以看到智能水表抄表系统的工程目录，如图 7.16 所示，系统的工程框架和停车收费管理系统的工程框架相同，详见 7.1.3 节。

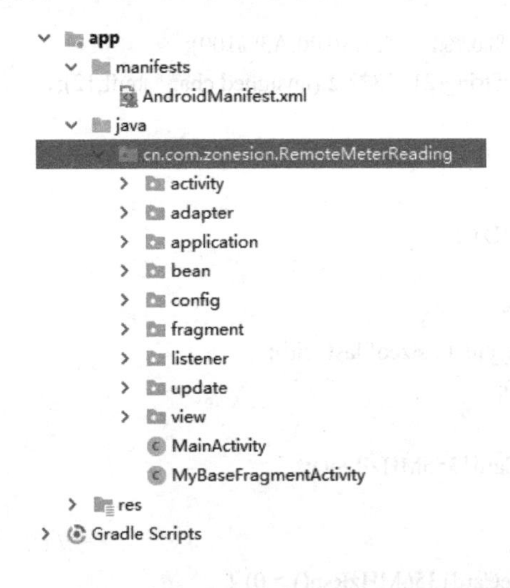

图 7.16 智能水表抄表系统的工程目录

2）软件设计

根据智云 Android 端应用程序接口的定义，系统的应用设计主要采用实时数据 API 接口，实时数据 API 接口的流程见 3.1.3 节的图 3.11。

（1）LCApplication.java 程序代码剖析。智能水表抄表系统中的 LCApplication.java 程序代码和城市环境信息采集系统的 LCApplication.java 程序代码相同，详见 3.1.3 节的相关内容。

（2）HomepageFragment.java 程序代码剖析。下面的代码通过(LCApplication) getActivity().getApplication()获取 LCApplication 类中的 WSNRTConnect 对象。

```
private void initViewAndBindEvent() {
    preferences = getActivity().getSharedPreferences("user_info", Context.MODE_PRIVATE);
    lcApplication = (LCApplication) getActivity().getApplication();
    wsnrtConnect = lcApplication.getWSNRConnect();
    lcApplication.registerOnWSNDataListener(this);
    editor = preferences.edit();
}
```

下面的代码通过按钮的事件监听器对远程设备进行手动控制。

```
@Override
    public void onActivityCreated(@Nullable Bundle savedInstanceState) {
        super.onActivityCreated(savedInstanceState);
        initSetting();
        openOrCloseLamp.setOnClickListener(new View.OnClickListener() {
            @Override
                public void onClick(View v) {
                    //TODO Auto-generated method stub
                    if (totalMac != null) {
                        if (openOrCloseLamp.getText().equals("开启")) {
                            new Thread(new Runnable() {
                                @Override
                                    public void run() {
                                        wsnrtConnect.sendMessage(totalMac,
                                        "{OD1=1,D1=?}".getBytes());
                                    }
                            }).start();
                        }
                        if (openOrCloseLamp.getText().equals("关闭")) {
                            new Thread(new Runnable() {
                                @Override
                                    public void run() {
                                        wsnrtConnect.sendMessage(totalMac,
                                        "{CD1=1,D1=?}".getBytes());
                                    }
                            }).start();
                        }
                    }else {
                        Toast.makeText(lcApplication, "请等待 MAC 地址上线", Toast.LENGTH_
                                    SHORT).show();
                    }
        }
    });
}
```

下面的代码通过复写 onMessageArrive 方法来处理节点接收到的无线数据包，实现了设备的 MAC 地址获取，并在当前页面显示设备的状态。

```
@Override
    public void onMessageArrive(String mac, String tag, String val) {
        if ("TYPE".equals(tag) && "605".equals(val.substring(2, val.length()))) {
            totalMac = mac;
        }
        if (mac.equals(totalMac) && "A0".equals(tag)) {
            enterState.setText("在线");
            enterState.setTextColor(getResources().getColor(R.color.line_text_color));
            cardFlag = true;
            if (!textNumber.getText().toString().equals(val)) {
                currentCardNo = val;
                textNumber.setText(val);
            }else {
                hasShowedChargeInfo = false;
                hasShowedConsumeInfo = false;
            }
        }
        if (tag.equalsIgnoreCase("D1") && mac.equalsIgnoreCase(totalMac)) {
            devState.setText("在线");
            devState.setTextColor(getResources().getColor(R.color.line_text_color));
            int numResult = Integer.parseInt(val);
            if ((numResult & 0X01) == 0x01) {
                imageDevState.setImageDrawable(getResources().getDrawable(R.drawable.waterpump_on));
                openOrCloseLamp.setText("关闭");
                openOrCloseLamp.setBackground(getResources().getDrawable(R.drawable.close));
            }else {
                imageDevState.setImageDrawable(getResources().getDrawable(R.drawable.waterwump_off));
                openOrCloseLamp.setText("开启");
                openOrCloseLamp.setBackground(getResources().getDrawable(R.drawable.open));
            }
        }
        if (mac.equals(totalMac) && "A3".equals(tag)) {
            textSurplus.setText(val);
            tableState.setText("在线");
            tableState.setTextColor(getResources().getColor(R.color.line_text_color));
        }
        if (mac.equals(totalMac) && "A5".equals(tag)) {
            textUsed.setText(val);
        }
        if (mac.equals(totalMac) && "D1".equals(tag)) {
            devState.setText("在线");
            devState.setTextColor(getResources().getColor(R.color.line_text_color));
            int numResult = Integer.parseInt(val);
        }
    }
```

3. Web 端应用设计

1）页面功能结构分析

智能水表抄表系统的 Web 端默认显示的是“运营首页”页面，在“运营首页”页面上设计了 3 个模块，分别是水表抄表显示模块、用户信息管理模块、水阀开关控制模块。智能水表抄表系统 Web 端的“运营首页”页面如图 7.17 所示。

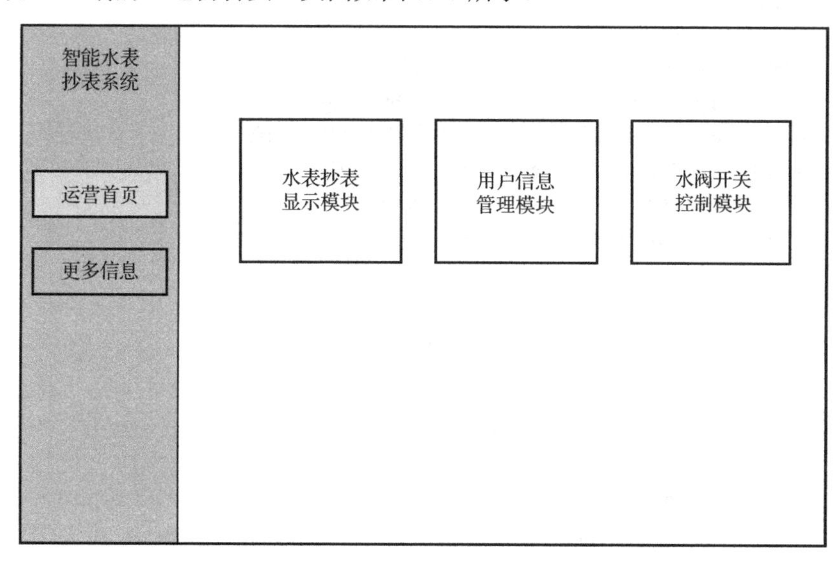

图 7.17　智能水表抄表系统 Web 端的“运营首页”页面

“更多信息”页面的主要功能是进行智云服务器的连接配置，和城市环境信息采集系统中的 Web 端“更多信息”页面类似，参见图 3.14。

2）软件设计

智能水表抄表系统 Web 端的 JS 开发逻辑与 Android 端的开发逻辑相似，首先通过配置 ID 和 KEY 与智云服务器进行连接，再通过实时监听数据的方法来获取相关传感器的数据并进行处理。JS 开发的部分代码如下。

在 getConnect()函数中定义了实时连接对象 rtc，连接成功回调函数是 rtc.onConnect，数据服务掉线回调函数是 rtc.onConnectLost，消息处理回调函数是 rtc.onmessageArrive。

```
function getConnect() {
    config["id"] = config["id"] ? config["id"] : $("#id").val();
    config["key"] = config["key"] ? config["key"] : $("#key").val();
    config["server"] = config["server"] ? config["server"] : $("#server").val();
    //创建数据连接服务对象
    rtc = new WSNRTConnect(config["id"], config["key"]);
    rtc.setServerAddr(config["server"] + ":28080");
    rtc.connect();
    rtc._connect = false;
    //连接成功回调函数
    rtc.onConnect = function() {
```

```
            $("#ConnectState").text("数据服务连接成功！ ");
            rtc._connect = 1;
            message_show("数据服务连接成功！ ");
            $("#idkeyInput").text("断开").addClass("btn-danger");
            $("#id,#key,#server").attr('disabled',true);
        };
        //数据服务掉线回调函数
        rtc.onConnectLost = function() {
            rtc._connect = 0;
            $("#ConnectState").text("数据服务连接掉线！ ");
            $("#idkeyInput").text("连接").removeClass("btn-danger");
            message_show("数据服务连接失败，检查网络或 ID、KEY");
            $("#RFIDLink").text("离线").css("color", "#e75d59");
            $("#doorLink").text("离线").css("color", "#e75d59");
            $("#id,#key,#server").removeAttr('disabled',true);
        };
        //消息处理回调函数
        rtc.onmessageArrive = function (mac, dat) {
            //console.log(mac+" >>> "+dat);
            if (dat[0]=='{' && dat[dat.length-1]=='}') {
                dat = dat.substr(1, dat.length-2);
                var its = dat.split(',');
                for (var i=0; i<its.length; i++) {
                    var it = its[i].split('=');
                    if (it.length == 2) {
                        process_tag(mac, it[0], it[1]);
                    }
                }
                if (!mac2type[mac]) { //如果没有获取到 TYPE 值，主动去查询
                    rtc.sendMessage(mac, "{TYPE=?,A0=?,A1=?,A2=?,A3=?,A4=?,A5=?,A6=?,A7=?,D1=?}");
                }
            }
        }
    }
```

下述 JS 开发代码的功能是根据设备连接情况，更新页面显示的水表信息，以及设备水阀的状态，实现了远程抄表功能。

```
var wsn_config = {
    "605" : {
        "online" : function() {
            $(".online_605").text("在线").css("color", "#96ba5c");
        },
        "pro" : function (tag, val) {
            if(tag=="D1"){
                if(val & 0x02){
                    $("#pumpBtn").text("关闭");
                    $("#pumpStatus").attr("src", "img/WaterPump-on.png");
                    state.pump = true;
```

```
            }else{
                $("#pumpBtn").text("打开");
                $("#pumpStatus").attr("src", "img/WaterPump-off.png");
                state.pump = false;
            }
        }
        else if(tag=="A0"){
            var msg ="";
            config.card.val = {};
            msg = "用户卡号: "+val;
            //显示当前卡号
            $("#ruc_rfid").text(val);
            message_show(msg);
        }
        else if(tag=="A3"){
            $("#parkNum").text("设备余量:"+val);
        }
        else if(tag=="A4"){
            message_show("扣费成功! "+val);
        }
        else if(tag=="A5"){
            $("#usedNum").text("累计用量:"+val);
        }
    }
  },
};
```

智能水表抄表系统 Web 端其他部分的代码请查看本书配套资料中的项目源文件。

7.2.4 开发验证

1. Web 端应用测试

在 Web 端打开智能水表抄表系统后，在“运营首页”下可以看到 Web 端的主页，如图 7.18 所示。

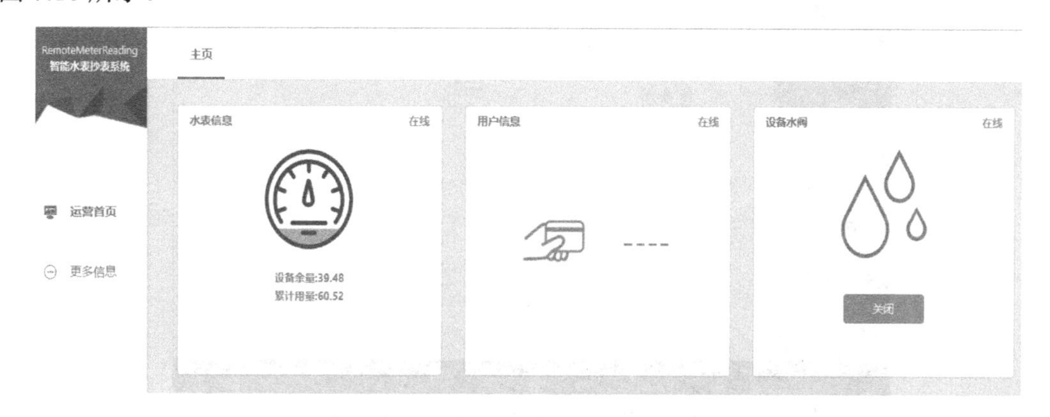

图 7.18 智能水表抄表系统的 Web 端主页

在 Sensor-EL 开发平台上使用用户卡片时，页面会显示读取到的信息，如图 7.19 所示。

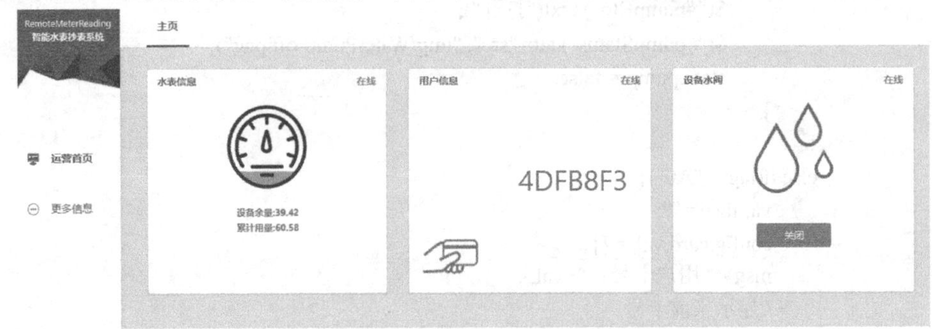

图 7.19　读取到用户卡片的信息

设备在线后可以通过 Sensor-EL 开发平台上的 K2 按键来模拟用量的减少，也可通过产生的随机数来定时减少用量，网页上会显示余量、累计用量等信息，在线测试如图 7.20 所示。

图 7.20　在线测试

2．Android 端应用测试

Android 端应用测试同 Web 端应用测试流程基本一致，可参考本系统的 Web 端应用测试进行操作。智能水表抄表系统的 Android 端“运营首页”页面如图 7.21 所示。

图 7.21　智能水表抄表系统的 Android 端“运营首页”页面

7.2.5　总结与拓展

本节基于 NB-IoT 实现了射频传感器的识别，通过 Android 和 HTML5 技术实现了 Android 端和 Web 端的应用设计，能通过 NB-IoT 实时获取射频传感器的数据，实现了基于 NB-IoT 的智能水表抄表系统。

第8章 LTE高级应用开发

LTE（Long Term Evolution，长期演进）技术是一种用于手机以及数据终端的高速数据通信标准。LTE 上行链路采用单载波频分多址技术（SC-FDMA），下行链路采用正交频分多址技术（OFDM），可支持更多系统带宽下的网络部署。有关更详尽的 LTE 内容请参考《物联网长距离无线通信技术应用与开发》。

本章通过基于LTE的仓库环境管理系统和自动化生产线计数系统这两个贴近生活的开发案例，详细地介绍 LTE 物联网系统的架构和软硬件开发，实现了识别类传感器、控制类传感器和安防类传感器的驱动程序，进行了 Android 端和 Web 端的应用开发。

8.1 基于 LTE 的仓库环境管理系统

仓库环境管理系统（见图 8.1）通过温湿度监控系统来监测仓库内的温湿度，根据监测的数据来控制风扇、空调等设备，从而实现无人化的管理。

图 8.1 仓库环境管理系统

8.1.1 系统开发目标

（1）熟悉温湿度传感器和继电器硬件原理和数据通信协议，实现基于 STM32F103 的温湿度传感器和继电器的驱动程序开发，通过温湿度传感器采集的数据来控制继电器开关，实现仓库环境管理系统的设计。

（2）实现仓库环境管理系统的 Android 端应用开发和 Web 端应用开发。

8.1.2 系统设计分析

1. 系统的功能设计

从系统功能的角度出发，仓库环境管理系统可以分为两个模块：设备采集和控制模块以及系统设置模块，如图 8.2 所示。

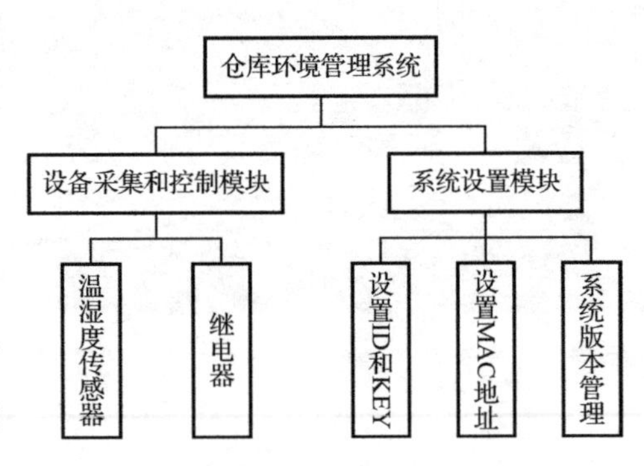

图 8.2 仓库环境管理系统的组成模块

仓库环境管理系统的功能需求如表 8.1 所示。

表 8.1 仓库环境管理系统的功能需求

功　能	功 能 说 明
采集数据显示	在上层应用页面中实时更新显示温湿度传感器采集的数据
除湿器实时控制	通过应用程序，对除湿器（空调）进行操作
智云连接设置	设置智云服务器的参数和设备的 MAC 地址

2. 系统的总体架构设计

仓库环境管理系统是基于物联网四层架构模型来设计的，其总体架构如图 8.3 所示。

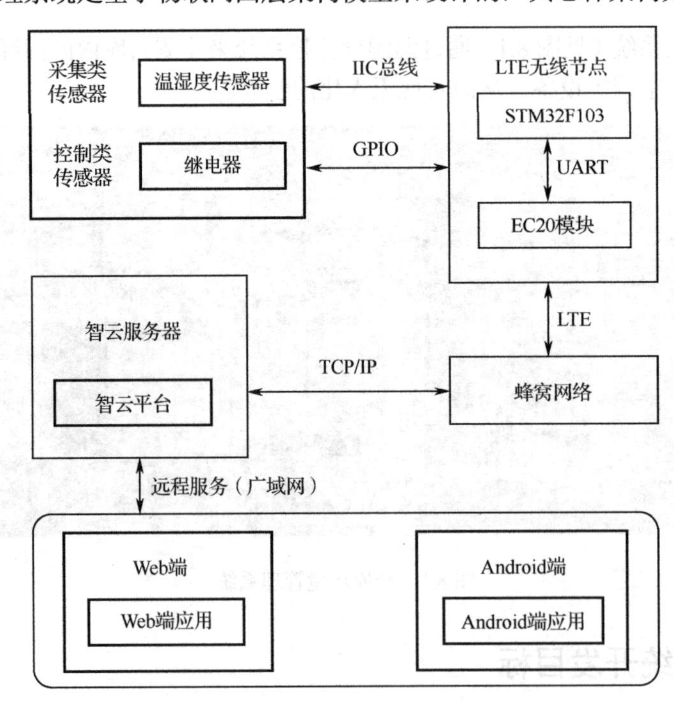

图 8.3 仓库环境管理系统的总体架构

3. 系统的数据传输

仓库环境管理系统的数据传输是在传感器节点、智云平台以及客户端（包括 Web 端和 Android 端）之间进行的，如图 8.4 所示，具体通信描述如下。

（1）传感器通过智能网关进行组网后与智云平台进行数据通信。

（2）传感器采集的数据通过 LTE 无线网络发送到智云平台，智云平台将数据发送给连接的客户端。

（3）客户端（Android 端和 Web 端）通过调用智云数据接口，实现数据实时采集功能。

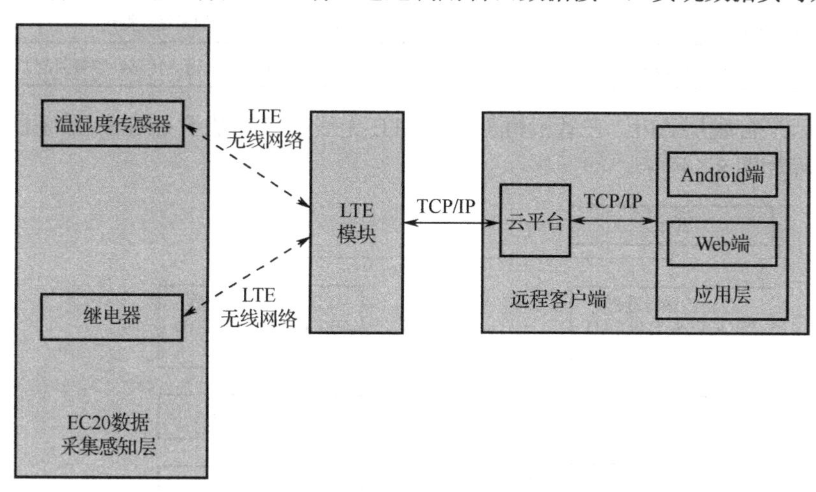

图 8.4　仓库环境管理系统的数据传输

8.1.3　系统的软硬件开发：仓库环境管理系统

1. 系统底层软硬件设计

1）感知层硬件设计

仓库环境管理系统感知层硬件包括 xLab 未来开发平台的 LTE 模块、增强型无线节点 ZXBeePlusB、采集类开发平台 Sensor-A 和控制类开发平台 Sensor-B。其中，LTE 模块负责汇集传感器采集的数据；LTE 无线节点（由增强型无线节点 ZXBeePlusB 实现）通过无线通信的方式向 LTE 模块发送传感器数据；采集类开发平台 Sensor-A 和控制类开发平台 Sensor-B 连接到 LTE 无线节点，对相关设备进行数据采集。本系统用到的传感器包括温湿度传感器和继电器，温湿度传感器的硬件接口电路如图 3.5 所示，继电器的硬件接口电路如图 4.18 所示。

2）系统底层开发

本系统是基于 LTE 无线网络进行开发的。

（1）LTE 智云开发框架。智云框架是在传感器应用程序接口和 SAPI 框架的基础上搭建起来的，通过合理调用这些接口函数，可以使 LTE 的开发形成一套系统的开发逻辑。传感器应用程序接口函数是在 sensor.c 文件中实现的，具体如表 8.2 所示。

表 8.2　传感器应用程序接口函数

函数名称	函数说明
sensorInit()	传感器初始化
sensorUpdate()	传感器数据定时上报
sensorControl()	传感器控制函数
sensorCheck()	传感器预警监测及处理函数
ZXBeeInfRecv()	处理节点接收到的无线数据包
PROCESS_THREAD(sensor, ev, data)	传感器进程（处理传感器上报、传感器预警监测）

（2）智云平台底层 API。在智云框架下，LTE 无线节点是基于 Contiki 系统框架开发的，详细程序流程如图 8.5 所示。

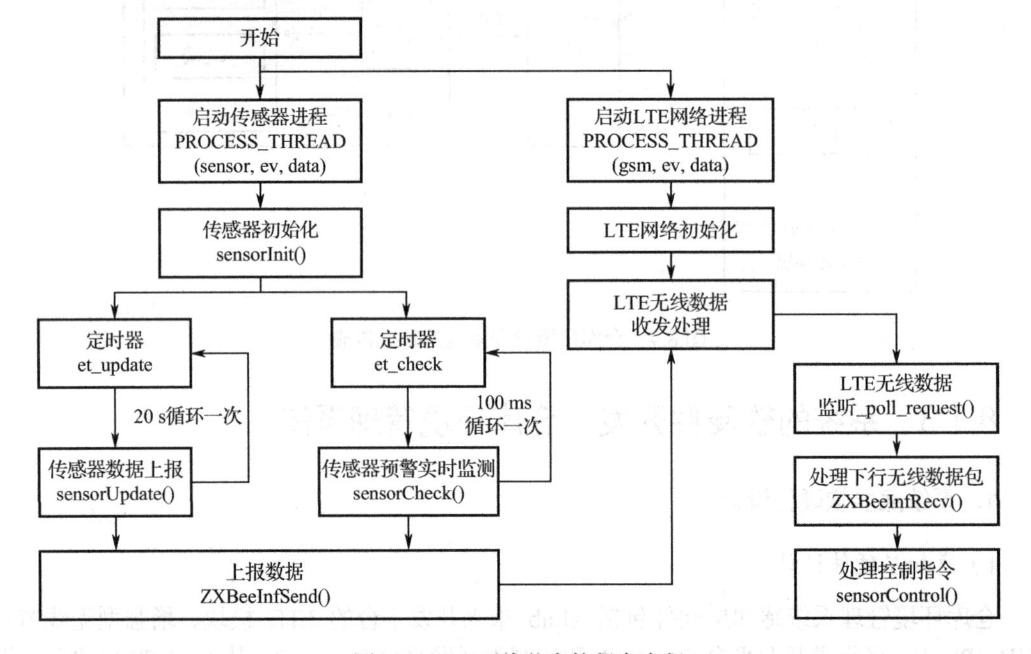

图 8.5　LTE 无线节点的程序流程

智云框架为 LTE 无线协议栈的上层应用提供分层的软件设计结构，将传感器的私有操作部分封装在 sensor.c 文件中，用户任务中的处理事件和节点类型选择在 sensor.h 文件中定义。sensor.h 文件中事件宏定义如下：

```
#define NODE_NAME "601"
```

sensor.h 文件中声明了智云框架下的传感器应用文件 sensor.c 中的函数，传感器进程用于启动传感器任务及定时器任务，相关代码如下：

```
PROCESS_THREAD(sensor, ev, data)
{
    static struct etimer et_update;
    PROCESS_BEGIN();
```

```
        sensorInit();
        etimer_set(&et_update, CLOCK_SECOND*10);
        while (1) {
            PROCESS_WAIT_EVENT_UNTIL(ev == PROCESS_EVENT_TIMER);
            if (etimer_expired(&et_update)) {
                printf("sensor->PROCESS_EVENT_TIMER: PROCESS_EVENT_TIMER trigger!\r\n");
                sensorUpdate();
                etimer_set(&et_update, CLOCK_SECOND*10);
            }
        }
        PROCESS_END();
}
```

sensorInit()函数用于初始化传感器，代码如下：

```
/*********************************************************************************************
*名称：sensorInit()
*功能：传感器初始化
*********************************************************************************************/
void sensorInit(void)
{
    printf("sensor->sensorInit(): Sensor init!\r\n");
    ……
}
```

sensorUpdate()函数用于更新传感器的数据，并将更新后的数据打包上报，相关代码如下：

```
/*********************************************************************************************
*名称：sensorUpdate()
*功能：处理主动上报的数据
*********************************************************************************************/
void sensorUpdate(void)
{
    char pData[32];
    char *p = pData;
    if (pData != NULL) {
        ZXBeeInfSend(p, strlen(p));                                    //将数据发送到智云平台
    }
    printf("sensor->sensorUpdate(): airPressure=%.1f\r\n", airPressure);
}
```

ZXBeeInfSend()函数用于发送数据，相关代码如下：

```
/*********************************************************************************************
*名称：ZXBeeInfSend()
*功能：节点将无线数据包发送到智云平台
*参数：*p—要发送的无线数据包；len—无线数据包的长度
*********************************************************************************************/
void ZXBeeInfSend(char *p, int len)
```

```
{
    leds_on(1);
    clock_delay_ms(50);
    zhiyun_send(p);
    leds_off(1);
}
```

ZXBeeInfRecv()函数用于处理节点接收到的无线数据包，相关代码如下：

```
/*********************************************************************************************
*名称：ZXBeeInfRecv()
*功能：处理节点接收到的无线数据包
*参数：*pkg—接收到的无线数据包；len—无线数据包的长度
*********************************************************************************************/
void ZXBeeInfRecv(char *buf, int len)
{
    uint8_t val;
    char pData[16];
    char *p = pData;
    char *ptag = NULL;
    char *pval = NULL;

    printf("sensor->ZXBeeInfRecv(): Receive LTE Data!\r\n");
    leds_on(1);
    clock_delay_ms(50);
    ptag = buf;
    p = strchr(buf, '=');
    if (p != NULL) {
        *p++ = 0;
        pval = p;
    }
    val = atoi(pval);

    //控制命令解析
    if (0 == strcmp("cmd", ptag)){
        sensorControl(val);
    }
    leds_off(1);
}
```

sensorControl()函数用于对传感器进行控制，相关代码如下：

```
/*********************************************************************************************
*名称：sensorControl()
*功能：传感器控制
*参数：cmd—控制命令
*********************************************************************************************/
void sensorControl(uint8_t cmd)
```

```
{
    //根据cmd参数执行对应的控制程序
}
```

通过在 sensor.c 文件中实现具体的函数即可快速地完成 LTE 项目开发。

3）传感器驱动设计

仓库环境管理系统中的传感器主要是采集类传感器和控制类传感器。

（1）数据通信协议的定义。本系统主要使用采集类可开发平台 Sensor-A 和控制类开发平台 Sensor-B，其 ZXBee 数据通信协议如表 8.3 所示。

表 8.3 采集类开发平台和控制类开发平台的 ZXBee 数据通信协议

传感器	属性	参数	权限	说明
Sensor-A（601）	温度	A0	R	温度值为浮点型数据，精度为0.1，-40～105，单位为℃
	湿度	A1	R	湿度值为浮点型数据，精度为0.1，范围为0～100，单位为%
	上报状态	D0(OD0/CD0)	R/W	D0 的 Bit0～Bit7 分别代表 A0～A7 传感器数据的上报状态，1 表示允许上报，D 表示不允许上报
	数据上报时间间隔	V0	R/W	循环上报的时间间隔
Sensor-B（602）	上报状态	D0(OD0/CD0)	R/W	D0 的 Bit0～Bit7 分别代表 A0～A7 传感器数据的上报状态，1 表示允许上报，D 表示不允许上报
	继电器	D1(OD1/CD1)	R/W	D1 的 Bit6、Bit7 分别代表继电器 K1、K2 的开关状态，0 表示断开，1 表示吸合
	数据上报时间间隔	V0	R/W	循环上报时间间隔

（2）驱动程序的开发。在智云框架下不仅可以很容易地实现传感器驱动程序的开发，还可以省略无线传感器节点的组网和用户任务的创建等烦琐过程。例如，直接调用 sensorInit()函数可以实现节点传感器的初始化；调用 ZXBeeInfRecv()函数可以处理节点接收到的无线数据包；ZXBeeInfSend()函数用于将无线数据包发送至智云平台；ZXBeeUserProcess()函数用于解析接收到的控制命令；设备状态的定时上报需要使用 PROCESS_THREAD()作为 sensorUpdate()函数和 sensorCheck()函数的定时进入接口以反馈设备状态信息。

在 sensor.c 中，需要在 sensorInit()函数中添加传感器初始化的内容，并通过定义上报事件和报警事件来实现设备工作状态的定时反馈，部分代码如下：

```
/*********************************************************************************************
*名称：sensorInit()
*功能：传感器初始化
*********************************************************************************************/
void sensorInit(void)
{
    //初始化传感器代码
    htu21d_init();                                  //HTU21D型温湿度传感器初始化
    relay_init();                                   //继电器初始化
}
```

温湿度传感器的初始化函数是htu21d_init()，代码如下：

```
/*********************************************************************************
*名称：htu21d_init()
*功能：HTU21D 型温湿度传感器初始化
*********************************************************************************/
void htu21d_init(void)
{
    iic_init();                                         //IIC 总线初始化
    iic_start();                                        //开启 IIC 总线
    iic_write_byte(HTU21DADDR&0xfe);                    //写 HTU21D 型温湿度传感器的 IIC 总线地址
    iic_write_byte(0xfe);
    iic_stop();                                         //停止 IIC 总线
    delay(600);                                         //短延时
}
/*********************************************************************************
*名称：htu21d_read_reg()
*功能：读取 HTU21D 型温湿度传感器寄存器的数据
*参数：cmd—寄存器地址
*返回：data—寄存器的数据
*********************************************************************************/
unsigned char htu21d_read_reg(unsigned char cmd)
{
    unsigned char data = 0;
    iic_start();                                        //开启 IIC 总线
    if(iic_write_byte(HTU21DADDR & 0xfe) == 0){         //写 HTU21D 型温湿度传感器的 IIC 总线地址
        if(iic_write_byte(cmd) == 0){                   //写寄存器地址
            do{
                delay(30);                              //延时 30 ms
                iic_start();                            //开启 IIC 总线
            }
            while(iic_write_byte(HTU21DADDR | 0x01) == 1);    //发送读信号
            data = iic_read_byte(0);                          //读取一个字节数据
            iic_stop();                                       //停止 IIC 总线
        }
    }
    return data;
}
/*********************************************************************************
*名称：htu21d_get_data()
*功能：HTU21D 型温湿度传感器采集的数据
*参数：order—指令
*返回：temperature—温度值  humidity—湿度值
*********************************************************************************/
int htu21d_get_data(unsigned char order)
{
    float temp = 0,TH = 0;
```

```
    unsigned char MSB,LSB;
    unsigned int humidity,temperature;
    iic_start();                                      //开启 IIC 总线
    if(iic_write_byte(HTU21DADDR & 0xfe) == 0){       //写 HTU21D 型温湿度传感器的 IIC 总线地址
        if(iic_write_byte(order) == 0){               //写寄存器地址
            do{
                delay(30);
                iic_start();
            }
            while(iic_write_byte(HTU21DADDR | 0x01) == 1);      //发送读信号
            MSB = iic_read_byte(0);                              //读取数据高 8 位
            delay(30);                                           //延时
            LSB = iic_read_byte(0);                              //读取数据低 8 位
            iic_read_byte(1);
            iic_stop();                                          //停止 IIC 总线
            LSB &= 0xfc;                                         //取出数据有效位
            temp = MSB*256+LSB;                                  //数据合并
            if (order == 0xf3){                                  //触发开启温度监测
                TH=(175.72)*temp/65536-46.85;                    //温度：T= -46.85 + 175.72 *ST/2^16
                temperature =(unsigned int)(fabs(TH)*100);
                if(TH >= 0)
                    flag = 0;
                else
                    flag = 1;
                return temperature;
            }else{
                TH = (temp*125)/65536-6;
                humidity = (unsigned int)(fabs(TH)*100);         //湿度：RH%= -6 + 125 *SRH/2^16
                return humidity;
            }
        }
    }
    iic_stop();
    return 0;
}
```

继电器的初始化函数是 relay_init()，代码如下：

```
/*********************************************************************************************
*名称：relay_init()
*功能：继电器初始化
*********************************************************************************************/
void relay_init(void)
{
    GPIO_InitTypeDef   GPIO_InitStructure;
    RCC_APB2PeriphClockCmd(RCC_APB2Periph_GPIOA, ENABLE);
    GPIO_InitStructure.GPIO_Pin =   GPIO_Pin_5 | GPIO_Pin_4;
    GPIO_InitStructure.GPIO_Mode = GPIO_Mode_Out_PP;
```

```
        GPIO_InitStructure.GPIO_Speed = GPIO_Speed_2MHz;
        GPIO_Init(GPIOA, &GPIO_InitStructure);
        relay_control(0x00);
}
/*********************************************************************************************
*名称：relay_control()
*功能：继电器控制
*********************************************************************************************/
void relay_control(unsigned char cmd)
{
        if(cmd & 0x01){
                GPIO_ResetBits(GPIOA,    GPIO_Pin_5);
        }else{
                GPIO_SetBits(GPIOA,    GPIO_Pin_5);
        }
        if(cmd & 0x02){
                GPIO_ResetBits(GPIOA,    GPIO_Pin_4);
        }else{
                GPIO_SetBits(GPIOA,    GPIO_Pin_4);
        }
}
```

2. Android 端应用设计

1）Android 工程设计框架

打开 Android Studio 开发环境，可以看到本系统的工程目录，如图 8.6 所示，系统的工程框架说明如表 8.4 所示。

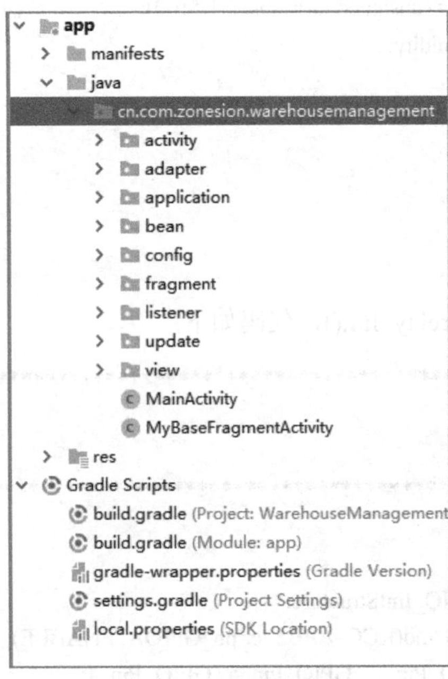

图 8.6 工程目录列表

表8.4 仓库环境管理系统的工程框架

类 名	说 明
activity	
IdKeyShareActivity.java	在IDKey页面单击“分享”按钮时，可弹出activity，用于分享二维码图片
adapter	
HdArrayAdapter.java	历史数据显示适配器
application	
LCApplication.java	LCApplication继承application类，使用单例模式（Singleton Pattern）创建WSNRTConnect对象
bean	
HistoricalData.java	历史数据的bean类，用于将从智云服务器获得的历史数据记录（JSON形式）转换成该类对象
IdKeyBean.java	IdKeyBean用来描述用户设备的ID、KEY，以及智云服务器的地址SERVER
config	
Config.java	config用于修改用户的ID、KEY，以及智云服务器的地址和MAC地址
fragment	
BaseFragment.java	页面基础Fragment定义类
HDFragment.java	历史数据页面
HistoricalDataFragment.java	历史数据显示页面
HomepageFragment.java	展示首页页面的Fragment
IDKeyFragment.java	IDKey选项的页面
MacSettingFragment.java	当用户设置被监测项的MAC地址时显示的页面
MoreInformationFragment.java	更多信息显示页面
RunHomePageFragment.java	运营首页显示页面
VersionInformationFragment.java	显示版本等相关信息的页面
listener	
IOnWSNDataListener.java	传感器数据监听器接口
update	
UpdateService.java	应用下载服务类
view	
APKVersionCodeUtils.java	获取当前本地apk的版本
CustomRadioButton.java	自定义按钮类
PagerSlidingTabStrip.java	自定义滑动控件类
MainActivity.java：主页面类	
MyBaseFragmentActivity.java：系统Fragment通信类	

2）软件设计

根据智云Android端应用程序接口的定义，系统的应用设计主要采用实时数据API接口，实时数据API接口的流程见3.1.3节的图3.11。

（1）LCApplication.java 程序代码剖析。仓库环境管理系统中的 LCApplication.java 程序代码和城市环境信息采集系统的 LCApplication.java 程序代码相同，详见 3.1.3 节的相关内容。

（2）HomepageFragment.java 程序代码剖析。下面的代码通过(LCApplication) getActivity().getApplication()获取 LCApplication 类中的 WSNRTConnect 对象。

```
private void initViewAndBindEvent() {
    preferences = getActivity().getSharedPreferences("user_info", Context.MODE_PRIVATE);
    lcApplication = (LCApplication) getActivity().getApplication();
    wsnrtConnect = lcApplication.getWSNRConnect();
    lcApplication.registerOnWSNDataListener(this);
    editor = preferences.edit();
}
```

下面的代码功能是当湿度超过设置的湿度上限时，打开除湿器。

```
private void temperatureTooHigh() {
    if (sensorAMac != null) {
        if (temperatureUpperLimit <= currentTemperature && isSecurityMode == true) {
            wsnrtConnect.sendMessage(sensorAMac, "{OD1=64,D1=?}".getBytes());
        } else {
            wsnrtConnect.sendMessage(sensorAMac, "{CD1=64,D1=?}".getBytes());
        }
    }else {
        Toast.makeText(lcApplication, "请等待 MAC 地址上线", Toast.LENGTH_SHORT).show();
    }
}
```

下面的代码通过复写 onMessageArrive 方法来处理节点接收到的无线数据包，实现了设备的 MAC 地址获取，并在当前的页面显示设备的状态。

```
@Override
    public void onMessageArrive(String mac, String tag, String val) {
        if (sensorAMac == null) {
            wsnrtConnect.sendMessage(mac, "{TYPE=?}".getBytes());
        }
        if ("TYPE".equals(tag) && "601".equals(val.substring(2, val.length()))) {
            sensorAMac = mac;
        }
        if (tag.equalsIgnoreCase("A0") && mac.equalsIgnoreCase(sensorAMac)) {
            textTemperatureState.setText("在线");
            textTemperatureState.setTextColor(getResources().getColor(R.color.line_text_color));
            temperatureText.setText(val + "℃");
            currentTemperature = Float.parseFloat(val);
        }
        if (tag.equalsIgnoreCase("A1") && mac.equalsIgnoreCase(sensorAMac)) {
            textHumidityState.setText("在线");
            textHumidityState.setTextColor(getResources().getColor(R.color.line_text_color));
            humidityText.setText(val + "%");
```

```
        }
        if (tag.equalsIgnoreCase("D1") && mac.equalsIgnoreCase(sensorAMac)) {
            textDehumidifierState.setText("在线");
            textDehumidifierState.setTextColor(getResources().getColor(R.color.line_text_color));
            int numResult = Integer.parseInt(val);
            if ((numResult & 0X40) == 0x40) {
                imageDehumidifierState.setImageDrawable(getResources().getDrawable(R.drawable.
                                                                    dehumidifier_on));
                openOrCloseLamp.setText("关闭");
                openOrCloseLamp.setBackground(getResources().getDrawable(R.drawable.close));
            }else {
                imageDehumidifierState.setImageDrawable(getResources().getDrawable(R.drawable.
                                                                    dehumidifier_off));
                openOrCloseLamp.setText("开启");
                openOrCloseLamp.setBackground(getResources().getDrawable(R.drawable.open));
            }
        }
    }
```

（3）HDFragment.java 程序代码剖析。HDFragment.java 程序中关键代码如下：

```
public class temperatureHDFragment extends BaseFragment implements IOnWSNDataListener{
    ......
    @Override
    public void initData() {
        //TODO Auto-generated method stub
        super.initData();
        initViewAndBindEvent();
    }
    private void initViewAndBindEvent() {
        config = Config.getConfig();
        lcApplication = (LCApplication) getActivity().getApplication();
        lcApplication.registerOnWSNDataListener(this);
        wsnrtConnect = lcApplication.getWSNRConnect();
        preferences = getActivity().getSharedPreferences("user_info", Context.MODE_PRIVATE);
        wsnHistory = new WSNHistory();                    //声明一个 WSNHistory 实例
        initSetting();
        //声明一个 ArrayAdapter 对象，用来对 Spinner 进行适配
        ......
    }
    /***初始化用户的 ID 和 KEY 以及智云服务器地址*/
     private void initSetting(){
        String id;
        String key;
        String serverAddress;
        id = preferences.getString("id",config.getUserId());
        key = preferences.getString("key",config.getUserKey());
        serverAddress = preferences.getString("server" + ":8080",config.getServerAddress()+ ":8080");
```

```
        wsnHistory.setIdKey(id,key);
        wsnHistory.setServerAddr(serverAddress);
    }
    /***通过 Java 反射来让 Spinner 选择同一个选项时会触发 onItemSelected 事件*/
        private void reflectChangeSpinnerPosition(int position){
        ……
    }
    /***该方法用于更新 UI*@param json*/
    private void mainThreadUpdateUI(final String json){
        getActivity().runOnUiThread(new Runnable() {
            @Override
                public void run() {
                    lineChartView.setVisibility(View.VISIBLE);
                    initLineChart(json);
                    textDataTip.setVisibility(View.GONE);
                }
        });
    }
    private void onSpinnerItemSelected(final int position) {
        hasRunned = true;
        if (position == 1) {
            new Thread(new Runnable() {
                @Override
                    public void run() {
                        try {
                            String result = wsnHistory.queryLast1H(airHumidityMac + "_A1");
                            mainThreadUpdateUI(result);
                        } catch (Exception e) {
                            e.printStackTrace();
                        }
                    }
            }).start();
        } else if (position == 2) {
            ……
        }
        @Override
            public void onStart() {
                super.onStart();
                reflectChangeSpinnerPosition(position);
            }
        @Override
            public void onDestroyView() {
                lcApplication.unregisterOnWSNDataListener(this);
                super.onDestroyView();
            }
        private void initLineChart(String json) {
            //数据拆线图的实现代码
```

```
        }
        /*当智云服务器信息到达时的回调方法*/
        @Override
            public void onMessageArrive(String mac, String tag, String val) {
                if (airHumidityMac == null) {
                    wsnrtConnect.sendMessage(mac, "{TYPE=?}".getBytes());
                }
                if (tag.equalsIgnoreCase("TYPE") && val.substring(2, val.length()).equals("601")){
                    airHumidityMac = mac;
                }
                if (airHumidityMac != null&&position!=0) {
                    if (!hasRunned) {
                        onSpinnerItemSelected(airHumiditySearch.getSelectedItemPosition());
                    }
                }
            }
    }
}
```

Android 端其余部分的代码请查看项目的源文件。

3. Web 端应用设计

1）页面功能结构分析

仓库环境管理系统的 Web 端默认显示的是“运营首页”页面，在“运营首页”页面上设计了 4 个模块，分别是当前温度显示模块、当前湿度显示模块、除湿器控制模块、湿度阈值设置模块。仓库环境管理系统 Web 端的“运营首页”页面如图 8.7 所示。

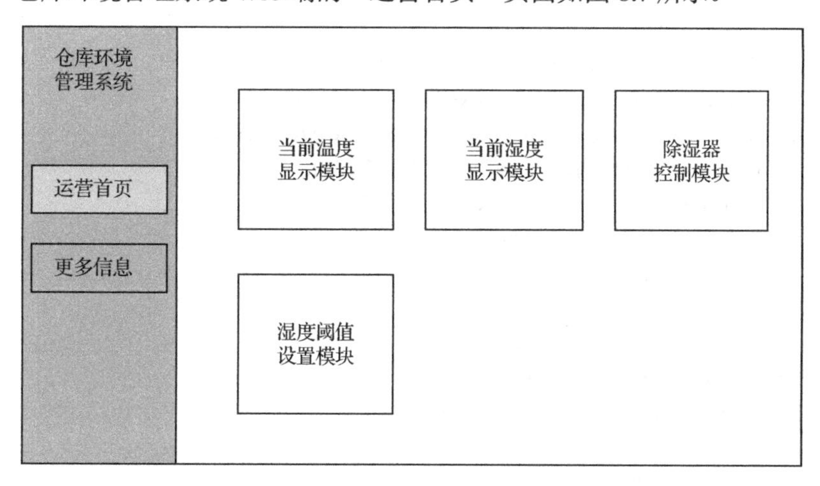

图 8.7　仓库环境管理系统 Web 端的“运营首页”页面

“更多信息”页面的主要功能是进行智云服务器的连接配置，和城市环境信息采集系统中的 Web 端“更多信息”页面类似，参见图 3.14。

2）软件设计

仓库环境管理系统 Web 端的 JS 开发逻辑与 Android 端的开发逻辑相似，首先通过配置 ID 和 KEY 与智云服务器进行连接，再通过实时监听数据的方法来获取相关传感器的数据并进行处理。JS 开发的部分代码如下。

在 getConnect()函数中定义了实时连接对象 rtc，连接成功回调函数是 rtc.onConnect，数据服务掉线回调函数是 rtc.onConnectLost，消息处理回调函数是 rtc.onmessageArrive。

```
function getConnect() {
    config["id"] = config["id"] ? config["id"] : $("#id").val();
    config["key"] = config["key"] ? config["key"] : $("#key").val();
    config["server"] = config["server"] ? config["server"] : $("#server").val();
    //创建数据连接服务对象
    rtc = new WSNRTConnect(config["id"], config["key"]);
    rtc.setServerAddr(config["server"] + ":28080");
    rtc.connect();
    rtc._connect = false;
    //连接成功回调函数
    rtc.onConnect = function() {
        $("#ConnectState").text("数据服务连接成功！");
        rtc._connect = 1;
        message_show("数据服务连接成功！");
        $("#idkeyInput").text("断开").addClass("btn-danger");
        $("#id,#key,#server").attr('disabled',true);
    };
    //数据服务掉线回调函数
    rtc.onConnectLost = function() {
        rtc._connect = 0;
        $("#ConnectState").text("数据服务连接掉线！");
        $("#idkeyInput").text("连接").removeClass("btn-danger");
        message_show("数据服务连接失败，检查网络或 ID、KEY");
        $("#RFIDLink").text("离线").css("color", "#e75d59");
        $("#doorLink").text("离线").css("color", "#e75d59");
        $("#id,#key,#server").removeAttr('disabled',true);
    };
    //消息处理回调函数
    rtc.onmessageArrive = function (mac, dat) {
        //console.log(mac+" >>> "+dat);
        if (dat[0]=='{' && dat[dat.length-1]=='}') {
            dat = dat.substr(1, dat.length-2);
            var its = dat.split(',');
            for (var i=0; i<its.length; i++) {
                var it = its[i].split('=');
                if (it.length == 2) {
                    process_tag(mac, it[0], it[1]);
                }
            }
```

```
                if (!mac2type[mac]) { //如果没有获取到 TYPE 值，主动去查询
                    rtc.sendMessage(mac, "{TYPE=?,A0=?,A1=?,A2=?,A3=?,A4=?,A5=?,A6=?,A7=?,D1=?}");
                }
            }
        }
    }
```

下述 JS 开发代码的功能是根据设备连接情况在页面更新设备的状态，并实现除湿器控制中的“打开”或“关闭”按钮功能。

```
var wsn_config = {
    "601" : {
        "online" : function() {
            $(".online_601").text("在线").css("color", "#96ba5c");
        },
        "pro" : function (tag, val) {
            if(tag=="A0"){
                thermometer('temp','℃','#2293ed', -20, 80, val);
            }
            if(tag=="A1"){
                thermometer('humi','%','#ff7850', 0, 100, val);
                if(config["mac_602"]!=""){
                    //超过较大值，打开除湿器
                    if(val>config["threshold"] && !state.Dehumidifier){
                        $("#Dehumidifier").text("关闭");
                        $("#doorImg").attr("src", "img/dehumidifier-on.png");
                        rtc.sendMessage(config["mac_602"],"{OD1=64,D1=?}");
                        message_show("超出最大湿度阈值，将自动打开除湿器");
                    }
                    //低于较小值，关闭除湿器
                    else if(val<config["threshold"] && state.Dehumidifier){
                        $("#Dehumidifier").text("打开");
                        $("#doorImg").attr("src", "img/dehumidifier-off.png");
                        rtc.sendMessage(config["mac_602"],"{CD1=64,D1=?}");
                        message_show("小于最大湿度阈值，将自动关闭除湿器");
                    }
                }
            }
        }
    },
    "602" : {
        "online" : function() {
            $(".online_602").text("在线").css("color", "#96ba5c");
        },
        "pro" : function (tag, val) {
            if(tag=="D1"){
                if(val & 64){
                    $("#Dehumidifier").text("关闭");
```

```
                            $("#doorImg").attr("src", "img/dehumidifier-on.png");
                            state.Dehumidifier = true;
                        }else{
                            $("#Dehumidifier").text("打开");
                            $("#doorImg").attr("src", "img/dehumidifier-off.png");
                            state.Dehumidifier = false;
                        }
                    }
                }
            }
};
```

下面的代码通过 rtc.sendMessage(config["mac_602"], cmd)方法对设备进行控制。

```
$("#Dehumidifier").on("click", function() {
    if(page.checkOnline() && page.checkMac("mac_602")){
        var state = $(this).text() == "关闭", cmd;
        if(state){
            cmd = "{CD1=64,D1=?}";
        }else{
            cmd = "{OD1=64,D1=?}";
        }
        console.log(cmd);
        rtc.sendMessage(config["mac_602"], cmd);
    }
});
```

仓库环境管理系统 Web 端其他部分的代码请查看本书配套资料中的项目源文件。

8.1.4 开发验证

1. Web 端应用测试

在 Web 端打开仓库环境管理系统后，在“运营首页”下可以看到 Web 端的主页，如图 8.8 所示。

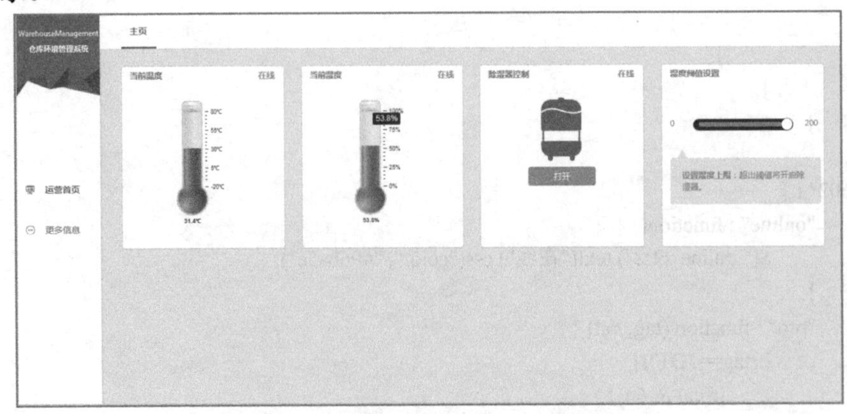

图 8.8 仓库环境管理系统 Web 端的主页

当设备在线后，可以通过“除湿器控制”中的“打开”或“关闭”按钮对设备进行手动控制，如图 8.9 所示。

当仓库中的湿度超过设置的湿度阈值时，本系统可以自动开启除湿器，如图 8.10 所示。

图 8.9　手动控制除湿器

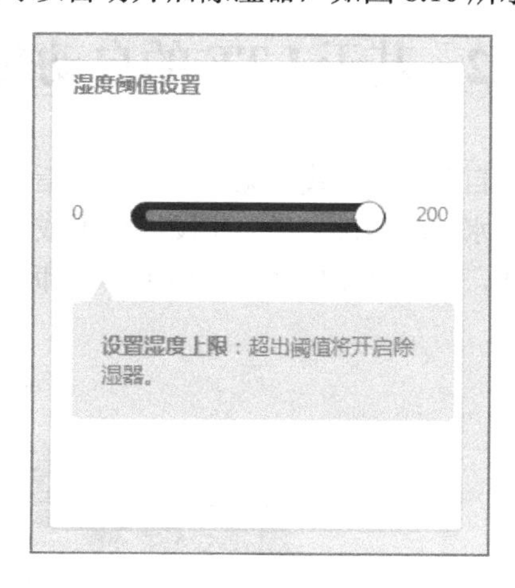

图 8.10　自动开启除湿器

2. Android 端应用测试

Android 端应用测试同 Web 端应用测试流程基本一致，可参考本系统的 Web 端应用测试进行操作。仓库环境管理系统的 Android 端“运营首页”页面如图 8.11 所示。

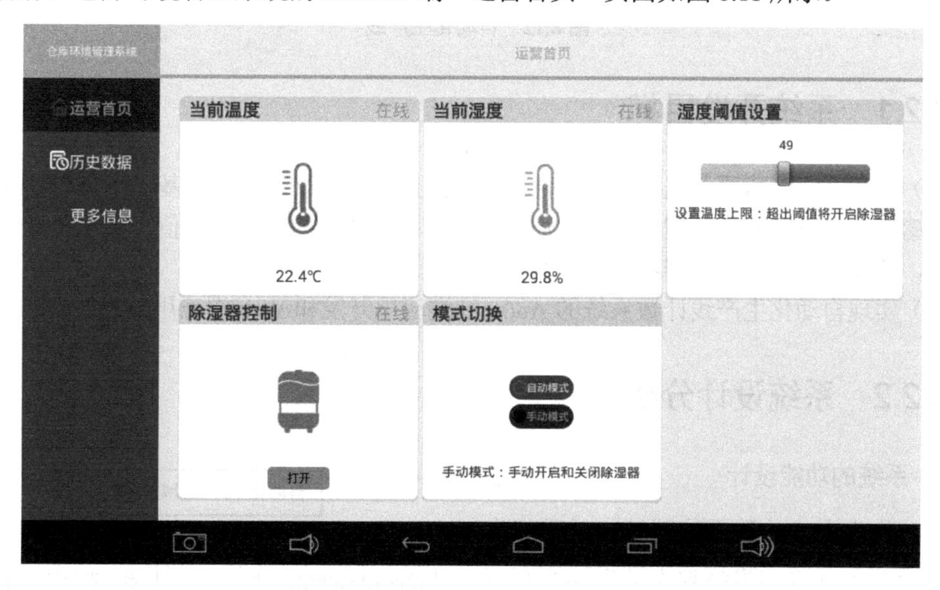

图 8.11　仓库环境管理系统的 Android 端“运营首页”页面

8.1.5　总结与拓展

本节基于 LTE 实现了温湿度传感器的数据采集和对继电器的控制，通过 Android 和 HTML5 技术实现了 Android 端和 Web 端的应用设计，能根据温湿度传感器实时采集的信息来对继电器（用继电器来模拟除湿器的开关）进行控制，实现了基于LTE的仓库环境管理系统。

8.2 基于LTE的自动化生产线计数系统

自动化生产线（见图8.12）是指由自动化机器实现产品生产过程的一种组织形式，它是在流水线的基础上进一步发展而形成的，其特点是加工对象自动地由一台机床传送到另一台机床，并由机床自动进行加工、装卸、检验等。

图8.12　自动化生产线

8.2.1 系统开发目标

（1）熟悉光栅传感器、继电器等硬件原理和数据通信协议，并实现基于STM32F103的光栅传感器、继电器的驱动程序开发，通过比较光栅传感器的次数与阈值，实现传送带的智能控制，从而实现自动化生产线计数系统的设计。

（2）实现自动化生产线计数系统的Android端应用开发和Web端应用开发。

8.2.2 系统设计分析

1．系统的功能设计

自动化生产线计数系统能够实时采集光栅传感器状态并将采集状态主动推送到智云平台，通过比较光栅传感器的次数与设置的阈值，能够控制传送带的启停，从而实现自动化生产线计数系统的设计。从系统功能的角度出发，自动化生产线计数系统可以分为两个模块：设备采集和控制模块以及系统设置模块，如图8.13所示。

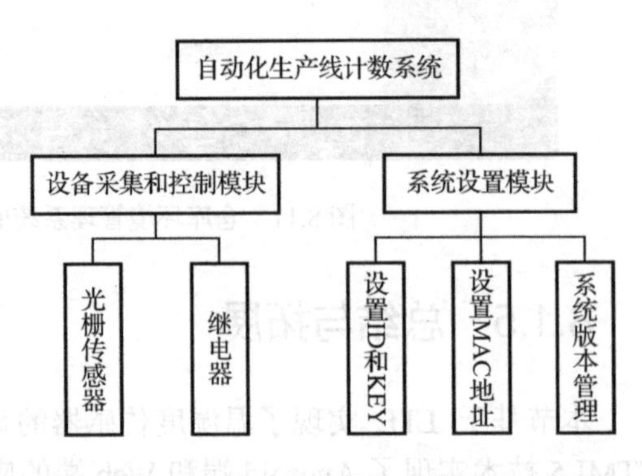

图8.13　自动化生产线计数系统的组成模块

自动化生产线计数系统的功能需求如表 8.5 所示。

表 8.5　自动化生产线计数系统的功能需求

功　　能	功 能 说 明
采集数据显示	在上层应用页面中实时更新显示光栅传感器的状态
传送带实时控制	通过应用程序，对传送带进行智能控制
智云连接设置	设置智云服务器的参数和设备的 MAC 地址

2．系统的总体架构设计

自动化生产线计数系统是基于物联网四层架构模型来设计的，其总体架构如图 8.14 所示。

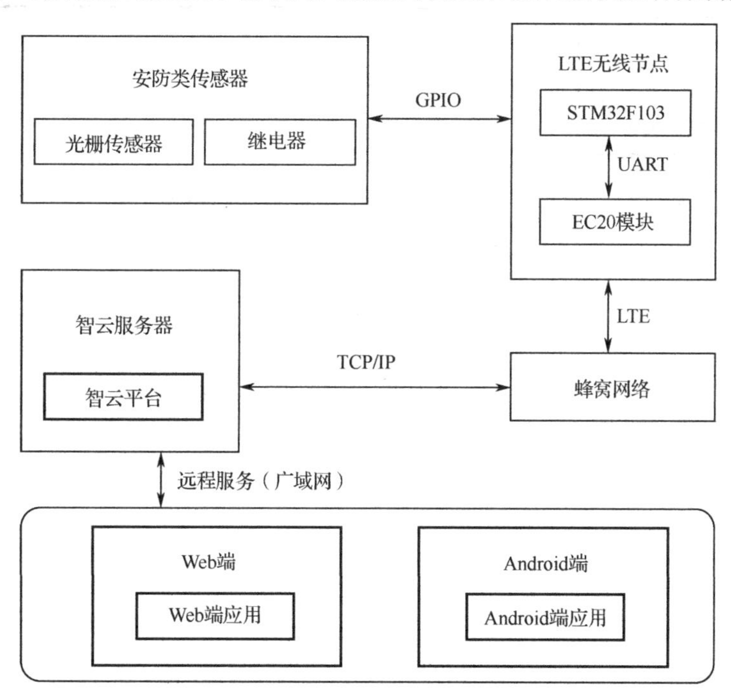

图 8.14　自动化生产线计数系统的总体架构

3．系统的数据传输

自动化生产线计数系统中的数据传输是在传感器节点、智云平台以及客户端（包括 Web 端和 Android 端）之间进行的，如图 8.15 所示。

8.2.3　系统的软硬件开发：自动化生产线计数系统

1．系统底层软硬件设计

1）感知层硬件设计

自动化生产线计数系统感知层的硬件主要包括 xLab 未来开发平台的 LTE 模块、LTE 无线节点、安防类开发平台 Sensor-C。其中 LTE 无线节点通过无线通信的方式向智云平台发送传感器数据，接收智云平台相关控制命令；安防类开发平台 Sensor-C 连接到 LTE 无线节点，对相关设备进行采集和控制。传感器包括光栅传感器和继电器。

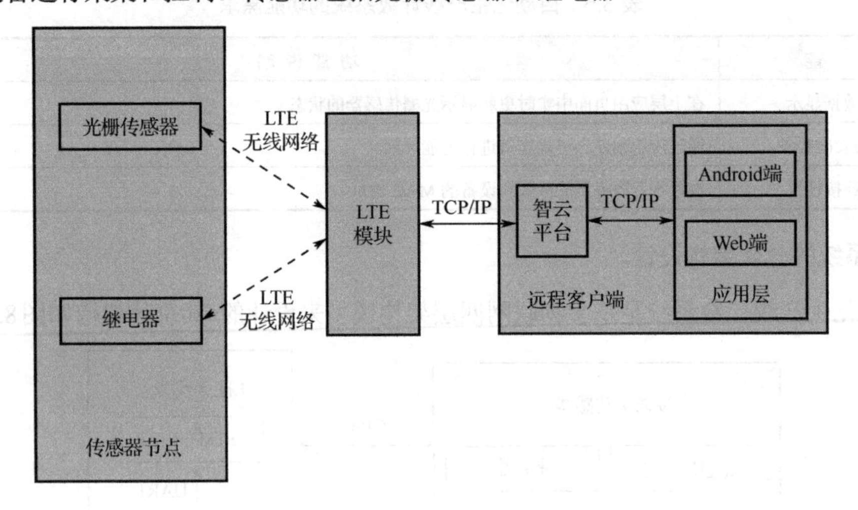

图 8.15　自动化生产线计数系统的数据传输

光栅传感器的硬件接口电路如图 8.16 所示，继电器的硬件接口电路如图 4.18 所示。

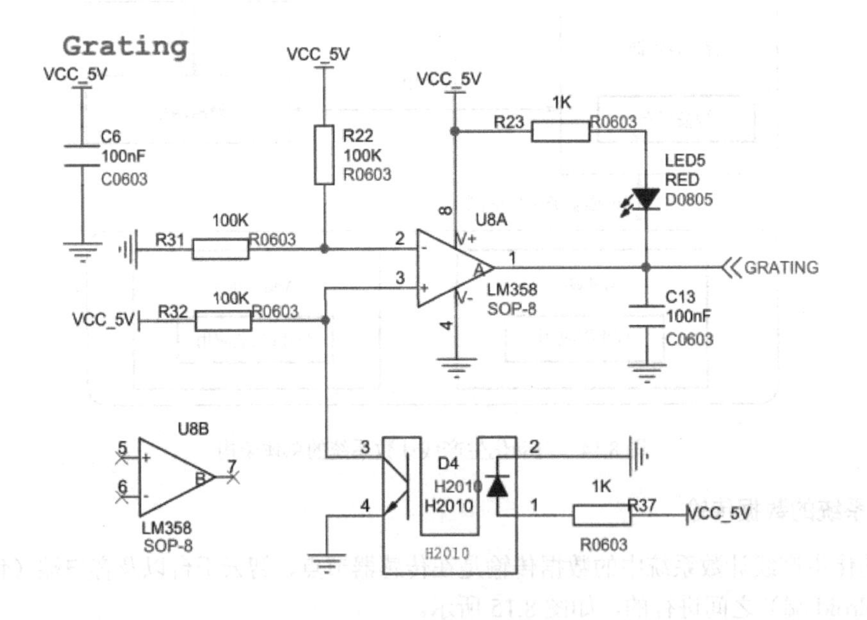

图 8.16　光栅传感器的硬件接口电路

2）系统底层开发

本系统是基于 LTE 无线网络进行开发的。

（1）LTE 智云开发框架。本系统采用的智云框架和仓库环境管理系统采用的智云框架相同，详见 8.1.3 节。

（2）智云平台应用接口分析。本系统的智云平台应用接口和仓库环境管理系统相同，详见 8.1.3 节。

3）传感器驱动设计

（1）数据通信协议的定义。本系统主要使用的是安防类开发平台Sensor-c，其ZXBee数据通信协议定义如表8.6所示。

表8.6 安防类开发平台的ZXBee数据通信协议

开发平台	属 性	参 数	权限	说 明
Sensor-C（603）	光栅（红外对射）状态	A5	R	光栅状态值，0或1变化，1表示监测到阻挡，0表示未监测到阻挡
	继电器	D1(OD1/CD1)	R/W	D1的Bit6～Bit7分别代表继电器K1、K2的开关状态，0表示断开，1表示吸合
	上报状态	D0(OD0/CD0)	R/W	D0的Bit0～Bit7分别代表A0～A7传感器数据的上报
	数据上报时间间隔	V0	R/W	循环上报时间间隔

（2）驱动程序的开发。在智云框架下不仅可以很容易地实现传感器驱动程序的开发，还可以省略无线传感器节点的组网和用户任务的创建等烦琐过程。例如，直接调用sensorInit()函数可以实现传感器的初始化；调用ZXBeeInfRecv()函数可以处理节点接收到的无线数据包；ZXBeeInfSend()函数用于将无线数据包发送至智云平台；ZXBeeUserProcess()函数用于解析接收到的控制命令；设备状态的定时上报需要使用PROCESS_THREAD()作为sensorUpdate()函数和sensorCheck()函数的定时进入接口来反馈设备状态信息。

在sensor.c中，需要在sensorInit()函数中添加传感器初始化的内容，并通过定义上报事件和报警事件来实现设备工作状态的定时反馈，部分代码如下：

```
/*********************************************************************************
*名称：sensorInit()
*功能：传感器初始化
*********************************************************************************/
void sensorInit(void)
{
    //初始化传感器代码
    grating_init();                                          //光栅传感器初始化
    relay_init();                                            //继电器初始化
}
```

光栅传感器的初始化函数是grating_init()，代码如下：

```
/*********************************************************************************
*名称：grating_init()
*功能：光栅传感器初始化
*********************************************************************************/
void grating_init(void)
{
    GPIO_InitTypeDef    GPIO_InitStructure;
    RCC_APB2PeriphClockCmd(RCC_APB2Periph_GPIOA, ENABLE);          //使能PA端口的时钟
    GPIO_InitStructure.GPIO_Pin = GPIO_Pin_7;
    GPIO_InitStructure.GPIO_Speed = GPIO_Speed_2MHz;
```

```
        GPIO_InitStructure.GPIO_Mode = GPIO_Mode_IN_FLOATING;
        GPIO_Init(GPIOA, &GPIO_InitStructure);
}
unsigned char get_grating_status(void)
{
        if(GPIO_ReadInputDataBit(GPIOA,GPIO_Pin_7))                //监测传感器引脚
            return 1;                                              //监测到信号返回 1
        else
            return 0;                                              //没有监测到信号返回 0
}
```

继电器的初始化函数是 relay_init()，代码如下：

```
/*********************************************************************************
*名称：relay_init()
*功能：继电器初始化
*********************************************************************************/
void relay_init(void)
{
        GPIO_InitTypeDef   GPIO_InitStructure;
        RCC_APB2PeriphClockCmd(RCC_APB2Periph_GPIOA, ENABLE);
        GPIO_InitStructure.GPIO_Pin =   GPIO_Pin_5 | GPIO_Pin_4;
        GPIO_InitStructure.GPIO_Mode = GPIO_Mode_Out_PP;
        GPIO_InitStructure.GPIO_Speed = GPIO_Speed_2MHz;
        GPIO_Init(GPIOA, &GPIO_InitStructure);
        relay_control(0x00);
}
/*********************************************************************************
*名称：relay_control()
*功能：继电器控制
*********************************************************************************/
void relay_control(unsigned char cmd)
{
        if(cmd & 0x01){
            GPIO_ResetBits(GPIOA,   GPIO_Pin_5);
        }else{
            GPIO_SetBits(GPIOA,   GPIO_Pin_5);
        }
        if(cmd & 0x02){
            GPIO_ResetBits(GPIOA,   GPIO_Pin_4);
        }else{
            GPIO_SetBits(GPIOA,   GPIO_Pin_4);
        }
}
```

2. Android 端应用设计

1）Android 工程设计框架

打开 Android Studio 开发环境，可以看到自动化生产线计数系统的工程目录，如图 8.17 所示，系统的工程框架和仓库环境管理系统的工程框架相同，详见 8.1.3 节。

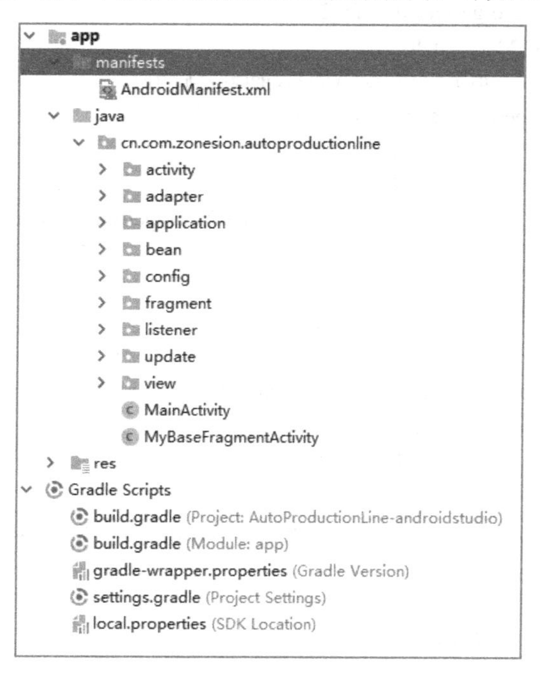

图 8.17　自动化生产线计数系统的工程目录

2）软件设计

根据智云 Android 端应用程序接口的定义，系统的应用设计主要采用实时数据 API 接口，实时数据 API 接口的流程见 3.1.3 节的图 3.11。

（1）LCApplication.java 程序代码剖析。自动化生产线计数系统中的 LCApplication.java 程序代码和城市环境信息采集系统的 LCApplication.java 程序代码相同，详见 3.1.3 节的相关内容。

（2）HomepageFragment.java 程序代码剖析。下面的代码通过(LCApplication) getActivity().getApplication()获取 LCApplication 类中的 WSNRTConnect 对象。

```
private void initInstance(){
    config = Config.getConfig();
    lcApplication = (LCApplication) getActivity().getApplication();
    lcApplication.registerOnWSNDataListener(this);
    wsnrtConnect = lcApplication.getWSNRConnect();
    preferences = getActivity().getSharedPreferences("user_info", Context.MODE_PRIVATE);
    editor = preferences.edit();
}
```

下面的代码通过传送带开关按钮的 setOnClickListener 监听器调用 wsnrtConnect.sendMessage 接口，从而实现对节点设备的控制。

```
openOrCloseBelt.setOnClickListener(new OnClickListener() {
    @Override
        public void onClick(View v) {
            //TODO Auto-generated method stub
            if (sensorCMac != null) {
                if (openOrCloseBelt.getText().equals("开启")) {
                    new Thread(new Runnable() {
                        @Override
                            public void run() {
                                wsnrtConnect.sendMessage(sensorCMac,
                                                        "{OD1=64,D1=?}".getBytes());
                            }
                    }).start();
                }
                if (openOrCloseBelt.getText().equals("关闭")) {
                    new Thread(new Runnable() {
                        @Override
                            public void run() {
                                wsnrtConnect.sendMessage(sensorCMac,
                                                        "{CD1=64,D1=?}".getBytes());
                            }
                    }).start();
                }
            }else {
                Toast.makeText(lcApplication, "请等待 MAC 地址上线", Toast.LENGTH_SHORT).show();
            }
        }
    });
}
```

下面的代码通过 onCheckedChanged 方法实现模式切换的功能。

```
@Override
    public void onCheckedChanged(CompoundButton buttonView, boolean isChecked) {
        //TODO Auto-generated method stub
        switch (buttonView.getId()) {
            case R.id.Infinite_module:
            if (isChecked) {
                isSecurityMode = true;
                openOrCloseBelt.setEnabled(false);
                InfiniteModuleTip.setVisibility(View.VISIBLE);
                finiteModuleTip.setVisibility(View.GONE);
            }
            break;
            case R.id.finite_module:
```

```
                if (isChecked) {
                    isSecuritуMode = false;
                    openOrCloseBelt.setEnabled(true);
                    InfiniteModuleTip.setVisibility(View.GONE);
                    finiteModuleTip.setVisibility(View.VISIBLE);
                }
                break;
            }
        }
```

下面的代码实现了在自动模式下对开关设备（传送带）的控制。

```
private void limitofilluminationTooHigh(){
    if (upperLimit <= conveyorBelt && isSecurityMode == true) {
        wsnrtConnect.sendMessage(sensorCMac, "{CD1=64,D1=?}".getBytes());
    } else {
        wsnrtConnect.sendMessage(sensorCMac, "{OD1=64,D1=?}".getBytes());
    }
}
```

下面的代码通过复写 onMessageArrive 方法来处理节点接收到的无线数据包，实现了设备 MAC 地址的获取，并在当前的页面显示设备的状态。

```
@Override
    public void onMessageArrive(String mac, String tag, String val) {
        if (sensorCMac == null) {
            wsnrtConnect.sendMessage(mac, "{TYPE=?}".getBytes());
        }
        if ("TYPE".equals(tag) && "603".equals(val.substring(2, val.length()))) {
            sensorCMac = mac;
        }
        if (tag.equals("A5") && mac.equals(sensorCMac)) {
            textBatchState.setText("在线");
            textBatchState.setTextColor(getResources().getColor(R.color.line_text_color));
            int numResult = Integer.parseInt(val);
            if ((numResult & 0X1) == 0x1) {
                conveyorBelt+=1;
            }else {
                conveyorBelt+=0;
            }
            textGasDisabled.setText(String.valueOf(conveyorBelt));
            if(seekBarThreshold.getProgress() != 0) {
                limitofilluminationTooHigh();
            }
        }
        if (tag.equals("D1") && mac.equals(sensorCMac)) {
            textConveyorBeltState.setText("在线");
            textConveyorBeltState.setTextColor(getResources().getColor(R.color.line_text_color));
```

```
                int numResult = Integer.parseInt(val);
                if ((numResult & 0X40) == 0x40) {
                    openOrCloseBelt.setText("关闭");
                openOrCloseBelt.setBackground(getResources().getDrawable(R.drawable.close));
                 }else {
                    openOrCloseBelt.setText("开启");
                openOrCloseBelt.setBackground(getResources().getDrawable(R.drawable.open));
                 }
            }
        }
```

3. Web端应用设计

1）页面功能结构分析

自动化生产线计数系统的Web端默认显示的是“运营首页”页面，在“运营首页”页面上设计了3个模块，分别是产品计数显示模块、传送带控制模块、计数阈值设置模块。自动化生产线计数系统的Web端“运营首页”页面如图8.18所示。

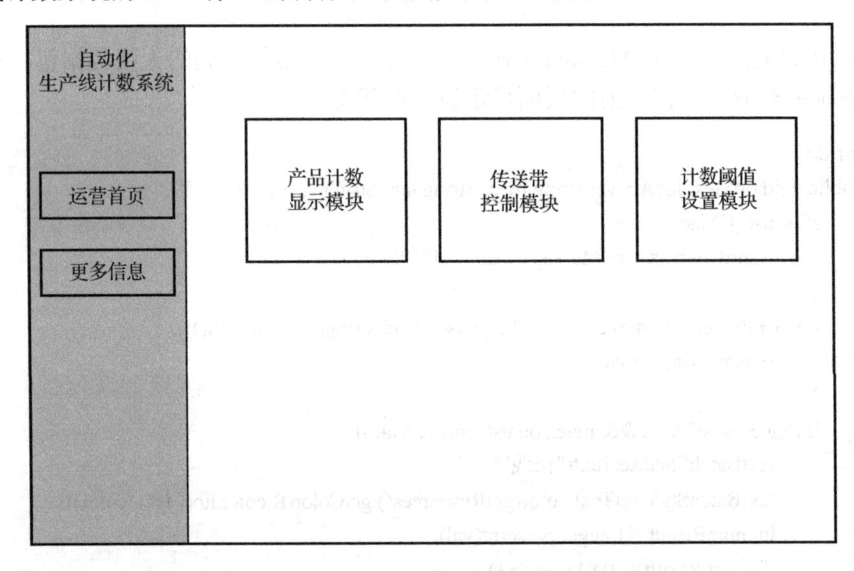

图8.18　自动化生产线计数系统的Web端“运营首页”页面

“更多信息”页面的主要功能是进行智云服务器的连接配置，和城市环境信息采集系统中的Web端“更多信息”页面类似，参见图3.14。

2）软件设计

自动化生产线计数系统Web端的JS开发逻辑与Android端的开发逻辑相似，首先通过配置ID和KEY与智云服务器进行连接，再通过实时监听数据的方法来获取相关传感器的数据并进行处理。JS开发的部分代码如下。

在getConnect()函数中定义了实时连接对象rtc，连接成功回调函数是rtc.onConnect，数据服务掉线回调函数是rtc.onConnectLost，消息处理回调函数是rtc.onmessageArrive。

```
function getConnect() {
    config["id"] = config["id"] ? config["id"] : $("#ID").val();
    config["key"] = config["key"] ? config["key"] : $("#KEY").val();
    config["server"] = config["server"] ? config["server"] : $("#server").val();
    //创建数据连接服务对象
    rtc = new WSNRTConnect(config["id"], config["key"]);
    rtc.setServerAddr(config["server"] + ":28080");
    rtc.connect();
    rtc._connect = false;
    //连接成功回调函数
    rtc.onConnect = function() {
        $("#ConnectState").text("数据服务连接成功！");
        rtc._connect = 1;
        message_show("数据服务连接成功！");
        $("#idkeyInput").text("断开").addClass("btn-danger");
        $("#id,#key,#server").attr('disabled',true)
    };
    //数据服务掉线回调函数
    rtc.onConnectLost = function() {
        rtc._connect = 0;
        $("#ConnectState").text("数据服务连接掉线！");
        $("#idkeyInput").text("连接").removeClass("btn-danger");
        message_show("数据服务连接失败，检查网络或 ID、KEY");
        $(".online_603").text("离线").css("color", "#e75d59");
    $("#id,#key,#server").removeAttr('disabled')
    };
    //消息处理回调函数
    rtc.onmessageArrive = function (mac, dat) {
        //console.log(mac+" >>> "+dat);
        if (dat[0]=='{' && dat[dat.length-1]=='}') {
            dat = dat.substr(1, dat.length-2);
            var its = dat.split(',');
            for (var i=0; i<its.length; i++) {
                var it = its[i].split('=');
                if (it.length == 2) {
                    process_tag(mac, it[0], it[1]);
                }
            }
            if (!mac2type[mac]) { //如果没有获取到 TYPE 值，主动去查询
                rtc.sendMessage(mac,
"{TYPE=?,A0=?,A1=?,A2=?,A3=?,A4=?,A5=?,A6=?,A7=?,D1=?}");
            }
        }
    }
}
```

下述 JS 开发代码的功能是根据设备连接情况，在页面上更新传送带的“开启”或“关闭”状态。

```
var wsn_config = {
    "603" : {
        "online" : function() {
            $(".online_603").text("在线").css("color", "#96ba5c");
        },
        "pro" : function (tag, val) {
            if(tag=="A5"){
                if(val == state.lastNum && state.Conveyor == true){
                    state.lastNum = val;
                    if(val){
                        console.log(val);
                        state.num ++;
                        $("#num").text(state.num);
                        if(state.num >= config["threshold"] && state.Conveyor){
                            message_show("超出传送上限，将关闭传送带！ ");
                            rtc.sendMessage(config["mac_603"], "{CD1=64,D1=?}");
                        }
                    }
                }
            }
            else if(tag=="D1"){
                if(val & 64){
                    $("#transporterImg").attr("src", "img/transporter.gif");
                    $("#transporterStatus").text("关闭");
                    state.Conveyor = true;
                }else{
                    $("#transporterImg").attr("src", "img/transporter.png");
                    $("#transporterStatus").text("开启");
                    state.Conveyor = false;
                }
            }
        }
    },
};
```

下面的代码通过 rtc.sendMessage(config["mac_603"], cmd)方法实现了对传送带的控制。

```
//传送带开关
$("#transporterStatus").click(function(){
    if (page.checkOnline() && page.checkMac("mac_603")){
        var curState = $(this).text(), cmd;
        console.log(curState);
        if(curState=="开启"){
            cmd = "{OD1=64,D1=?}";
        }else{
```

```
                cmd = "{CD1=64,D1=?}";
            }
            console.log("cmd="+cmd);
            rtc.sendMessage(config["mac_603"], cmd);
        }
    });
```

自动化生产线计数管理系统 Web 端其他部分的代码请查看本书配套资料中的项目源文件。

8.2.4　开发验证

1．Web 端应用测试

在 Web 端打开自动化生产线计数系统后，在“运营首页”下可以看到 Web 端的主页，如图 8.19 所示。

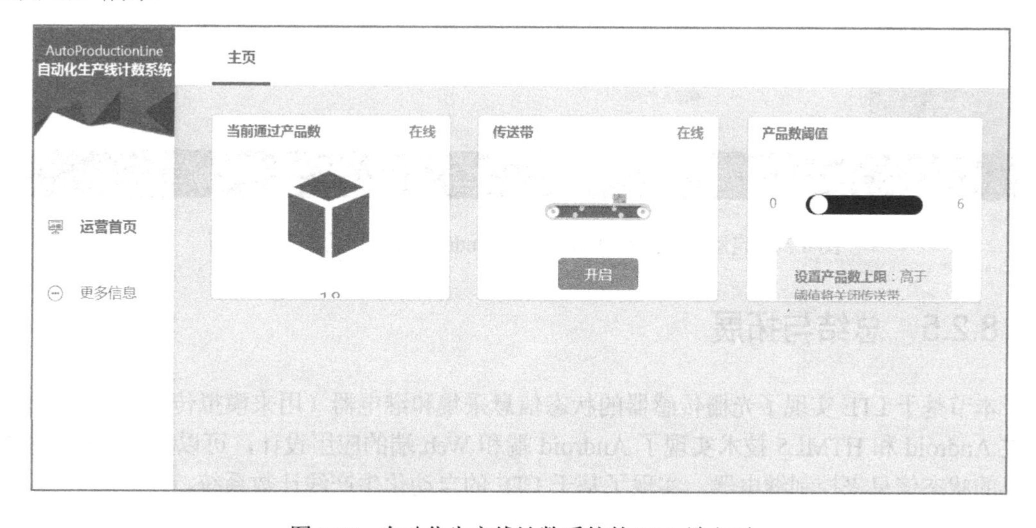

图 8.19　自动化生产线计数系统的 Web 端主页

在设备在线后，可以通过“开启”或“关闭”按钮来控制传送带，如图 8.20 所示。

在自动化生产线计数系统的 Web 端主页中，还可以根据实际情况来设置产品数阈值，当实际产品数高于设置的阈值时将关闭传送带，如图 8.21 所示。

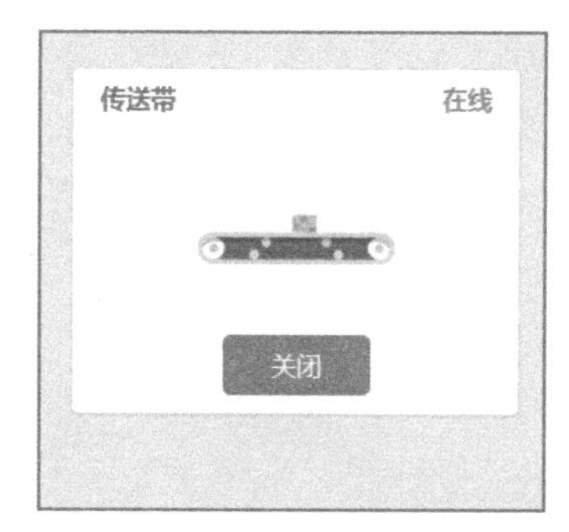

图 8.20　通过“开启”或“关闭”按钮来控制传送带

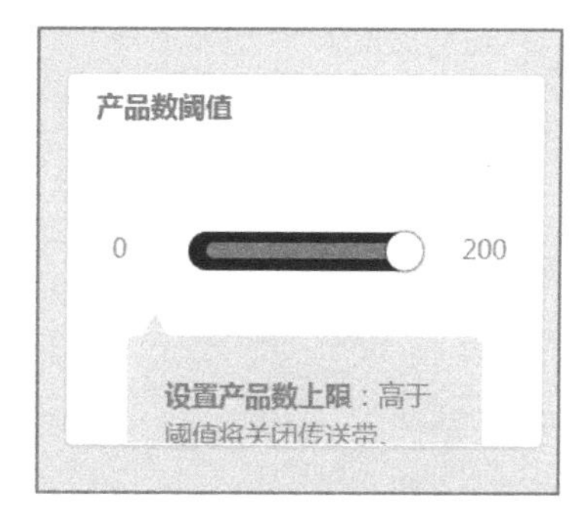

图 8.21　产品数阈值设置

2．Android 端应用测试

Android 端应用测试同 Web 端应用测试流程基本一致，可参考本系统的 Web 端应用测试进行操作。自动化生产线计数系统的 Android 端“运营首页”页面如图 8.22 所示。

图 8.22　自动化生产线计数系统的 Android 端“运营首页”页面

8.2.5　总结与拓展

本节基于 LTE 实现了光栅传感器的状态信息采集和继电器（用来模拟传送带）的控制，通过 Android 和 HTML5 技术实现了 Android 端和 Web 端的应用设计，可以根据光栅传感器采集的状态信息来控制继电器，实现了基于 LTE 的自动化生产线计数系统。

参 考 文 献

[1] 刘云山. 物联网导论. 北京：科学出版社，2010.

[2] 工业和信息化部. 信息化和工业化深度融合专项行动计划（2013—2018）. 工信部信[2013]317 号.

[3] 工业和信息化部. 工业和信息化部关于印发信息通信行业发展规划（2016—2020 年）的通知. 工信部规[2016]424 号.

[4] 工业和信息化部. 物联网“十二五”发展规划，2011.

[5] 国务院关于积极推进“互联网＋”行动的指导意见[J]. 中华人民共和国国务院公报，2015(20):20-22.

[6] 廖建尚. 面向物联网的 CC2530 与传感器应用开发. 北京：电子工业出版社，2018.

[7] 廖建尚，等. 面向物联网的嵌入式系统开发——基于 CC2530 和 STM32 微处理器. 北京：电子工业出版社，2019.

[8] LoRa 技术特点和系统架构. https://www.eefocus.com/communication/392976/r0.

[9] Lora 技术用语解析. https://blog.csdn.net/qq_33658067/article/details/78059774.

[10] NB-IoT 网络架构. https://blog.csdn.net/simon_csx/article/details/79106789.

[11] 王洪亮. 基于无线传感器网络的家居安防系统研究[D]. 河北科技大学，2012.

[12] 沈寿林. 基于 ZigBee 的无线抄表系统设计与实现[D]. 南京邮电大学，2016.

[13] 姚兵兵. 基于 BLE 智能车位锁的设计与实现[D]. 东南大学，2017.

[14] 赵晓伟. 基于蓝牙 BLE 的智能体温测量系统的设计与实现[D]. 南京邮电大学，2015.

[15] 林海龙. 基于 Wi-Fi 的位置指纹室内定位算法研究[D]. 华东师范大学，2016.

[16] 吴奇. 基于 WIFI 的手机签到考勤系统开发[D]. 中国地质大学（北京），2018.

[17] 曾丽丽. 基于 NB-IoT 数据传输的研究与应用[D]. 安徽理工大学，2018.

[18] 常云果. 基于 NB-IoT 的飞行动物远程监测系统[D]. 郑州大学，2018.

[19] 陈钇安. 基于 LoRa 全无线智能水表抄表应用的研究[D]. 湖南大学，2018.

[20] 陈伦斌. 无线 LoRa 在输电线路监测中的组网设计与实现[D]. 西安理工大学，2017.

[21] LoRa 技术特点和系统架构. https://www.eefocus.com/communication/392976/r0.

[22] 黄琦敏. LTE 系统中多播业务的吞吐量和公平性研究[D]. 南京邮电大学，2018.

[23] 马晓慧. LTE 下行链路关键技术的研究与实现[D]. 西安电子科技大学，2009.

[24] 廖建尚，等. 物联网短距离无线通信技术应用与开发. 北京：电子工业出版社，2019.

[25] 廖建尚，等. 物联网长距离无线通信技术应用与开发. 北京：电子工业出版社，2019.

[26] https://developer.android.google.cn/.

[27] http://developer.android.com/.